T0296774

NICOTINIC ACETYLCHOLINE RECEPTORS IN HEALTH AND DISEASE

NICOTINIC ACETYLCHOLINE RECEPTORS IN HEALTH AND DISEASE

R. THOMAS BOYD

Department of Neuroscience
The Ohio State University College of Medicine
Wexner Medical Center
Columbus, OH, United States

ACADEMIC PRESS

An imprint of Elsevier

ELSEVIER

Academic Press is an imprint of Elsevier
125 London Wall, London EC2Y 5AS, United Kingdom
525 B Street, Suite 1650, San Diego, CA 92101, United States
50 Hampshire Street, 5th Floor, Cambridge, MA 02139, United States
The Boulevard, Langford Lane, Kidlington, Oxford OX5 1GB, United Kingdom

Notices
Knowledge and best practice in this field are constantly changing. As new research and experience broaden our understanding, changes in research methods, professional practices, or medical treatment may become necessary.

Practitioners and researchers must always rely on their own experience and knowledge in evaluating and using any information, methods, compounds, or experiments described herein. In using such information or methods they should be mindful of their own safety and the safety of others, including parties for whom they have a professional responsibility.

To the fullest extent of the law, neither the Publisher nor the authors, contributors, or editors, assume any liability for any injury and/or damage to persons or property as a matter of products liability, negligence or otherwise, or from any use or operation of any methods, products, instructions, or ideas contained in the material herein.

ISBN: 978-0-12-819958-9

For information on all Academic Press publications
visit our website at https://www.elsevier.com/books-and-journals

Publisher: Nikki Levi
Acquisitions Editor: Anna Valutkevich
Editorial Project Manager: Kyle Gravel
Production Project Manager: Swapna Srinivasan
Cover Designer: Mark Rogers

Typeset by STRAIVE, India

Working together
to grow libraries in
developing countries

www.elsevier.com • www.bookaid.org

To: Dr. Dennis McKay, my good friend, colleague, and mentor.

Contents

About the author

Dr. R. Thomas Boyd graduated from the University of Illinois at Urbana-Champaign in 1977 with a BS in Physiology. He received a PhD in Molecular Biology from the University of Texas at Austin in 1986. After working as a postdoctoral researcher at the University of California, San Diego, and the Salk Institute for 4 years, he moved to Ohio State in 1990, where he is a Professor in the Department of Neuroscience. Dr. Boyd has also been involved in undergraduate and graduate education for most of his career. He served on the Neuroscience Graduate Studies Program Graduate Studies Committee for many years and currently is on the Biomedical Sciences Graduate Programs (BSGP) Graduate Studies Committee. Dr. Boyd also served for 4 years as the Associate Director of the BSGP. He was involved in the development of the Neuroscience major at Ohio State and has taught a large undergraduate neuroscience class for more than 15 years. He is also Graduate Studies Chair of the new Applied Master's in Neuroscience at Ohio State.

Research activities in the Boyd laboratory have emphasized a molecular biological analysis of neuronal nicotinic acetylcholine receptors (nAChRs). Dr. Boyd studied transcriptional regulation of nAChR subunit genes as well identifying a new allosteric site on the $\alpha 4\beta 2$ neuronal nAChR. This site was explored for its potential as a binding site for new drugs that modulate nAChR activity and could be used to aid smoking cessation. His lab was the first to clone the zebrafish neuronal nAChR genes and determined the expression pattern of zebrafish neuronal nAChR genes during development. The information derived from these studies is being used to develop zebrafish as a model for studying the role of nAChRs in normal development of the nervous system and the mechanisms by which nicotine perturbs vertebrate nervous system development.

Preface

Acetylcholine is a neurotransmitter with widespread expression in the central nervous system, autonomic nervous system, skeletal muscle, immune system, and even the skin. Many organisms, including fungi and bacteria, express acetylcholine. Acetylcholine was one of the first neurotransmitters discovered. Acetylcholine signals through two major neurotransmitter receptor families, nicotinic acetylcholine receptors (nAChRs) and muscarinic acetylcholine receptors (mAChRs). While the widespread effects of acetylcholine are mediated by both of these receptor families, this book will focus on nAChRs. nAChRs are so named because nicotine also interacts with these receptors. Many effects of smoking and nicotine addiction are mediated by nAChRs. nAChRs are involved in numerous diseases, including Alzheimer's, Parkinson's, schizophrenia, cancer, and autism. nAChRs are important potential translational targets for treatment of these diseases as well as therapy for addiction. A greater understanding of the basic neurobiology and clinical roles of nAChRs will provide important insights into future clinical treatment of many major disorders. Translational aspects of nAChRs will be woven throughout this book.

There is a rich literature on muscle nAChRs and a number of disorders involve nAChRs at the neuromuscular junction (NMJ). However, there is much to cover regarding the various roles of neuronal nAChRs, and I will try to do them justice. The chapters will not be exhaustive reviews of the literature, but will highlight important areas of research and focus on important concepts in each area. A recent search of PubMed using the term "nicotinic receptor" identified over 26,000 papers. I apologize in advance to anyone whose papers I have not included. This was not due to any lack of regard for their important contributions to the field, but rather a need to focus the chapters. This book can serve as an introduction to this important field for medical students and neuroscience graduate students.

nAChR structure

1.1 Early structural studies on muscle and Torpedo nAChRs

All neuronal nAChRs share a common structure, but small changes in the subunit composition allow for differential function and localization. While we will focus on the overall structure of the nAChR from both muscle and neurons, function will be described in more detail in the context of nAChR pharmacology in another chapter. However, function can't be separated from structure, so some elements will be summarized here.

The nAChR was first chemically characterized from electric organs and subsequently from muscle. The first biochemistry was done using the Electrophorus electricus electric organ (Changeux, 2012). The ability to isolate electroplaques and the development of compounds that bound the presumptive receptors were key. The isolated electroplaques showed ion flux in response to nicotinic agonists, supporting that ion channels may be present in the tissue. The use of the snake venom toxin alpha bungarotoxin (α-Bgt) to bind the nAChR was an important step to isolate nAChRs from the electric organ, since it was shown that α-Bgt blocked neuromuscular junction (NMJ) transmission and also blocked the flux of cations in the electroplaques (Changeux, 2012). The use of α-Bgt and other toxins with affinity chromatography allowed the isolation of the nAChR proteins. Subsequently, these methods were applied to the electroplaques of *Torpedo marmorata*, which also had a high concentration of nAChRs.

The receptor was shown to be composed of five subunits, designated α1, β1, γ1, δ1 (two copies of the α1 subunit were present) forming a pentamer (Figs. 1.1–1.3). The two ACh-binding sites are located between the α and γ and α and δ subunits. A fifth subunit ε is also present in mature neuromuscular junction nAChRs, replacing the γ subunit and thus part of the α-ε ACh interface that forms an ACh-binding site (Papke, 2014). The sequences of a small number of amino acids in the receptor were obtained, and the advent of molecular cloning in the 1980s opened the door for

Nicotinic Acetylcholine Receptors in Health and Disease
https://doi.org/10.1016/B978-0-12-819958-9.00003-7

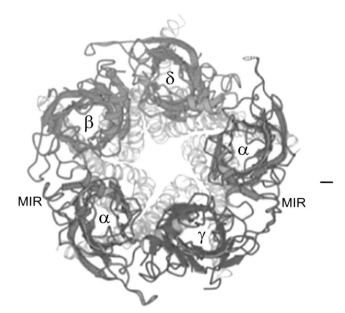

Fig. 1.1 Pentameric structure of the NMJ nAChR. Top view of the embryonic form of the muscle receptor with the ACh-binding site at the interface between αγ and αδ obtained from high-resolution microscopy. ACh binding sites represented in gold. *(From Hurst, R., Rollema, H., & Bertrand, D. (2013). Nicotinic acetylcholine receptors: From basic science to therapeutics. Pharmacology and Therapeutics, 137, 22–54.)*

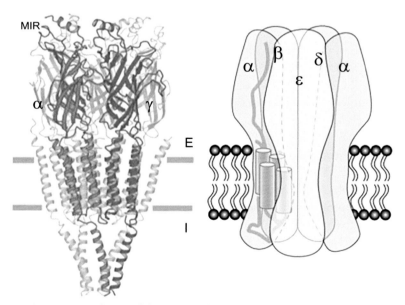

Fig. 1.2 Cross-sectional view of the NMJ nAChR. Cross-sectional view of the NMJ nAChR as a ribbon diagram and schematic. MIR: main immunogenic region. Binding sites are at the α-γ and α-δ interfaces in the embryonic form. During development, there is a switch and the α-γ-binding site is replaced with an α-ε interface. *(From Hurst, R., Rollema, H., & Bertrand, D. (2013). Nicotinic acetylcholine receptors: From basic science to therapeutics. Pharmacology and Therapeutics, 137, 22–54.)*

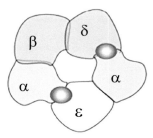

Fig. 1.3 ACh-binding sites at subunit interfaces. Binding of ACh to the adult form of the NMJ nAChR (α-ε and α-δ). *(From Hurst, R., Rollema, H., & Bertrand, D. (2013). Nicotinic acetylcholine receptors: From basic science to therapeutics. Pharmacology and Therapeutics, 137, 22–54.)*

characterization of the individual subunits at a molecular level (Noda et al., 1983). Each subunit has common structural features (Fig. 1.4) when aligned, and the locations of the structural features are generally conserved in each subunit, if not the exact sequences. These include an extracellular domain (ECD), transmembrane regions (M1–M4), an amphipathic region (MA), A, B, C loops, a Cys loop, and main immunogenic region (MIR) (Unwin, 2005). Many of the basic features of the muscle nAChR that are common to the neuronal nAChRs were determined in these early studies, such as the basic subunit structure, ligand-binding sites, and identification of the ion channel. As gene sequencing and structural studies progressed, the nAChRs took their place as a member of the Cys-loop ligand-gated family of receptors along with glycine, serotonin 5-HT3, and GABA A/C receptors.

The first electron micrographic images of the receptor were obtained by Nigel Unwin and resolved a membrane-associated Torpedo nAChR at 4 Angstroms (Unwin, 2005). The total length was shown to be 160 Angstroms with a 20 Angstrom extracellular facing vestibule and the binding pockets for ACh located 40 Angstroms above the membrane and on opposite sides of the vestibule (Unwin, 2005). The vestibule is formed by the N terminal regions of the five subunits (Dani, 2015). The binding pockets are formed at α-δ or α-γ or ε interfaces in the ECD (Papke, 2014). The α-helical M1–M4 regions form the membrane part of the structure. TM2s form a symmetrical inner ring of the channel while TM1s, TM3s, and TM4s form an outer shell between the inner ring and the membrane lipids (Fig. 1.5) (Miyazawa et al., 2003). The TM2s interact in a closed channel to form a hydrophobic barrier that doesn't allow ion flow (Unwin, 2003). Binding

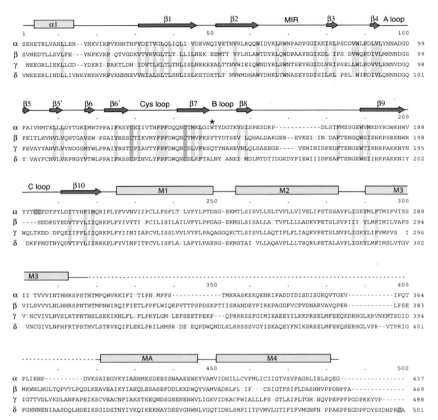

Fig. 1.4 Common features of nAChRs. Aligned amino acid sequences of the four ACh receptor polypeptide chains. The sequences are from *T. marmorata*, which differ in 48 places (*cyan* lettering) from those of *T. californica* (including the absence of the first residue of γ). Locations of the MIR (critical segment in *red*), named loops, αTrp149 (star), and some key cysteine residues (*green* background) are indicated. Conserved residues forming the hydrophobic cores of the subunits in the ligand binding domain and at the boundary between this domain and the membrane-spanning domain are shown with *pink* and *orange* background, respectively. Elements of secondary structure, for the α subunits, are indicated above the sequences (*yellow*, α-helix; *blue* and *red*, β-strands composing the inner and outer sheets of the β-sandwich). The exact extents of the α-helices and β-strands are not accurately represented, given the limited resolution, but are similar for all four polypeptides. *(From Unwin, N. (2005). Refined structure of the nicotinic acetylcholine receptor at 4A resolution. Journal of Molecular Biology, 346(4), 967–989. https://doi.org/10.1016/j.jmb.2004.12.031.)*

of ACh initiates movement in the binding sites, closing of the C-loop and a linked rotational conformational change in the TM2 regions that widens the channel by 3 Angstroms and changes interactions between ions and the receptor allowing cations to move through the channel (Fig. 1.6, Unwin, 2003). The closed channel presents a more hydrophobic environment,

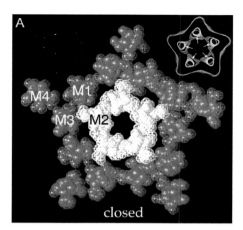

Fig. 1.5 Closed-channel structure. Cross sections at the gate in middle of the membrane, showing van der Waal's surfaces of the atoms encircling the closed pore. Inset (*blue*: closed channel; *white* and *brown*: open channel). The view is from the synaptic cleft; individual helices, M1–M4, are identified on one of the subunits. *(From Unwin, N. (2003). Structure and action of the nicotinic acetylcholine receptor explored by electron microscopy. FEBS Letters, 555 (1), 91–95. https://doi.org/10.1016/S0014-5793(03)01084-6.)*

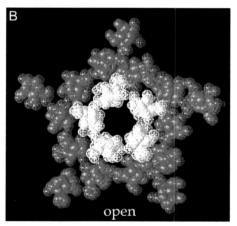

Fig. 1.6 Open-channel structure. Cross sections at the gate in middle of the membrane, showing van der Waal's surfaces of the atoms encircling the open pore. The open pore is modeled by applying 15 degrees clockwise rotations to each of the inner helices (colored *white*). The change in structure, involving a widening of the pore by 3 Angstroms, is consistent with the changes observed experimentally at 9 Angstrom resolution. The view is from the synaptic cleft; individual helices, M1–M4, are identified on one of the subunits. *(From Unwin, N. (2003). Structure and action of the nicotinic acetylcholine receptor explored by electron microscopy. FEBS Letters, 555 (1), 91–95. https://doi.org/10.1016/S0014-5793(03)01084-6.)*

while the open is more hydrophilic. The intracellular loop formed by the TM3-TM4 region is disordered except for an α-helical region designated MA. Intracellular sequences contribute to the walls of the vestibule inside the neuron. The smaller TM1-TM2 and TM2-TM3 loops are also important for function (Unwin, 2005). The modeled MA regions of the subunits are predicted to form a structure with five windows of about 8 Angstrom in diameter (with contributions from the TM1-TM2 loop, C terminal of TM3, and the N-terminal of TM4), which would allow movement of cations and repel anions (Unwin, 2005). The intracellular and extracellular vestibules contain negative charges that would allow cations to be focused near the openings and contribute to selectivity of the channel.

1.2 Diversity of neuronal nAChRs

The overall pentameric structure of neuronal nAChR is very similar to that of the muscle/Torpedo receptors, but with refinements due to more variation in subunit combinations. The structures and sequences of the neuronal subunits are also related to the NMJ/Torpedo receptors, and much of the early functional work on NMJ receptors will apply in general to neuronal nAChRs. The first mammalian neuronal nAChR subunit was cloned from rat in 1986 (Boulter et al., 1986) and others quickly followed. Since the muscle subunits were designated α1, β1, γ, δ, and ε, the neuronal subunits were also designated with Greek letters. Twelve neuronal nAChR genes have been cloned from mammals and other vertebrates and placed into two groups (more about diversity in a subsequent chapter). Nine alpha subunit genes (most homologous to the neuromuscular junction α1 subunit) are designated α2–α10. Three beta subunit genes (most homologous to the non-α subunits of the neuromuscular receptor) are designated β2–β4. The alpha subunits also contain characteristic adjacent vicinal cysteines in the extracellular region, while the beta subunits do not, same as in the NMJ nAChR (Fig. 1.4) (Dani, 2015). Neuronal nAChR subtypes are defined as a specific combination of subunits (Fig. 1.7). The properties of each subtype of nAChR are determined by the combination of subunits. Subtypes are designed as α4β2, α7, or α3β4α5, for example. Some descriptions also are written α4β2*, α6β2*, α7*, and α3β4* with the asterisk indicating that other subunits known or unknown might be present in that subtype (Lukas et al., 1999).

While all nAChRs are pentamers, some are heteromeric and some homomeric (α7, α9). The heteromeric nAChRs are composed of multiple alpha subunits and multiple beta subunits (Fig. 1.8) (Zoli et al., 2015). The

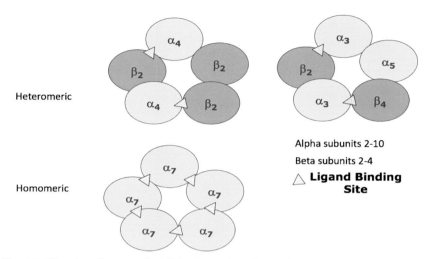

Fig. 1.7 Diversity of neuronal nAChRs. Examples of some heteromeric and homomeric nAChR subtypes.

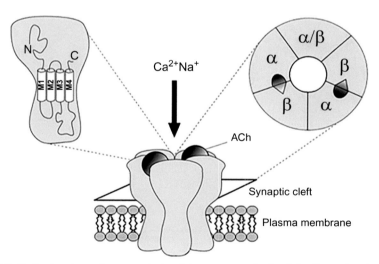

Fig. 1.8 Basic structure of neuronal nicotinic receptors. Each of the five subunits contains an extracellular amino terminal portion followed by three hydrophobic transmembrane domains (M1–M3), a large intracellular loop, and then a fourth hydrophobic transmembrane domain (M4). Neuronal nAChR subunits are assembled as pentamers, same as for the NMJ nAChRs. Ca^{2+} and Na^{+} enter the cell through the channels. nACh-binding sites are located at α-β interfaces in a heteromeric receptor. *(From Zoli, M., Pistillo, F., & Gotti, C. (2015). Diversity of native nicotinic receptor subtypes in mammalian brain.* Neuropharmacology, 96, *302–311. https://doi.org/10.1016/j.neuropharm.2014.11. 003.)*

five subunits are arranged around a central channel or pore, with the overall structure of the neuromuscular nAChR. Heteromeric nAChRs have two ACh-binding sites as in NMJ nAChRs, while more can be present in homomeric nAChRs. Each of these nAChRs is defined as a subtype. Although there are nine alpha and three beta subunit proteins, not all mathematical combinations exist. However, more than a dozen have been detected by expression in vitro or in vivo (see Chapter 2). Specific subunits contribute to the binding site for acetylcholine, but others play an accessory role and are important in determining the properties of the receptor, but do not contribute to the structure of the binding pocket directly (Fig. 1.9). Concatameric constructs have been used in vitro to express combinations of subunits in specific orders. These have been useful to study the pharmacology and binding site structures of many subtypes.

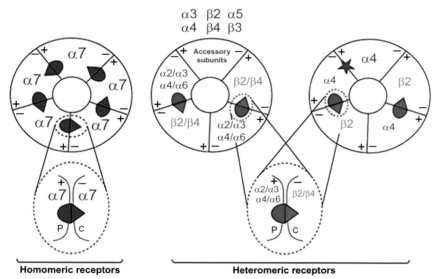

Fig. 1.9 ACh-binding pockets vary depending on the nAChR subtype. The pentameric arrangement of nAChR subunits in an α7 homopentameric subtype (left), heteromeric receptor subtype (middle), and the (α4)3(β2)2 subtype (right). The localization of the subunit interfaces of the orthosteric-binding sites is indicated, together with the primary component P (+) carried by the α subunits and the complementary component C (−) carried by an α or non-α subunit. In addition to the two orthosteric sites, the (α4)3(β2)2 subtype has a binding site at the α4/α4 interface (star). *(From Zoli, M., Pistillo, F., & Gotti, C. (2015). Diversity of native nicotinic receptor subtypes in mammalian brain.* Neuropharmacology, 96, *302–311. https://doi.org/10.1016/j.neuropharm.2014.11.003.)*

Subtypes vary in biophysical and pharmacological properties that are determined by diversity in the binding sites and channel structure. All of the neuronal nAChR subtypes are cation channels gating Na^+, K^+, and Ca^{2+}, with a permeability to each, especially Ca^{2+}, specific to a subtype. Each subtype varies in regard to affinity for nicotine, acetylcholine, and other cholinergic agonists and antagonists. Each subtype varies with regard to the ability to be desensitized, channel open time, probability of opening, single-channel conductance, and other biophysical properties. A more detailed description of the subtypes and their distribution and the detailed function and pharmacology of neuronal nAChRs will be presented in other chapters.

1.3 Structure of individual nAChR subunits

Each of the individual neuronal nAChR subunit proteins is homologous and has conserved features. Each has a long N-terminal or ECD of approximately 210–250 amino acids (Giastas et al., 2018). The N-terminal domains of the five subunits form the ligand-binding domain (Tsetlin et al., 2011). This is followed by four transmembrane domains (TM1–TM4) of about 18–27 amino acids in length (Görne-Tschelnokow et al., 1994; Tsetlin et al., 2011). The TMs have a high level of sequence identity to each other and to the same domains in other nAChRs, including those of the Torpedo/NMJ nAChR subtypes. Hydrophobicity blots predicted the presence of the TM domains. A large intracellular cytoplasmic domain runs between the TM3 and TM 4 (varies the most between neuronal subunits in length and sequence), and a short C terminal extracellular domain follows the last TM domain. There is a short extracellular loop (ECL) between TM2 and TM3. The ion channel is formed by association of the five TM2 domains, one TM2 contributed by each subunit as in the NMJ receptor. The other transmembrane regions surround the pore (Dani, 2015). Both the N terminus and the C terminus are located extracellularly (Fig. 1.10). All of these features of neuronal nAChRs are similar to those found in the Torpedo NMJ nACh (Fig. 1.4). The next sections will describe these features of the nAChR in more detail. This will focus on refining what was learned from Torpedo and NMJ nAChRs to neuronal nAChRs using crystal structure of neuronal nAChRs and mutated versions of the acetylcholine-binding protein (AChBP).

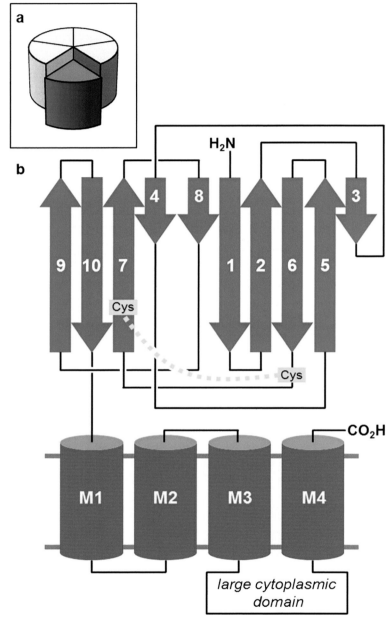

Fig. 1.10 Overall structure of neuronal nAChR subunits. (A) Schematic of a Cys-loop receptor pentamer with one monomeric unit highlighted and (B) topological map of a typical Cys-loop receptor monomer with notable features highlighted. The N-terminal sequence is formed by a complex series of loops important for ligand binding and channel function. *(From Sparling, B. A., & DiMauro, E. F. (2017). Progress in the discovery of small molecule modulators of the Cys-loop superfamily receptors. Bioorganic and Medicinal Chemistry Letters, 27, 3207–3218.)*

1.3.1 Extracellular domain (ECD)

The extracellular domain is formed by components of both alpha and beta subunits. This domain contains the binding sites for ACh as well as forming an entryway or vestibule extending into the synaptic cleft. The snake venom toxin α–Bgt used to characterize the Torpedo nAChR bound to the dissociated alpha subunit, and it was thought this subunit alone formed the acetylcholine-binding site. However, numerous studies, some described here, determined that the binding pocket for acetylcholine is formed by residues from two subunits of the ECD. Heteromeric nAChRs are activated by ACh occupying two binding sites, while homomeric receptors are maximally activated by binding at three sites (Rayes et al., 2009; Zoli et al., 2015). The number of possible combinations of binding interfaces is much greater than for the Torpedo/NMJ nAChRs due to the possible combinations of α and β subunits available.

One of the first studies on the structure of the ECD of neuronal-like nAChRs was done using an acetylcholine-binding protein (AChBP) from *Lymnaea stagnalis*, a freshwater snail (Smit et al., 2003). This protein was shown to be released by glial cells. Structurally and pharmacologically, it is most similar to the ECD of the homomeric α7 nAChRs and is similar to other nAChRs (overall 20%–24% sequence identity). The AChBP is 210 amino acids long, and this length is consistent with the length of vertebrate nAChR ECDs, and the overall structure was similar to mouse muscle α1 or Torpedo nAChRs (Rucktooa et al., 2009). This is only a binding protein that assembles as homopentamers (5 protomers) and doesn't have TM domains or an ion channel. It is most similar to α7 homomeric nAChRs in that it has a relatively low affinity for ACh and a higher affinity for nicotine (Celie et al., 2004). Nevertheless, it has been an important first model for studying the structure of vertebrate nAChRs (Smit et al., 2003).

Since the AChBP was able to be crystalized, with and without ligands, much was learned about the structure of the nAChR-binding domains (Fig. 1.11, Celie et al., 2004). From previous work on muscle and Torpedo receptors, it was known that binding sites were at subunit interfaces. As the AChBP is close in homology to the neuronal α7 nAChR, this work also showed this to be the case for the AChBP with five binding sites in the AChBP homopentamer. Six loops were defined for the muscle nAChR (A–F) (A–C provided by one subunit, and D-F provided by the complementary subunit), and these were also identified in the AChBP (Figs. 1.12 and 1.13). Five sites were seen as with an α7 homomeric neuronal

A   B

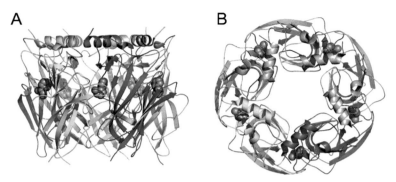

Fig. 1.11 Binding pockets of the ACh binding protein (AChBP). The pentameric AChBP is shown bound to nicotine. (A) Schematic representation of AChBP with nicotine (*pink*) bound. One subunit in *yellow*, one in *blue*, view with membrane at the bottom in nAChRs. (B) Orthogonal view of (A), toward the membrane in nAChRs. *(From Celie, P. H. N., Van Rossum-Fikkert, S. E., Van Dijk, W. J., Brejc, K., Smit, A. B., & Sixma, T. K. (2004). Nicotine and carbamylcholine binding to nicotinic receptors as studied in AChB crystal structures. Neuron, 41 (6), 907–914. https://doi.org/10.1016/S0896-6273(04)00115-1.)*

nAChR. When nAChRs are compared with the AChBP, principle subunit (α) residues are more conserved than residues on the complementary subunit of the binding site (Celie et al., 2004). Binding is most similar to alpha subunits, but the determinants of gating and the role of complementary subunits were difficult to determine with the AChBP. When HEPES bound and nicotine bound structures were compared, the C loop moved as was seen in NMJ nAChRs in the ligand-bound state (Celie et al., 2004). The affinity for nicotinic ligands is not identical between nAChRs and the AChBP, most likely due to differences on the complementary face (Rucktooa et al., 2009).

To analyze the structure of other nAChR subtypes other than α7, further studies are conducted to mutate the AChBP sequence toward that in other nAChR subunits (Giastas et al., 2018). The overall structure of the AChBP, while different than nAChRs, shares the main structural features of other nAChRs ECDs and thus helped define the important structural elements of the ECD of neuronal nAChRs although no Cys loop was present. The ability to crystalize the AChBP led to an understanding of the structure of the nAChR ECD and showed many overall structural similarities to the NMJ receptor structure determined by Unwin.

The structure of the α4β2 nAChR (Morales-Perez et al., 2016) allowed a comparison to be made between the Torpedo/NMJ nAChRs and the heteromeric neuronal nAChRs, Fig. 1.14. This structure showed the great similarity between NMJ and neuronal nAChR structures. This work tied

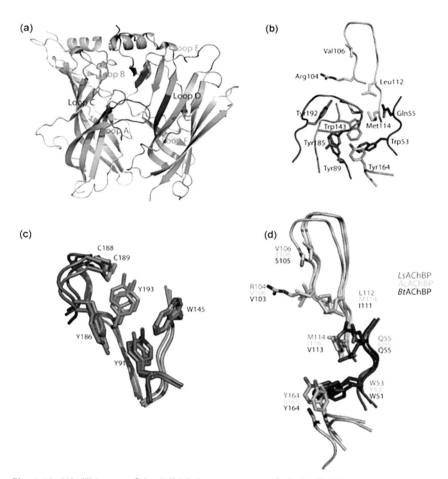

Fig. 1.12 (A)–(F) Loops of the AChBP. *Lymnaea stagnalis* (Ls) AChBP protomer-protomer interface making up the ligand-binding site, with loops contributing to the binding interface highlighted (A). Residues involved in contacts with nicotine in LsAChBP are shown in detail in a blown-up view (B). Superposition of the *Aplysia californica* (Ac) (PDB: 2BR7), Ls (PDB: 1UX2), and *Bulinas truncatus* (Bt) AChBP (PDB: 2BJ0) residues contributing to the principal (C) or complementary (D) face of the ligand-binding site. Principal face residues are highly conserved while the complementary face displays more variability. *(From Rucktooa, P., Smit, A. B., & Sixma, T. K. (2009). Insight in nAChR subtype selectivity from AChBP crystal structures.* Biochemical Pharmacology, *78 (7), 777–787. https://doi.org/10.1016/j.bcp.2009.06.098.)*

together the studies with the AChBP and neuronal nAChRs. The ECD are followed by six loops that form the binding pockets, Fig. 1.15 (Papke, 2014). The binding pocket is formed by a primary or principal surface (+) (usually alpha subunit residues) and a complementary surface (−). Three loops, designated A, B, and C on the alpha subunit form the primary surface, while D,

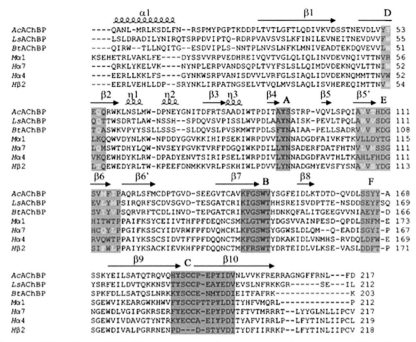

Fig. 1.13 Homology of AChBP sequence to other muscle and neuronal nAChR ECDs. 3D-coffee-based sequence alignment of *Aplysia* (Ac), *Lymnea* (Ls) and *Bulinus* (Bt) AChBPs with the mouse α1, human α7, α4, and β2 nAChR subunits, taking into account available structural information from the Ac, Ls, and Bt AChBPs and mouse α1 subunit. Loops A–F, contributing to the binding site, are highlighted in blue blocks for principal site residues and beige blocks for complementary site residues. Principal face residues involved in agonist binding are colored *red*, whereas those from the complementary face are colored *yellow*. (From Rucktooa, P., Smit, A. B., & Sixma, T. K. (2009). Insight in nAChR subtype selectivity from AChBP crystal structures. Biochemical Pharmacology, 78 (7), 777–787. https://doi.org/10.1016/j.bcp.2009.06.098.)

E, and F loops form the complementary surface. Two loops most important for binding and enclosing the bound ligand are respectively the Cys-loop and the C-loop. The Cys loop is 13–14 conserved amino acids linked by disulfide bonds and is a feature of alpha subunits (Dani, 2015). The sequences of these sites are critical for determining the pharmacology of the various nAChR subtypes. Two binding pockets are formed, often with differential sensitivity to ligand binding. In heteromeric receptors α2, α3, α4, or α6 subunits form the principal component, and the complementary surface is provided mainly by the β2 or β4 subunits, but also by α4 and α2 in some subtypes (Zoli et al., 2015). Other subunits are accessory and contribute

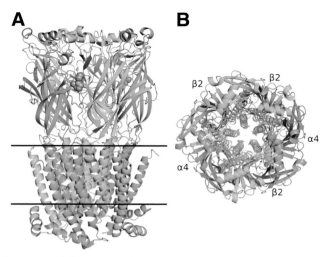

Fig. 1.14 Structure of the neuronal α4β2 nAChR. Overall architecture of the α4β2 nAChR. (A) View of α4β2 parallel to the plasma membrane (PDB ID: 5KXI) (Morales-Perez et al., 2016). α subunits are shown in *green* and β in *cyan*, while nicotine is in orange spheres. Solid lines indicate the approximate limits of the membrane. (B) View of α4β2 along the channel axis. Color coding as in (A). The coordinates of all the structures depicted were retrieved from Protein Data Bank (www.wwpdb.org), and PyMol (www.pymol.org) was used to generate the figures. *(From Giastas, P., Zouridakis, M., & Tzartos, S. J. (2018). Understanding structure-function relationships of the human neuronal acetylcholine receptor: Insights from the first crystal structures of neuronal subunits. British Journal of Pharmacology, 175 (11), 1880–1891. https://doi.org/10.1111/bph.13838.)*

to the overall structure, but not to the orthosteric-binding site such as α5 and β3. α3, α4, β2, and β4 subunits can also occupy accessory positions (Zoli et al., 2015). It was later shown in many studies that the binding pocket of neuronal nAChRs is formed by components of both alpha (not including α5) and beta subunits (Papke, 2014) in heteromeric receptors, by alpha subunits in homomeric receptors (α7), and by forming unorthodox interfaces such as α4-α4, α2-α2, and α9α10 nAChRs (Giastas et al., 2018).

The α4β2 receptor composed of 2 α and 3 β subunits structure co-crystallized with nicotine was obtained at 3.9A resolution (Morales–Perez et al., 2016). The structure determined here was most likely of the desensitized, nonconducting state. Nicotine bound at the α/β interface as ACh did for interfaces in the Torpedo/NMJ receptor and the AChBP. The α4 subunit was part of the (+) or primary side, and the β subunit was part of the (−) or complementary side of the binding pocket. The ABC loops were part of + side and DEF loops (part of − side, also as is seen

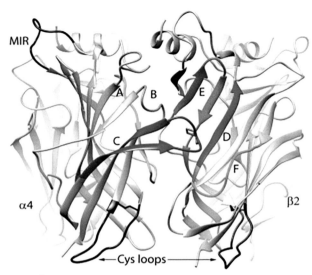

Fig. 1.15 Loops of the neuronal α4β2 nAChR. Homology model of the interface between α4 and β2 subunits highlighting the subdomains important for agonist binding that are located on the primary (Loops A, B, and C in the α4 subunit) or complementary surface (Loops D, E, and F in the β2 subunit) of the ligand-binding domain. Only the contours of the backbone are shown except for the disulfide-linked vicinal cysteines on the C-loop of α4. *MIR*, main immunogenic region. *(The figure was prepared in UCSF Chimera by Dr. Nicole Horenstein. From Papke, R. L. (2014). Merging old and new perspectives on nicotinic acetylcholine receptors. Biochemical Pharmacology, 89(1), 1–11. https://doi. org/10.1016/j.bcp.2014.01.029.)*

in NMJ or Torpedo receptors. A large extracellular vestibule is also seen with a channel lined by 5 TM2 α-helices (Morales-Perez et al., 2016). The pore diameter at the narrowest point was determined to be 3.8A. An MX domain, similar to the MA region in the NMJ receptors, was also resolved. Most of the intracellular loop was not included in the crystalized receptor.

The opening of the channel is initiated when agonist-binding sites are occupied, and a conformational change in the ECD is induced. This change ultimately leads to the opening of the channel. To examine the conformational change, the structure of the AChBP was determined when bound to acetylcholine and in an unbound state. Other agonists may interact with residues in other loops (Miller & Smart, 2010). The C loop moves after agonist binding in such a way as to trap the agonist, with the C loop less accessible after agonist binding. In the presence of antagonists or no agonist, the C loop is in an uncapped conformation (Miller & Smart, 2010). Unwin (2005) showed for Torpedo nAChRs without agonist the C loop is uncapped and that agonist bound subunits are rotated compared with those without agonist. The move-ment of the binding site must be communicated to the TM domains. In the region where the ECD is close to the membrane, alpha subunit loops are close to the TM2-TM3 loop (Unwin, 2005). The inner beta sheet of the β1–β2 loop and the Cys loop connecting the β6 and β7 strands are close to the TM2 channel (Miller & Smart, 2010). These loops in beta subunits are further from the channel (Miller & Smart, 2010). Movement of these loops commu-nicates the ligand binding to changes in channel structure.

1.3.2 Channel structure

Since the AChBP doesn't contain a channel, studies were done using Torpedo, muscle, and more recently on α4β2 crystals (Morales-Perez et al., 2016). The overall structure of the α4β2 nAChR was similar to that shown for NMJ nAChR (compare Fig. 1.1 with Fig. 1.14). Early work on Torpedo nAChRs revealed features of the channel as it did for other parts of the structure. The channel is formed by the five TM2 alpha helices. The channel was shown to be wider at each end with a narrowing in the middle (Miyazawa et al., 2003). Five TM regions line the channel with the 15 TM1, TM3, and TM4s (5 subunits × 3 TMs) surrounding the channel and separating the chan-nel from the membrane lipids (Fig. 1.5). Hydrophobic side chains stabilize the TM1, TM3, and TM4 interactions with the membrane, while TM2 lines the channel and is water-facing (Miyazawa et al., 2003). TM2 is thought to have minimal contact with the rest of the TMs. Labeling studies of specific residues have supported the overall structure of lipid and water-facing surfaces.

The TM2 helices are 40 Angstroms in length with rings of homologous residues bestowing distinct properties to various parts of the channel. The TM2 spans about 23 amino acids and the amino acids at the cytoplasmic end noted as $0'$ and the residues at the extracellular end as $20'$ (Miller & Smart, 2010). A negatively charged ring of aspartates is in the ECD vesti-bule about 15 A from the membrane that may concentrate cations (Miller & Smart, 2010). Two rings contain negatively charged groups, which would be expected to focus on cations present at both the intracellular and extracellular ends of the channel (Miyazawa et al., 2003). The extra-cellular ring at $20'$ is important for current rectification properties while the intracellular ring at $-1'$ is important for determining permeability to cat-ions vs anions (Miller & Smart, 2010). Another negative ring is just inside the pore on the intracellular side. When the channel is closed (not bound by ligand), hydrophobic residues (leucine at 251 and valine at 255) near the middle of the channel form a 3A opening that doesn't allow ion flow

(Albuquerque et al., 2009). This gate is located then at $9'$-$14'$ (Miller & Smart, 2010). Changes at the $9'$ leucine ring to more hydrophilic residues increase channel opening supporting the role of the hydrophobic $9'$ ring in closing the channel (Miller & Smart, 2010). This narrow opening presents an energy barrier to flow of Na^+ or K^+ ions. Binding of agonist leads to opening of the channel within microseconds. Thus, there needs to a physical connection through which changes in binding site structure such as closing of the C-loop are transmitted to the TM2 domains. As described previously, binding of agonist induces movement of the C-loop of about 11 Angstroms; the movement essentially traps the ligand in the binding site (Albuquerque et al., 2009). This movement is transmitted to TM2 via the Cys-loop and $\beta1$-$\beta2$ loop and results in a 15-degree rotation of TM2 (Miyazawa et al., 2003). Multiple studies have shown that the rigid movement of the Cys-loop is required for this allosteric effect linking ligand binding and channel opening. This rotation of the TM2 moves the hydrophobic ring to the side exposing a more hydrophilic environment (Miyazawa et al., 2003). Alpha subunit TM2s move first (Miller & Smart, 2010). Binding of two ligands required for opening would move two TM2s and provide greater destabilization of the gate than binding at just one site (Miller & Smart, 2010). This channel opening or widening was imaged by Unwin (1995) after a brief agonist exposure. This work showed that upon agonist binding, the hydrophobic residues are moved tangentially allowing polar residues to be exposed.

1.3.3 Intracellular domains

The intracellular domain between TM3 and TM4 is the most diverse in sequence and length and varies between each subunit in a given species. However, the sequence of a specific neuronal nAChR subunit (i.e., $\alpha4$) is conserved across species (Papke, 2014). This indicates that the variation in sequence of the intracellular domain of each subunit may have a defined and distinct role from that of other subunits, and that these specific functions have been maintained during evolution. A smaller cytoplasmic loop is present between TM1 and TM2 also. An alpha helix (MA) from each subunit of the cytoplasmic loop between TM3–TM4 contributes to the intracellular vestibule of the Torpedo nAChR (Unwin, 2005). The rest of the TM3–TM4 loop appears to be disordered (Unwin, 2005). The MA sequences contribute to five openings of about 8 A that are part of the intracellular vestibule, forming a type of filter allowing cations but not anions to enter (Unwin, 2005). The crystal structure of the $\alpha4\beta2$ nAChR was solved

(Morales-Perez et al., 2016), but most of the intracellular loop was deleted to form the crystal, so little information about the TM3-TM4 loop was available. A sequence similar to MA, MX is present in the α4β2 neuronal nAChR (Morales-Perez et al., 2016) and is predicted to form an α-helix as in Torpedo nAChRs. Most of the TM3-TM4 loop is disordered and hydrophilic in the α4β2 structure (Giastas et al., 2018).

The cytoplasmic loop of muscle nAChRs is important for clustering due to interactions with Rapsyn (Tsetlin et al., 2011), and multiple roles were shown for neuronal nAChR intracellular loops. When the cytoplasmic loop of the of α3 was substituted for the α7 loop of α7 nAChRs, the receptor protein was localized to the postsynaptic membrane in embryonic ciliary ganglion as occurs for α3 nAChR subunits, but with reduced surface expression (Williams et al., 1998). α7 nAChRs are normally localized perisynaptically in ciliary ganglion neurons. Substitution of the α5 loop in the α7 nAChR resulted in receptors that didn't localize to the postsynaptic membrane or have reduced the surface levels, although normally α5 nAChRs subunits in CG are found postsynaptically (Williams et al., 1998).

Mutations at specific serine residues S336A and S470A in mouse α4 subunits (analogous to the same sites in the human α4 nAChR subunits) reduced the normal increase in surface expression seen after nicotine exposure in subcortical and hippocampal neurons (Zambrano et al., 2019). A S530A mutation, also in the cytoplasmic loop, affected upregulation in subcortical neurons (diencephalon and hindbrain). Some of the mutations affected clustering as well (Zambrano et al., 2019).

Kracun et al. (2008) used loop chimeras, each containing a different loop from α1-α10 and β1-β4 subunits, with common ECD (α7) and TM (5-HT3A) domains to examine the effects of the cytoplasmic loop on function and localization. The receptors would therefore be α7 homomers, and surface expression could be monitored by using α-Bgt binding. These subunits were expressed in transfected tsA201 cells, not neurons, but revealed that levels of surface expression and intracellular assembly varied between loops with α1, α4, α7, and α8 having the highest surface expression (Kracun et al., 2008). The presence of distinct cytoplasmic loops also affected the channel properties and desensitization.

A novel role of the loop is in linking the ionotropic α7 nAChRs to a G-protein–coupled signaling pathway. Multiple proteins including G proteins have been found to interact with the TM3-TM4 cytoplasmic loop (Kabbani & Nichols, 2018). Multiple proteins have also been detected that bind to the β2 nAChR cytoplasmic loop (Kabbani et al., 2007). A specific sequence, G-protein-binding cluster (GPBC), was identified in the α7 loop and those of other nAChRs (Kabbani & Nichols, 2018; King et al., 2015). Conditions of agonist exposure that produce ionotropic α7 nAChR desensitization lead to activation of G protein signaling with subsequent IP3-induced Ca^{2+} release (Kabbani & Nichols, 2018). While the sequence of the loops in various subunits is known, the structure of this region is not known due to lack of ability to form crystals (Papke, 2014). The most we can say is that specific sequences described briefly above determine features of the subunit protein/nAChR including trafficking, assembly, and function of the channel for some subtypes.

References

Albuquerque, E. X., Pereira, E. F. R., Alkondon, M., & Rogers, S. W. (2009). Mammalian nicotinic acetylcholine receptors: From structure to function. *Physiological Reviews, 89*(1), 73–120. https://doi.org/10.1152/physrev.00015.2008.

Boulter, J., Evans, K., Goldman, D., Martin, G., Treco, D., Heinemann, S., & Patrick, J. (1986). Isolation of a cDNA clone coding for a possible neural nicotinic acetylcholine receptor alpha subunit. *Nature, 319*(6052), 368–374. https://doi.org/10.1038/319368a0.

Celie, P. H. N., Van Rossum-Fikkert, S. E., Van Dijk, W. J., Brejc, K., Smit, A. B., & Sixma, T. K. (2004). Nicotine and Carbmylcholine Binding to Nicotinic Receptors as Studied in AChB Crystal Structures. *Neuron, 41*(6), 907–914. https://doi.org/10.1016/S0896-6273(04)00115-1.

Changeux, J. P. (2012). The nicotinic acetylcholine receptor: The founding father of the pentameric ligand-gated Ion channel superfamily. *Journal of Biological Chemistry, 287*(48), 40207–40215. https://doi.org/10.1074/jbc.R112.407668.

Dani, J. A. (2015). Neuronal nicotine acetylcholine receptor structure and function and response to nicotine. *International Review of Neurobiology, 124*, 3–19. https://doi.org/10.1016/bs.irn.2015.07.001.

Giastas, P., Zouridakis, M., & Tzartos, S. J. (2018). Understanding structure-function relationships of the human neuronal acetylcholine receptor: Insights from the first crystal structures of neuronal subunits. *British Journal of Pharmacology, 175*(11), 1880–1891. https://doi.org/10.1111/bph.13838.

Görne-Tschelnokow, U., Strecker, A., Kaduk, C., Naumann, D., & Hucho, F. (1994). The transmembrane domains of the nicotinic acetylcholine receptor contain alpha-helical and beta structures. *EMBO Journal, 13*(2), 338–341. https://doi.org/10.1002/j.1460-2075.1994.tb06266.x.

Kabbani, N., & Nichols, R. A. (2018). Beyond the channel: Metabotropic signaling by nicotinic receptors. *Trends in Pharmacological Sciences, 39*(4), 354–366. https://doi.org/10.1016/j.tips.2018.01.002.

Kabbani, N., Woll, M., Levenson, R., Lindstrom, J., & Changeux, J.-P. (2007). Intracellular complexes of the β2 subunit of the nicotinic acetylcholine receptor in brain identified by proteomics. *Proceedings of the National Academy of Sciences of the United States of America*, 104.

King, J. R., Nordman, J. C., Bridges, S. P., Lin, M. K., & Kabbani, N. (2015). Identification and characterization of a G protein-binding cluster in α7 nicotinic acetylcholine receptors. *Journal of Biological Chemistry, 290*(33), 20060–20070. https://doi.org/10.1074/jbc.M115.647040.

Kracun, S., Harkness, P. C., Gibb, A. J., & Millar, N. S. (2008). Influence of the M3-M4 intracellular domain upon nicotinic acetylcholine receptor assembly, targeting and function. *British Journal of Pharmacology*, *153*(7), 1474–1484. https://doi.org/10.1038/sj. bjp.0707676.

Lukas, R. J., Changeux, J. P., Le Novère, N., Albuquerque, E. X., Balfour, D. J. K., Berg, D. K., … Wonnacott, S. (1999). International Union of Pharmacology. XX. Current status of the nomenclature for nicotinic acetylcholine receptors and their subunits. *Pharmacological Reviews*, *51*(2), 397–401. http://www.pharmrev.org.

Miller, P. S., & Smart, T. G. (2010). Binding, activation and modulation of Cys-loop receptors. *Trends in Pharmacological Sciences*, *31*(4), 161–174. https://doi.org/10.1016/j. tips.2009.12.005.

Miyazawa, A., Fujiyoshi, Y., & Unwin, N. (2003). Structure and gating mechanism of the acetylcholine receptor pore. *Nature*, *423*(6943), 949–955. https://doi.org/10.1038/ nature01748.

Morales-Perez, C. L., Noviello, C. M., & Hibbs, R. E. (2016). X-ray structure of the human α4β2 nicotinic receptor. *Nature*, *538*(7625), 411–415. https://doi.org/10.1038/ nature19785.

Noda, M., Furutani, Y., Takahashi, H., Toyosato, M., Tanabe, T., Shimizu, S., … Numa, S. (1983). Cloning and sequence analysis of calf cDNA and human genomic DNA encoding alpha-subunit precursor of muscle acetylcholine receptor. *Nature*, *305* (5937), 818–823. https://doi.org/10.1038/305818a0.

Papke, R. L. (2014). Merging old and new perspectives on nicotinic acetylcholine receptors. *Biochemical Pharmacology*, *89*(1), 1–11. https://doi.org/10.1016/j.bcp.2014.01.029.

Rayes, D., De Rosa, M. J., Sine, S. M., & Bouzat, C. (2009). Number and locations of agonist binding sites required to activate homomeric cys-loop receptors. *Journal of Neuroscience*, *29*(18), 6022–6032. https://doi.org/10.1523/JNEUROSCI.0627-09.2009.

Rucktooa, P., Smit, A. B., & Sixma, T. K. (2009). Insight in nAChR subtype selectivity from AChBP crystal structures. *Biochemical Pharmacology*, *78*(7), 777–787. https://doi.org/ 10.1016/j.bcp.2009.06.098.

Smit, A. B., Brejc, K., Syed, N., & Sixma, T. K. (2003). Structure and function of AChBP, homologue of the ligand-binding domain of the nicotinic acetylcholine receptor. *Annals of the New York Academy of Sciences*, *998*, 81–92. https://doi.org/10.1196/ annals.1254.010.

Tsetlin, V., Kuzmin, D., & Kasheverov, I. (2011). Assembly of nicotinic and other Cys-loop receptors. *Journal of Neurochemistry*, *116*(5), 734–741. https://doi.org/10.1111/j.1471-4159.2010.07060.x.

Unwin, N. (1995). Acetylcholine receptor channel imaged in the open state. *Nature*, *373* (6509), 37–43. https://doi.org/10.1038/373037a0.

Unwin, N. (2003). Structure and action of the nicotinic acetylcholine receptor explored by electron microscopy. *FEBS Letters*, *555*(1), 91–95. https://doi.org/10.1016/S0014-5793(03)01084-6.

Unwin, N. (2005). Refined structure of the nicotinic acetylcholine receptor at 4A resolution. *Journal of Molecular Biology*, *346*(4), 967–989. https://doi.org/10.1016/j.jmb.2004.12.031.

Williams, B. M., Temburni, M. K., Levey, M. S., Bertrand, S., Bertrand, D., & Jacob, M. H. (1998). The long internal loop of the α3 subunit targets nAChRs to subdomains within individual synapses on neurons in vivo. *Nature Neuroscience*, *1*(7), 557–562. https://doi. org/10.1038/2792.

Zambrano, C. A., Escobar, D., Ramos-Santiago, T., Bollinger, I., & Stitzel, J. (2019). Serine residues in the α4 nicotinic acetylcholine receptor subunit regulate surface α4β2* receptor expression and clustering. *Biochemical Pharmacology*, *159*, 64–73. https://doi.org/ 10.1016/j.bcp.2018.11.008.

Zoli, M., Pistillo, F., & Gotti, C. (2015). Diversity of native nicotinic receptor subtypes in mammalian brain. *Neuropharmacology*, *96*, 302–311. https://doi.org/10.1016/j. neuropharm.2014.11.003.

CHAPTER TWO

Neuronal nAChR localization, subtype diversity, and evolution

2.1 Subtype diversity

As was demonstrated for the neuromuscular nAChRs, those expressed in the nervous system are all pentamers and members of the Cys-loop family. Eight alpha subunit genes designated α2–7, α9, and α10 have been isolated from humans and several other species. α8 has been detected in chick and zebrafish. Three beta subunits β2–β4 have also been similarly isolated. Similar related subunit genes have been detected in various species (Pedersen et al., 2019). All nAChRs except for homomeric receptors express distinct combinations of alpha and beta subunits (except for α9α10 heteromers), while homomeric nAChRs are composed of only alpha subunits (α7, α8, and α9). Each pentameric combination is defined as a subtype. Each subtype can express a unique pharmacology and biophysical properties of the channel (covered in more detail elsewhere). Each subtype has a distinct pattern of expression and also overlaps with other subtypes. In addition, various subtypes are also associated with specific functions. The major subtypes of neuronal nAChRs initially defined were the widely expressed α4β2 nAChR, a major nicotine-binding receptor, the α7 nAChRs that bind alpha bungarotoxin (α-Bgt) and the α3β4 nAChR subtype considered as a "ganglionic" receptor. With the development of molecular cloning, binding, knockout mice (KO), and immunochemistry techniques, numerous neuronal subtypes were identified. As more genes were identified, it was realized that more subtypes may be present than was previously thought.

2.2 Subunit genes

Table 2.1 summarizes the known genes expressed in humans. For more details of the evolution of these genes and other vertebrate nAChR genes, see the excellent paper of Pedersen et al. (2019). The human genes

Table 2.1 Human neuronal nAChR genes.

Gene	Chromosome	Exons
α2	8	6
α3	15	6
α4	20	6
α5	15	6
α6	8	6
α7	15	10
α9	4	5
α10	11	5
β2	1	6
β3	8	6
β4	15	6

Localization and number of exons in human nAChR genes.

evolved from primordial ancestors as a result of two tetraploidization events. The most closely related genes share the same number of exons, which often have similar lengths (Pedersen et al., 2019). An exception occurs with the α4 cytoplasmic region encoded in exon 5 that is significantly longer than in other subunits. Subtypes that express homologous subunits such as α2/ α4, β3/α5, α6/β3, or β2/β4 express one of the pair, but usually not both. An alignment of human nAChR proteins (Table 2.2) using ClustalM (Madeira et al., 2019) shows a high level of sequence identity for all subunits, but especially for some pairs such as α2/α4, β3/α5, or α6/α3.

The genes for nAChRs are highly conserved in vertebrates, Fig. 2.1. Even across species as diverse as human, cat, bat, zebrafish, turtle, lizard,

Table 2.2 Protein identity among human neuronal nAChR genes.

	α9	α10	α7	β2	β4	α5	β3	α2	α4	α3	α6
α9	100	55	38	33	34	35	35	36	35	36	36
α10	55	100	41	36	37	35	36	37	38	37	35
α7	38	41	100	37	37	36	35	40	39	39	37
β2	33	36	37	100	65	42	44	51	53	49	48
β4	34	37	37	65	100	43	45	51	50	50	46
α5	35	35	36	42	43	100	68	52	53	48	48
β3	35	36	35	44	45	68	100	56	54	48	48
α2	36	37	40	51	51	52	56	100	69	59	55
α4	35	38	39	53	50	53	54	69	100	58	55
α3	36	37	39	49	50	48	48	59	58	100	66
α6	36	35	37	48	46	48	48	55	55	66	100

% Amino acid identity.

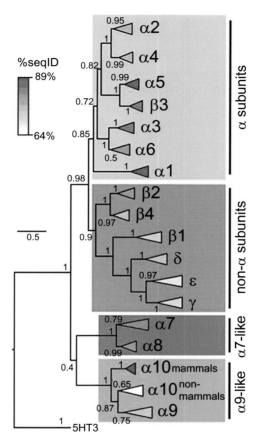

Fig. 2.1 Subtype genes phylogenetic relationships. Phylogenetic relationships between vertebrate nicotinic subunits. The branches corresponding to the same subunits of different species were collapsed up to the node at which one subunit separates from its closest neighbor. Triangle lengths denote the divergence on sequence identity from the subunit node. Triangles were colored according to the average percentage of sequence identity between all pairs of sequences within the branch. Shadings denote the different groups of subunits. Light greens, α subunits; dark green, non-α subunits; purple, o7-like subunits; orange, α9-like subunits. Numbers in branches indicate the bootstrap value obtained after 1000 replicates. Scale bar indicates the number of amino acid substitutions per site. *(From Marcovich, I., Moglie, M. J., Freixas, A. E. C., Trigila, A. P., Franchini, L. F., Plazas, P. V., Lipovsek, M., & Elgoyhen, A. B. (2019). Distinct evolutionary trajectories of neuronal and hair cell nicotinic acetylcholine receptors. bioRxiv. https://doi.org/10.1101/621342.)*

and alligators, the amino sequence identities are high, usually greater than 70% for any given subunit protein. Given that many combinations are seen in widespread locations, there appears to be a strong negative selection pressure to prevent major sequence changes (Marcovich et al., 2019; Pedersen et al., 2019). An exception maybe the α9 and α10 subunits expressed in hair cells (Marcovich et al., 2019) which show divergence from other vertebrate nAChRs. This may be a reflection of the limited expression of nAChRs composed of α9 and α10 subunits and divergent functional and pharmacological properties.

Many of these genes are also present in rat, mice, primates, and chick as shown above, while some genes (such those containing α8 or its orthologs) may be present only in other species such as chick or zebrafish. More subtypes could be as yet found as the exact combination of some subtypes (those with*) may contain unidentified subunits (Lukas et al., 1999). β4-containing nAChRs may be less common in humans than in mice or rats (Zoli et al., 2015). Not all species have the same distribution of all subtypes, which should be a consideration when choosing an animal model for research involving nAChRs.

2.3 Subtype assemblies

Given the potential combinations of subunit adding up to five, thousands of subtypes are mathematically possible. Over the years, immunochemistry, radioactive ligand binding, functional assays, and PCR have been used to identify and isolate nAChR subtypes (Zoli et al., 2015). These studies have been conducted using receptors isolated in vivo or reconstructed in oocytes or cell lines and give clues as to how many nAChR subtypes are actually found in the nervous system. It should be noted that less receptors may actually be expressed in vivo than have been found in in vitro systems (Zoli et al., 2015). The subunits must be expressed in the same cell and have sequences that allow assembly and trafficking of a specific subunit. Identification has been complicated by the poor subunit specificity of many anti-nAChR antibodies (Moser et al., 2007). In general, there are few subtype-specific ligands. Some combinations have not been defined as additional subunits may be present, which have not been identified. These are often presented in the literature as α4β2* or α3β4* to indicate that other subunits may be present (Lukas et al., 1999). While many subunits RNAs/proteins have been identified to be co-expressed in various brain regions and even individual neurons, it is difficult to know for sure whether all of the subtypes expressed in vitro from these subunits are present in high amounts in these neurons. Table 2.3 summarizes combinations assembled

Table 2.3 Pentameric nAChR subtypes.

Subtype	Method(s) of analysis	Reference
Chick α7	Oocyte electrophysiology	Couturier et al. (1990)
Chick α8	Oocyte physiology, α-Bgt binding	Gerzanich et al. (1994)
Rat α9	Oocyte electrophysiology	Elgoyhen et al. (1994)
Chick α10	Oocyte electrophysiology	Lipovsek et al. (2014)
Rat α2β2	Oocyte electrophysiology	Deneris et al. (1988)
Rat α2β4	Oocyte electrophysiology	Duvoisin et al. (1989)
Rat α3β2	Oocyte electrophysiology	Boulter et al. (1987)
Rat α3β4	Oocyte electrophysiology	Duvoisin et al. (1989)
Rat α4β2	Oocyte electrophysiology	Boulter et al. (1987)
Rat α4β4	Oocyte electrophysiology	Duvoisin et al. (1989)
Chick α6β2,α6β4	BOSC 23 cells, electrophysiology, Ca^{2+} imaging	Fucile et al. (1998)
Human α6β4, α6β2, α6β4β3	HEK cells, immunoprecipitation, binding	Tumkosit et al. (2006)
Rat α7β2	Oocytes and HEK cells, electrophysiology, co-precipitation	Khiroug et al. (2002)
Chick α7β3	Oocyte electrophysiology, α-Bgt binding	Palma et al. (1999)
Human α7β4	Oocyte electrophysiology, α-Bgt binding, immunopurification	Criado et al. (2012)
Rat α9α10, α10	Oocyte electrophysiology	Elgoyhen et al. (2001)
Chick α2α5β2	Retina, optic tectum extracts BOSC23 cells, binding, co-precipitation	Vailati et al. (2003)
Rat α2α6β2	Rat LGN, SC binding, KO mice, immunoprecipitation	Gotti et al. (2005)
Human α2β2β3	Oocyte electrophysiology	Dash et al. (2012)
Human α2β3β4	Oocyte electrophysiology	Dash et al. (2012)
Rat α3α4β2	Rat cerebellum, immunoprecipitation binding, RNA probes	Turner and Kellar (2005)
Rat α3α4β4	Rat cerebellum, immunoprecipitation binding, RNA probes	Turner and Kellar (2005)
α4β2β4		

Continued

Table 2.3 Pentameric nAChR subtypes—cont'd

Subtype	Method(s) of analysis	Reference
Human α3α5β2	Oocyte electrophysiology, SH-SY5Y surface labeling, immunoprecipitation	Wang et al. (1996)
Human α3α5β4	Oocyte electrophysiology, SH-SY5Y surface labeling, immunoprecipitation	Wang et al. (1996)
Human α3α6β2	Oocyte electrophysiology, binding, immunoprecipitation	Kuryatov et al. (2000)
Chick α3α6β4	BOSC23 cells, patch clamp, Ca^{2+} imaging	Fucile et al. (1998)
Rat α3β2β4	Rat cerebellum, immunoprecipitation binding, RNA probes	Turner and Kellar (2005)
Human α3β2β3	Oocyte electrophysiology	Dash et al. (2012)
Human α3β3β4	Oocyte electrophysiology	Groot-Kormelink et al. (1998)
Chick α4α5β2	Oocyte electrophysiology SCG antisense knockdown	Ramirez-Latorre et al. (1996)
Human α4β2β3	cDNAs stably expressed in HEK cells epibatidine binding, Ca^{2+} flux assay	Kuryatov et al. (2008)
Human α4β3β4	Oocyte electrophysiology	Dash et al. (2012)
Human α5α6β2	Oocyte electrophysiology binding, immunoprecipitation	Kuryatov et al. (2000)
α6β2β4, α6β2β3α5		
Human α6β2β3	HEK cells epibatidine binding, immunoprecipitation	Tumkosit et al. (2006)
Human α6β3β4	Oocyte electrophysiology binding, immunoprecipitation	Kuryatov et al. (2000)
Rat α3α4α6β2 α4α6β2β3, α3α6β2 α3β2β3, α6β2β3	Optic nerve, immunoprecipitation	Cox et al. (2008)
Chick α3α5β2β4	Ciliary ganglion, co-immunoprecipitation	Conroy and Berg (1995)
Human α4α6β2β3	Oocyte, electrophysiology, concatamer expression, antibody binding	Kuryatov and Lindstrom (2011)
Human α5α6β3β4	Cell line expression in vivo lesioning, binding	Grinevich et al. (2005)

Adapted from Marcovich, I., Moglie, M. J., Freixas, A. E. C., Trigila, A. P., Franchini, L. F., Plazas, P. V., Lipovsek, M., & Elgoyhen, A. B. (2019). Distinct evolutionary trajectories of neuronal and hair cell nicotinic acetylcholine receptors. *bioRxiv*. https://doi.org/10.1101/621342.

in vitro, detected in vivo by binding or immunoprecipitation or show function (noted in Table 2.3). Some of these were also co-immunoprecipitated to show assembly in vivo. Human, rat, and chick subtypes are noted in the table. Many of these subtypes are also detected in vivo in various brain regions in numerous species.

2.4 Subtypes present in vivo

Many of the subtypes assembled in vitro have also been detected in vivo. α3β4 nAChRs are now found not only in autonomic ganglia but also in medial habenula (MHb), interpeduncular nucleus (IPN), retina, spinal cord, hippocampus, and thalamus (Gotti & Clementi, 2004). Some of these also contain α5. α6β3* nAChRs are found in locus coeruleus, ventral tegmental area (VTA), substantia nigra (SN), and retina (Gotti & Clementi, 2004). α7 and α4β2* nAChRs are widely expressed, while α9 and α9/α10 nAChRs are present in cochlea and a few other areas (Gotti & Clementi, 2004). Additional subtypes can be formed of varied stoichiometry of the same subunits. For example, α4β2 nAChRs can contain (α4)2 (β2)3 or (α4)3(β2)2 or (α4)2(β2)2α5 (Albuquerque et al., 2009). Each of these has different biophysical properties. One stoichiometry may be present in one cell type, the other in another ,or both may even be present in the same cells.

Many subtypes are present in specific cells or brain regions while others demonstrate co-expression of the genes in the same cells, making it possible or even likely that the subtypes are expressed in these neurons. Table 2.4 summarizes assembled subtypes detected in vivo. While this list is extensive,

Table 2.4 Assembled nAChRs detected in vivo.

Receptor	Localization	Reference
Mouse α3β4α5	MHb-IPN pathway	Fowler, Lu, Johnson, Marks, and Kenny (2011)
Rat α3β2*	Retinal afferents in LGN	Gotti et al. (2005)
Rat α3β2*	Retinal afferents in SC	Gotti et al. (2005)
Rat α4α6β2β3	Retinal afferents in LGN	Gotti et al. (2005)
Rat α4α6β2β3	Retinal afferents in SC	Gotti et al. (2005)
Rat α6β2β3	Retinal afferents in LGN	Gotti et al. (2005)
Rat α6β2β3	Retinal afferents in SC	Gotti et al. (2005)
Rat α2α6β2	Retinal afferents in LGN	Gotti et al. (2005)
Rat α4β2	Retinal afferents in LGN	Gotti et al. (2005)
Rat α4β2	Neurons in LGN	Gotti et al. (2005)

Continued

Table 2.4 Assembled nAChRs detected in vivo—cont'd

Receptor	Localization	Reference
Rat α4β2	Neurons in SC	Gotti et al. (2005)
Rat α4α5β2	Neurons in LGN	Gotti et al. (2005)
Rat α4α5β2	Neurons in SC	Gotti et al. (2005)
Rat α7	Neurons in LGN	Gotti et al. (2005)
Rat α7	Neurons in SC	Gotti et al. (2005)
Mouse α4β2*	Widely expressed multiple brain regions	Baddick and Marks (2011)
Mouse α7	Widely expressed multiple brain regions	Baddick and Marks (2011)
Rodent α3β4*	Autonomic ganglia thalamus, hippocampus, spinal cord, retina	Gotti and Clementi (2004)
Rat α4β2*	Striatum	Gotti and Clementi (2004)
Rat α4α5β2	Striatum	Gotti and Clementi (2004)
Rat α6α4β2β3	Striatum	Gotti and Clementi (2004)
Rat α6β2β3	Striatum	Gotti and Clementi (2004)
Chick α3β4	CG	Conroy and Berg (1995)
Chick α3α5β4	CG	Conroy and Berg (1995)
Chick α3α5β2β4	CG	Conroy and Berg (1995)
Chick α7	CG	Conroy and Berg (1995)
Rat α4β2	Cerebellum	Turner and Kellar (2005)
Rat α3β2β4	Cerebellum	Turner and Kellar (2005)
Rat α3β4	Cerebellum	Turner and Kellar (2005)
Rat α3α4β2	Cerebellum	Turner and Kellar (2005)
Rat α4β2β4	Cerebellum	Turner and Kellar (2005)
Rat α3α4β4	Cerebellum	Turner and Kellar (2005)
Rat α7	Hippocampus	Albuquerque et al. (2009)
Human α7	Cortex	Albuquerque et al. (2009)
Rat α7	Striatum	Albuquerque et al. (2009)
Rat α7	Olfactory bulb	Albuquerque et al. (2009)
Rat α7	VTA	Albuquerque et al. (2009)
Rat α7	Medial habenula	Albuquerque et al. (2009)
Rat α7	Frontal cortex	Albuquerque et al. (2009)
Rat α9α10	Olfactory epithelium, cochlea	Gotti and Clementi (2004)
Human α9,α9/α10	Cochlear hair cells	Marcovich et al. (2019)

CG, ciliary ganglion; *IPN*, interpeduncular nucleus; *LGN*, lateral geniculate nucleus; *MHb*, medial habenula; *SC*, superior colliculus; *VTA*, ventral tegmental area.

it is not exhaustive. Detection methods used included binding with subtype-selective compounds, in situ hybridization, PCR, immunoprecipitation, and slice electrophysiology. Figs. 2.2 and 2.3 illustrate nAChRs RNAs co-expressed in the same regions and the potential subtypes present (from Marcovich et al., 2019).

This list of in vivo subtypes in Tables 2.3 and 2.4 and Figs. 2.2 and 2.3 may not be exhaustive, but makes the point that multiple varied subtypes shown to have diverse functional and pharmacological properties are present throughout the nervous system. Subtypes found in rat, mice, or primates are likely to be expressed in analogous regions in humans. Many of these combinations shown in Table 2.4 are functional in vivo as in vitro (see Table 2.3). nAChRs present in the VTA and nucleus accumbans (NAc) are described in the nicotine addiction chapter. The nAChRs may not function exactly as in vitro due to associations with unknown subunits or additional endogenous modulation by Ca^{2+} or allosteric molecules such as Lynx proteins.

2.5 Regional localization

While nAChR subunits are widely expressed such as α4β2 and α7, some nAChR subtypes are expressed in a specific regional pattern. Other regions such as the MHb-IPN pathway express all known heteromeric receptors to some degree or other (Zoli et al., 2015). Multiple methods have been used to identify subtypes, and Table 2.3 shows some that have been identified. Subtype expression patterns have been determined for primates, mice, rats, and humans in some detail. A comparison of rodent, human, and primate brain distribution is shown Fig. 2.4, and Fig. 2.5 from Gotti, Zoli, & Clementi (2006); Zoli et al. (2015). Some subtype localization has been conserved. In rat cerebellum, 35% of the nAChRs are α4β2 subtype, 20% α3β4 subtype, and 10% contain α3α4β2, α3β2β4, and α3α4β4 (Zoli et al., 2015). Mice have a similar distribution, but β4 has not been detected so far in human cerebellum. Monkeys, humans, and rodents all share expression of α4β2* and α6β3β2* in the striatum (Zoli et al., 2015). A major difference occurs in α2 distribution, which in rodents is highly expressed in the IPN, but low elsewhere. However, α2 is highly and widely expressed in rhesus macaques. Human expression of α2 may be higher than in rodents (Zoli et al., 2015). α7β2 nAChRs are present in mice and human basal forebrain and may be involved in the pathology of Alzheimer's Disease, Fig. 2.5.

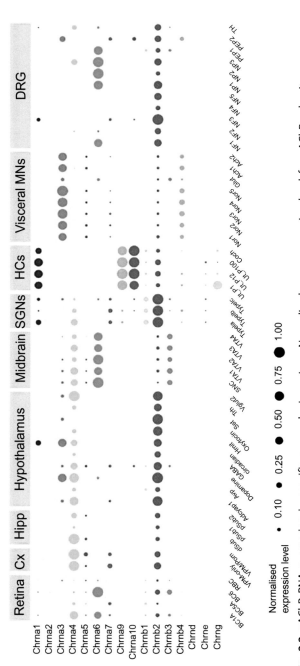

Fig. 2.2 nAChR RNA expression in specific mouse brain regions. Normalized mean expression level for nAChR subunits across mouse neuronal and sensory cell types. Circle sizes indicate the mean expression level for each cell type, normalized to the highest value observed within each data set. A large number of neuronal populations and brain regions were analyzed (along top with specific cells and regions along bottom). (From Marcovich, I., Moglie, M. J., Freixas, A. E. C., Trigila, A. P., Franchini, L. F., Plazas, P. V., Lipovsek, M., & Elgoyhen, A. B. (2019). *Distinct evolutionary trajectories of neuronal and hair cell nicotinic acetylcholine receptors. bioRxiv. https://doi.org/10.1101/621342.*)

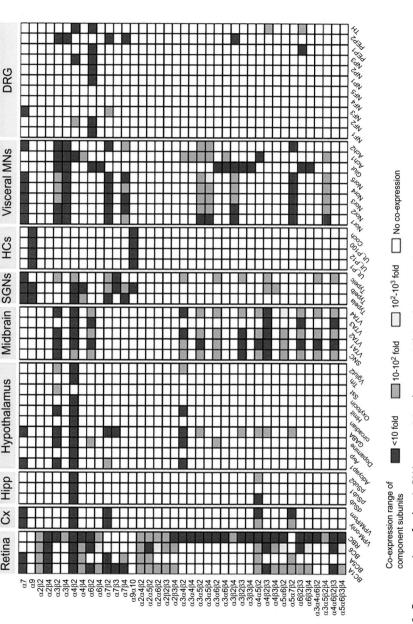

Fig. 2.3 Co-expression of subunit RNAs comprising known nAChR subtypes. Dark red squares, all component subunits are co-expressed within a 10-fold range of expression level. Light red squares, all component subunits are co-expressed within a 100-fold range of expression level. Pink squares, all component subunits are expressed within a 1000-fold range of expression level. White squares, at least one subunit of that receptor assembly in not expressed in that cell type. *(From Marcovich, I., Moglie, M. J., Freixas, A. E. C., Trigila, A. P., Franchini, L. F., Plazas, P. V., Lipovsek, M., & Elgoyhen, A. B. (2019). Distinct evolutionary trajectories of neuronal and hair cell nicotinic acetylcholine receptors. bioRxiv. https://doi.org/10.1101/621342.)*

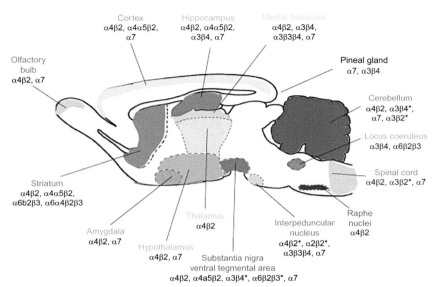

Fig. 2.4 nAChR subtype expression in rodent brain. The subtypes present in the cortex, cerebellum, hippocampus, interpeduncular nucleus, medial habenula, and pineal gland have been identified by binding, immunoprecipitation, and/or immunopurification assays in tissue from rat and/or wild-type and/or receptor subunit knockout mice. The subtypes present in the amygdala, hypothalamus, locus coeruleus, olfactory bulb, raphe nuclei, spinal cord, substantia nigra-ventral tegmental area, and thalamus have been deduced from in situ hybridization, single-cell PCR, and binding studies of tissues obtained from rat and/or wild-type and/or knockout mice. *(From Gotti, C., Zoli, M., & Clementi, F. (2006). Brain nicotinic acetylcholine receptors: Native subtypes and their relevance. Trends in Pharmaceutical Sciences, 27(9).)*

2.6 Cellular localization

nAChRs are located in multiple places on neurons, Fig. 2.6. These include soma, dendrites, presynaptic axon terminals, and other locations on axons. Some are at synapses and others extrasynaptically. nAChRs can be found in multiple locations on the same cell. A neuron may express multiple subtypes, which may be localized to one or more locations. Numerous electrophysiological studies were done to localize functional nAChRs on neurons (Albuquerque et al., 2009). These include outside–out patch recording to localize α7 nAChRs to the soma, expression of labeled subunits, and release of caged transmitters to localize to dendrites (Albuquerque et al., 2009). Slice recordings of various brain regions

(A)

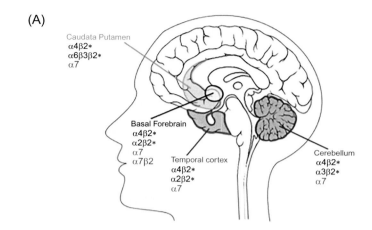

Caudata Putamen
α4β2*
α6β3β2*
α7

Basal Forebrain
α4β2*
α2β2*
α7
α7β2

Temporal cortex
α4β2*
α2β2*
α7

Cerebellum
α4β2*
α3β2*
α7

(B)

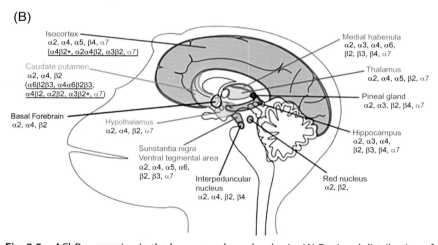

Isocortex
α2, α4, α5, β4, α7
(α4β2*, α2α4β2, α3β2, α7)

Caudate putamen
α2, α4, β2
(α6β2β3, α4α6β2β3,
α4β2, α2β2, α3β2*, α7)

Basal Forebrain
α2, α4, β2

Hypothalamus
α2, α4, β2, α7

Sunstantia nigra
Ventral tegmental area
α2, α4, α5, α6,
β2, β3, α7

Interpeduncular
nucleus
α2, α4, β2, β4

Medial habenula
α2, α3, α4, α6,
β2, β3, β4, α7

Thalamus
α2, α4, α5, β2, α7

Pineal gland
α2, α3, β2, β4, α7

Hippocampus
α2, α3, α4,
β2, β3, β4, α7

Red nucleus
α2, β2,

Fig. 2.5 nAChR expression in the human and monkey brain. (A) Regional distribution of nAChR subtypes identified in human brain. The subtypes in the temporal cortex, cerebellum, striatum and basal forebrain have been identified by means of quantitative immunoprecipitation studies using nAChRs obtained from postmortem brains, radiolabeled with 3H-Epibatine or 125I-α-Bungarotoxin, and subunit specific antibodies. (B) Regional distribution of monkey nAChR subunit mRNAs. The localization of the mRNA of the subunits is based on the results of in situ hybridization or single-cell PCR experiments. The striatum and cortex subtypes identified by means of quantitative immunoprecipitation studies are also shown (underlined). *(From Zoli, M., Pistillo, F., & Gotti, C. (2015). Diversity of native nicotinic receptor subtypes in mammalian brain. Neuropharmacology, 96, 302–311. https://doi.org/10.1016/j.neuropharm.2014.11.003.)*

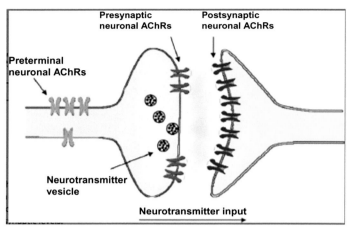

Fig. 2.6 Cellular Localization of nAChRs. nAChRs are located presynaptically, preterminally, or postsynaptically on cell bodies or dendrites. *(From Kalamida, D., Poulas, K., Avramopoulou, V., Fostieri, E., Lagoumintzis, G., Lazaridis, K., Sideri, A., Zouridakis, M., & Tzartos, S. J. (2007). Muscle and neuronal nicotinic acetylcholine receptors. FEBS Journal, 274(15), 3799–3845. https://doi.org/10.1111/j.1742-4658.2007.05935.x.)*

including the hippocampus have been used. Many somatodendritic nAChRs are in fact extrasynaptic. Somatic and dendritic nAChRs can mediate fast synaptic transmission. nAChRs can also be activated by volume transmission.

Presynaptic axon (terminal) expresses nAChRs of multiple subtypes. A major role of nAChRs in the CNS is via presynaptic signaling controlling the release of neurotransmitters (Nashmi & Lester, 2006). For example, α7 nAChRs are expressed on glutamatergic terminals in multiple brain regions and α6-containing nAChRs on dopaminergic terminals (Albuquerque et al., 2009). nAChRs are also expressed in preterminal regions on neurons including those releasing serotonin, GABA, norepinephrine, and acetylcholine. Presynaptic nAChRs mediate neurotransmitter release through mediating Ca^{2+} influx or local depolarization (Nashmi & Lester, 2006). Neurons expressing α3, β3, and β4 subunits are usually involved in ACh release (ChAT +) while neurons expressing α4 and α6 subunits are involved in the release of other neurotransmitters (Zoli et al., 2015). An example of the complex distribution of nAChRs in seen in the VTA and SN (Fig. 2.4, Albuquerque et al., 2009) where multiple subtypes are observed presynaptically as well as somatodendritically. nAChRs are even expressed in myelinated axons of retinal ganglion cells (Albuquerque et al., 2009).

2.7 Summary

Multiple combinations of subunits are present in many brain areas and locations on neurons. Many functional subtypes have been identified and localized in vivo and functionally analyzed in vitro and in vivo. The multiplicity of subtypes and locations can seem overwhelming, but even this brief review demonstrates the complexity of nAChR expression. This diversity provides for unique signaling possibilities that promote flexibility and changing responses to the demands of the cholinergic system.

References

Albuquerque, E. X., Pereira, E. F. R., Alkondon, M., & Rogers, S. W. (2009). Mammalian nicotinic acetylcholine receptors: From structure to function. *Physiological Reviews, 89*(1), 73–120. https://doi.org/10.1152/physrev.00015.2008.

Baddick, C. G., & Marks, M. J. (2011). An autoradiographic survey of mouse brain nicotinic acetylcholine receptors defined by null mutants. *Biochemical Pharmacology, 82*(8), 828–841. https://doi.org/10.1016/j.bcp.2011.04.019.

Boulter, J., Connolly, J., Deneris, E., Goldman, D., Heinemann, S., & Patrick, J. (1987). Functional expression of two neuronal nicotinic acetylcholine receptors from cDNA clones identifies a gene family. *Proceedings of the National Academy of Sciences of the United States of America, 84*(21), 7763–7767. https://doi.org/10.1073/pnas.84.21.7763.

Conroy, W. G., & Berg, D. K. (1995). Neurons can maintain multiple classes of nicotinic acetylcholine receptors distinguished by different subunit compositions. *Journal of Biological Chemistry, 270*(9), 4424–4431. https://doi.org/10.1074/jbc.270.9.4424.

Couturier, S., Bertrand, D., Matter, J. M., Hernandez, M. C., Bertrand, S., Millar, N., Valera, S., Barkas, T., & Ballivet, M. (1990). A neuronal nicotinic acetylcholine receptor subunit (α7) is developmentally regulated and forms a homo-oligomeric channel blocked by α-BTX. *Neuron, 5*(6), 847–856. https://doi.org/10.1016/0896-6273(90)90344-F.

Cox, B. C., Marritt, A. M., Perry, D. C., & Kellar, K. J. (2008). Transport of multiple nicotinic acetylcholine receptors in the rat optic nerve: High densities of receptors containing α6 and β3 subunits. *Journal of Neurochemistry, 105*(5), 1924–1938. https://doi.org/10.1111/j.1471-4159.2008.05282.x.

Criado, M., Valor, L. M., Mulet, J., Gerber, S., Sala, S., & Sala, F. (2012). Expression and functional properties of α7 acetylcholine nicotinic receptors are modified in the presence of other receptor subunits. *Journal of Neurochemistry, 123*(4), 504–514. https://doi.org/10.1111/j.1471-4159.2012.07931.x.

Dash, B., Bhakta, M., Chang, Y., & Lukas, R. J. (2012). Modulation of recombinant, α2*, α3* or α4*-nicotinic acetylcholine receptor (nAChR) function by nAChR β3 subunits. *Journal of Neurochemistry, 121*(3), 349–361. https://doi.org/10.1111/j.1471-4159.2012.07685.x.

Deneris, E. S., Connolly, J., Boulter, J., Wada, E., Wada, K., Swanson, L. W., Patrick, J., & Heinemann, S. (1988). Primary structure and expression of β2: A novel subunit of neuronal nicotinic acetylcholine receptors. *Neuron, 1*(1), 45–54. https://doi.org/10.1016/0896-6273(88)90208-5.

Duvoisin, R. M., Deneris, E. S., Patrick, J., & Heinemann, S. (1989). The functional diversity of the neuronal nicotinic acetylcholine receptors is increased by a novel subunit: β4. *Neuron, 3*(4), 487–496. https://doi.org/10.1016/0896-6273(89)90207-9.

Elgoyhen, A. B., Johnson, D. S., Boulter, J., Vetter, D. E., & Heinemann, S. (1994). α9: An acetylcholine receptor with novel pharmacological properties expressed in rat cochlear hair cells. *Cell*, *79*(4), 705–715. https://doi.org/10.1016/0092-8674(94)90555-X.

Elgoyhen, A. B., Vetter, D. E., Katz, E., Rothlin, C. V., Heinemann, S. F., & Boulter, J. (2001). α10: A determinant of nicotinic cholinergic receptor function in mammalian vestibular and cochlear mechanosensory hair cells. *Proceedings of the National Academy of Sciences of the United States of America*, *98*(6), 3501–3506. https://doi.org/10.1073/pnas.051622798.

Fowler, C. D., Lu, Q., Johnson, P. M., Marks, M. J., & Kenny, P. J. (2011). Habenular α5 nicotinic receptor subunit signalling controls nicotine intake. *Nature*, *471*(7340), 597–601. https://doi.org/10.1038/nature09797.

Fucile, S., Matter, J., Erkman, L., Ragozzino, D., Barabino, B., Grassi, F., Alemà, S., Ballivet, M., & Eusebi, F. (1998). The neuronal α 6 subunit forms functional heteromeric acetylcholine receptors in human transfected cells. *European Journal of Neuroscience*, *10*(1), 172–178. https://doi.org/10.1046/j.1460-9568.1998.00001.x.

Gerzanich, V., Anand, R., & Lindstrom, J. (1994). Homomers of α8 and α7 subunits of nicotinic receptors exhibit similar channel but contrasting binding site properties. *Molecular Pharmacology*, *45*(2), 212–220.

Gotti, C., & Clementi, F. (2004). Neuronal nicotinic receptors: From structure to pathology. *Progress in Neurobiology*, *74*(6), 363–396. https://doi.org/10.1016/j.pneurobio.2004.09.006.

Gotti, C., Moretti, M., Zanardi, A., Gaimarri, A., Champtiaux, N., Changeux, J. P., Whiteaker, P., Marks, M. J., Clementi, F., & Zoli, M. (2005). Heterogeneity and selective targeting of neuronal nicotinic acetylcholine receptor (nAChR) subtypes expressed on retinal afferents of the superior colliculus and lateral geniculate nucleus: Identification of a new native nAChR subtype α3β(α5 or β3) enriched in retinocollicular afferents. *Molecular Pharmacology*, *68*(4), 1162–1171. https://doi.org/10.1124/mol.105.015925.

Gotti, C., Zoli, M., & Clementi, F. (2006). Brain nicotinic acetylcholine receptors: Native subtypes and their relevance. *Trends in Pharmaceutical Sciences*, *27*(9).

Grinevich, V. P., Letchworth, S. R., Lindenberger, K. A., Menager, J., Mary, V., Sadieva, K. A., Buhlman, L. M., Bohme, G. A., Pradier, L., Benavides, J., Lukas, R. J., & Bencherif, M. (2005). Heterologous expression of human α6β4β3α5 nicotinic acetylcholine receptors: Binding properties consistent with their natural expression require quaternary subunit assembly including the α5 subunit. *Journal of Pharmacology and Experimental Therapeutics*, *312*(2), 619–626. https://doi.org/10.1124/jpet.104.075069.

Groot-Kormelink, P. J., Luyten, W. H. M. L., Colquhoun, D., & Sivilotti, L. G. (1998). A reporter mutation approach shows incorporation of the "orphan" subunit β3 into a functional nicotinic receptor. *Journal of Biological Chemistry*, *273*(25), 15317–15320. https://doi.org/10.1074/jbc.273.25.15317.

Khiroug, S. S., Harkness, P. C., Lamb, P. W., Sudweeks, S. N., Khiroug, L., Millar, N. S., & Yakel, J. L. (2002). Rat nicotinic ACh receptor α7 and β2 subunits co-assemble to form functional heteromeric nicotinic receptor channels. *Journal of Physiology*, *540*(2), 425–434. https://doi.org/10.1113/jphysiol.2001.013847.

Kuryatov, A., & Lindstrom, J. (2011). Expression of functional human α6β2β3* acetylcholine receptors in *Xenopus laevis* oocytes achieved through subunit chimeras and concatamers. *Molecular Pharmacology*, *79*(1), 126–140. https://doi.org/10.1124/mol.110.066159.

Kuryatov, A., Olale, F., Cooper, J., Choi, C., & Lindstrom, J. (2000). Human α6 AChR subtypes: Subunit composition, assembly, and pharmacological responses. *Neuropharmacology*, *39*(13), 2570–2590. https://doi.org/10.1016/S0028-3908(00)00144-1.

Kuryatov, A., Onksen, J., & Lindstrom, J. (2008). Roles of accessory subunits in α4β2 * nicotinic receptors. *Molecular Pharmacology*, *74*(1), 132–143. https://doi.org/10.1124/mol.108.046789.

Lipovsek, M., Fierro, A., Pérez, E. G., Boffi, J. C., Millar, N. S., Fuchs, P. A., Katz, E., & Elgoyhen, A. B. (2014). Tracking the molecular evolution of calcium permeability in a nicotinic acetylcholine receptor. *Molecular Biology and Evolution, 31*(12), 3250–3265. https://doi.org/10.1093/molbev/msu258.

Lukas, R. J., Changeux, J. P., Le Novère, N., Albuquerque, E. X., Balfour, D. J. K., Berg, D. K., Bertrand, D., Chiappinelli, V. A., Clarke, P. B. S., Collins, A. C., Dani, J. A., Grady, S. R., Kellar, K. J., Lindstrom, J. M., Marks, M. J., Quik, M., Taylor, P. W., & Wonnacott, S. (1999). International union of pharmacology. XX. Current status of the nomenclature for nicotinic acetylcholine receptors and their subunits. *Pharmacological Reviews, 51*(2), 397–401. http://www.pharmrev.org.

Madeira, F., Park, Y. M., Lee, J., Buso, N., Gur, T., Madhusoodanan, N., Basutkar, P., Tivey, A. R. N., Potter, S. C., Finn, R. D., & Lopez, R. (2019). The EMBL-EBI search and sequence analysis tools APIs in 2019. *Nucleic Acids Research, 47*(W1), W636–W641. https://doi.org/10.1093/nar/gkz268.

Marcovich, I., Moglie, M. J., Freixas, A. E. C., Trigila, A. P., Franchini, L. F., Plazas, P. V., Lipovsek, M., & Elgoyhen, A. B. (2019). Distinct evolutionary trajectories of neuronal and hair cell nicotinic acetylcholine receptors. *bioRxiv.* https://doi.org/10.1101/621342.

Moser, N., Mechawar, N., Jones, I., Gochberg-Sarver, A., Orr-Urtreger, A., Plomann, M., Salas, R., Molles, B., Marubio, L., Roth, U., Maskos, U., Winzer-Serhan, U., Bourgeois, J.-P., Le Sourd, A.-M., De Biasi, M., Schröder, H., Lindstrom, J., Maelicke, A., Changeux, J.-P., & Wevers, A. (2007). Evaluating the suitability of nicotinic acetylcholine receptor antibodies for standard immunodetection procedures. *Journal of Neurochemistry, 102*(2), 479–492. https://doi.org/10.1111/j.1471-4159.2007.04498.x.

Nashmi, R., & Lester, H. A. (2006). CNS localization of neuronal nicotinic receptors. *Journal of Molecular Neuroscience, 30*(1–2), 181–184. https://doi.org/10.1385/JMN:30:1:181.

Palma, E., Maggi, L., Barabino, B., Eusebi, F., & Ballivet, M. (1999). Nicotinic acetylcholine receptors assembled from the α7 and β3 subunits. *Journal of Biological Chemistry, 274*(26), 18335–18340. https://doi.org/10.1074/jbc.274.26.18335.

Pedersen, J. E., Bergqvist, C. A., & Larhammar, D. (2019). Evolution of vertebrate nicotinic acetylcholine receptors. *BMC Evolutionary Biology, 19*(1). https://doi.org/10.1186/s12862-018-1341-8.

Ramirez-Latorre, J., Yu, C. R., Qu, X., Perin, F., Karlin, A., & Role, L. (1996). Functional contributions of α5 subunit to neuronal acetylcholine receptor channels. *Nature, 380* (6572), 347–351. https://doi.org/10.1038/380347a0.

Tumkosit, P., Kuryatov, A., Luo, J., & Lindstrom, J. (2006). β3 subunits promote expression and nicotine-induced up-regulation of human nicotinic α6 nicotinic acetylcholine receptors expressed in transfected cell lines. *Molecular Pharmacology, 70*(4), 1358–1368. https://doi.org/10.1124/mol.106.027326.

Turner, J. R., & Kellar, K. J. (2005). Nicotinic cholinergic receptors in the rat cerebellum: Multiple heteromeric subtypes. *Journal of Neuroscience, 25*(40), 9258–9265. https://doi.org/10.1523/JNEUROSCI.2112-05.2005.

Vailati, S., Moretti, M., Longhi, R., Rovati, G. E., Clementi, F., & Gotti, C. (2003). Developmental expression of heteromeric nicotinic receptor subtypes in chick retina. *Molecular Pharmacology, 63*(6), 1329–1337. https://doi.org/10.1124/mol.63.6.1329.

Wang, F., Gerzanich, V., Wellst, G. B., Anand, R., Peng, X., Keyser, K., & Lindstrom, J. (1996). Assembly of human neuronal nicotinic receptor α5 subunits with α3, β2, and β4 subunits. *Journal of Biological Chemistry, 271*(30), 17656–17665. https://doi.org/10.1074/jbc.271.30.17656.

Zoli, M., Pistillo, F., & Gotti, C. (2015). Diversity of native nicotinic receptor subtypes in mammalian brain. *Neuropharmacology, 96*, 302–311. https://doi.org/10.1016/j.neuropharm.2014.11.003.

Further reading

Kalamida, D., Poulas, K., Avramopoulou, V., Fostieri, E., Lagoumintzis, G., Lazaridis, K., Sideri, A., Zouridakis, M., & Tzartos, S. J. (2007). Muscle and neuronal nicotinic acetylcholine receptors. *FEBS Journal*, *274*(15), 3799–3845. https://doi.org/10.1111/j.1742-4658.2007.05935.x.

CHAPTER THREE

Function and pharmacology of neuronal nAChRs

3.1 Introduction

nAChRs are involved in a number of disorders and are widely expressed as described elsewhere in this book. Although nAChRs are ligand-gated channels that influence excitability, due to their locations and properties of gating Ca^{2+}, they also influence release of other neurotransmitters, synaptic plasticity, and gene expression. The use of specific nAChR compounds has allowed for some of the functions to be dissected. Much of what we know about nAChR localization and function comes in part from using specific agonists, antagonists, and allosteric modulators for various nAChR subtypes. In the disease chapter, the roles of nAChRs and specific compounds that might be used for treatment have been described. The major subtypes focused on were the α4β2*, α6β2*, and α7 nAChRs. Drugs targeted against various nAChR subtypes may have antipsychotic, neuroprotection, cognitive-enhancing properties, or may be useful for smoking cessation. In addition to summarizing nAChR pharmacology, we will describe some aspects of channel function that have been discovered using various compounds with activity at the nAChRs, such as differences in ion channel physiology and biophysics and the ability of specific subtypes to be upregulated or desensitized. All of the nAChRs function as ligand-gated ion channels, with ligand binding stabilizing the receptor in the open state. However, allosteric modulators (both positive and negative) affect channel function. Toxins, silent desensitizers, inverse agonists, and partial agonists also reveal aspects of nAChR function and regulation.

3.2 Function

In a previous chapter describing the structure of nAChRs, the basic function of the receptor was outlined. The simplest description is that upon

Nicotinic Acetylcholine Receptors in Health and Disease
https://doi.org/10.1016/B978-0-12-819958-9.00002-5

binding of ACh, the nAChR is stabilized in an open state, and cations such as Na^+, Ca^{2+}, and K^+ pass through the channel. The channel is opened due to allosteric interactions passed from the binding site in the

Extracellular domain (ECD) to the channel domain formed by the four transmembrane (TM) regions present in each subunit. There are two binding sites for ACh with positive agonist binding cooperativity. Heterotropic allosteric modulation of the channels also comes about through binding of molecules at allosteric sites, some in the channel domain (Papke & Lindstrom, 2020).

However, nAChRs exist in multiple states and make transitions between these. The resting state with no ligand bound is the most stable state, with very few spontaneous openings (Papke, 2014). While binding of ligand stabilizes a transition from a closed or resting state to an open state, continued exposure to ACh, nicotine, or other ligands causes the channel to enter a desensitized state that still binds agonist, but is now not activatable Fig. 3.1. Agonists bind to the open state with relatively low affinity (K_d of ACh for the open state is $50\,\mu M$), but ACh binds with a K_d for the desensitized state of $1-5\,\mu M$ (Wonnacott & Barik, 2007). The low affinity for the active state allows for ACh to bind and dissociate and allow for repetitive signaling (Changeux, 2018). Thus, agonists bind the desensitized nAChR with higher affinity, but the channel doesn't conduct ions. Desensitized heteromeric nAChRs have a higher increase in affinity compared with the resting state than homomeric $\alpha7$ nAChRs. nAChRs exist in multiple desensitized states. An equilibrium exists between the open and closed states, and the nAChR can move between these states (Cecchini & Changeux, 2022). Channels move between states over a time course, Fig. 3.2. Initially most nAChRs would be bound by agonist (AR with large red spot) after ACh exposure (Papke & Lindstrom, 2020). $100\,\mu s$ after agonist binding, the majority of the nAChRs would be in a ligand bound and open state due to the low-energy barrier between the resting closed and open state with a small amount in the agonist-bound/desensitized state. After $100\,ms$, most would be in a ligand-bound desensitized state (large red dot) (Papke & Lindstrom, 2020). Some would still be functional, and this activity is referred to as "smoldering" (Campling et al., 2013). Nicotine is effective at producing "smoldering." Nicotine exposure would produce less activation due to a higher-energy barrier. However, with nicotine, a steady-state current would be produced due to a lower-energy barrier between the desensitized state and the open state (Papke, 2014). This can account for some of nicotine effects during smoking, the ability to maintain some signaling in the presence of continuous levels of nicotine.

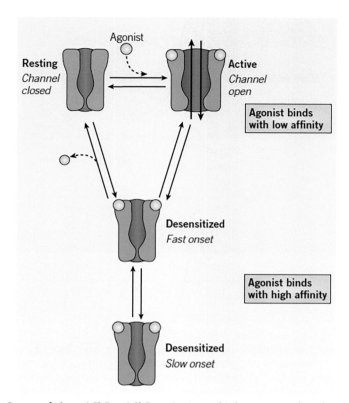

Fig. 3.1 States of the nAChR. nAChRs exist in multiple states and make transitions between these. The resting state with no ligand bound is the most stable state (resting), with very few spontaneous openings. Binding of ligand stabilizes a transition to an open state (active). Continued exposure to ACh, nicotine, or other ligands causes the channel to enter a desensitized state that still binds agonist, but is now not activatable. Agonists bind to the open state with relatively low affinity, but ACh binds with higher affinity to the desensitized states. *(From Wonnacott, S., & Barik, J. (2007). Nicotinic ACh receptors. Tocris Reviews, 28.)*

Heteromeric nAChRs have two orthosteric ligand-binding sites and exist in two open states, one brief and one long. It was thought that this was due to one agonist bound producing the short state and two bound for the long state. This was demonstrated to not be true, in that one binding site occupied could account for both short and long open states (Williams, Stokes, Horenstein, & Papke, 2011). In general, nAChRs with one agonist bound had the highest association with short-lived opening, and nAChRs with two ligands bound existed in the longer open state (Williams et al., 2011). One ligand could, however, produce nAChRs in the longer open

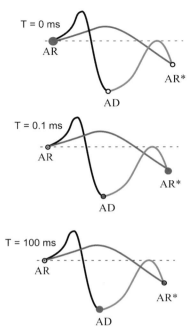

Fig. 3.2 Transitions of nAChRs between states. Channels move between states over a time course. Initially most nAChRs would be bound by agonist (AR with large *red* spot) after ACh exposure. In total, 100 μs after agonist binding, the majority of the nAChRs would be in a ligand-bound and open state (AR*, large *red* dot) due to the low-energy barrier between the resting closed and open state with a small amount of receptors in the agonist-bound/desensitized state (AD). After 100 ms, most would be in a ligand-bound desensitized state (AD, large *red* dot). *(From Papke, R. L., & Lindstrom, J. M. (2020). Nicotinic acetylcholine receptors: Conventional and unconventional ligands and signaling.* Neuropharmacology, 168, *108021. https://doi.org/10.1016/j.neuropharm.2020. 108021.)*

state. The probability of an nAChR moving to a desensitized state is generally associated with how long the channel is open. Two desensitization states exist for nAChRs, a fast and a slow component (Fig. 3.1). Desensitized states develop in receptors with both ligand–binding sites occupied more frequently. In muscle nAChRs, some open spontaneously, and a certain percentage are found in a desensitized state without ligand exposure (Changeux, 2018). Since the open state has lower affinity, ACh can bind and unbind quickly and allow for repeated fast stimulation. A model for nAChR transition between states is shown in Fig. 3.3 (Papke &

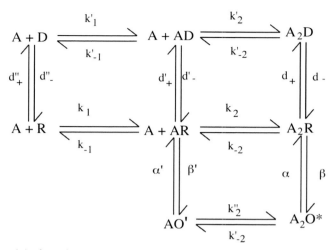

Fig. 3.3 Model of nAChR transitions between multiple open and desensitized states. This model represents transitions between states consistent with the view that there exist two open states AO′ and A2O*, the two binding sites, and multiple desensitized states. These transitions are based in part on agonist concentrations, and receptors are seen to move between the two open states, between states with 0, 1, or 2 ligands bound and between resting (ligand bound or not) and two desensitized states. $k′$ represents the rate constant for agonist association and $k′$-for agonist dissociation. α and $\alpha′$ are rate constants for channel closing and β and $\beta′$ rate constants for channel opening. d+ and d− represent rate constants for moving to and from desensitized states. *(From Papke, R. L., & Lindstrom, J. M. (2020). Nicotinic acetylcholine receptors: Conventional and unconventional ligands and signaling.* Neuropharmacology, 168, 108021. https://doi.org/10.1016/j.neuropharm.2020.108021.)

Lindstrom, 2020). This model represents two open states AO′ and A2O*, the two binding sites, and multiple desensitized states. These transitions are based in part on agonist concentrations, and receptors are seen to move between the two open states, between states with 0, 1, and 2 ligands bound and between resting (ligand bound or not) and two desensitized states. Equilibrium is achieved between the states allowing for a sometimes long macroscopic currents at many agonist concentrations (Williams et al., 2011).

Heteromeric α4β2 (low sensitivity) nAChRs can mediate signaling in that after rapid exposure to agonist, each receptor has an 80% probability of opening (Papke, 2014). This can generate a large synchronous current (Li & Steinbach, 2010) that also decays quickly due to desensitization and breakdown of ACh by AChE. Nicotine and other full agonists also produce this response (Li & Steinbach, 2010). There seems to be a disconnect between the number of receptors detected by binding and the actual number of "activatable" nAChRs, wherein most of the nAChRs are not activatable (Li & Steinbach, 2010). In contrast, the primordial homomeric α7 nAChRs have a lower probability of opening and a much shorter open time than heteromeric nAChRs (Papke, 2014). α7 nAChRs desensitize rapidly but recover quickly after agonist removal (Williams, Wang, & Papke, 2011b).

What is the physical basis for the desensitized state? It is known that the channel opens upon ligand binding through an allosteric transition, but what blocks ion flow in a ligand–bound desensitized receptor? A desensitization gate at the cytoplasmic end of the glycine receptor (GlyR) α1 was proposed that would produce desensitization by gate closure (Gielen & Corringer, 2018). This indicates that in a pentameric ligand-gated ion channel, the activation and desensitization gates are unique at least for slow dissociation (Gielen & Corringer, 2018).

nAChRs are also thought to possess a desensitization gate. Imaging of desensitized α4β2 and α3β4 nAChRs shows that nicotine remains bound to the orthosteric sites, and the activation gate is open (Papke, 2014). However, a desensitization gate at the cytoplasmic end (as for the ligand–gated glycine receptor) is closed (Papke, 2014). The α3β4 nAChR channel is narrow at the cytoplasmic side, and functional evidence indicates that this is the location of the nAChR desensitization gate (Gharpure et al., 2019).

Drugs have been used to probe the function of nAChRs. Type II PAMs (described more below) slow desensitization and can reactivate desensitized receptors (Norleans et al., 2019). One particular PAM, (R)-7bromo-N-(piperidin-3-yl)benzo[b]thiophene-2-carboxamide (Br-PBTC), was shown to potentiate the opening of α2* and α4* nAChRs and also to reactivate desensitized α4* nAChRs (Norleans et al., 2019). It was most able to reactivate at the (α4)3(β2)2 (low sensitivity) stoichiometry that contained an α4/α4 interface. Br-PBTC binds in a pocket between the α4 TM1, TM3, and TM4 domains (Norleans et al., 2019). This binding is proposed to slow the closing of the desensitization gate in this region of the structure and also to promote reopening of the gate (Norleans et al., 2019). When nAChR channels are conducting ions, both activation gate and desensitization gate are open, but when desensitized, the activation gate is open but the desensitization gate is shut (Norleans et al., 2019).

3.3 Pharmacology overview

As summarized elsewhere in this book, nAChRs are present in different combinations of subunits that define subtypes. These have diverse patterns of expression in the brain and elsewhere in the body as well as diverse cellular location (presynaptic, postsynaptic, and extrasynaptic). These combinations exhibit different functional characteristics. Diverse subtypes with specific subunit combinations and stoichiometries vary in response to specific cholinergic compounds. The nAChR channels are opened by binding of agonists at the orthosteric site. In addition to ACh, other agonists bind at the orthosteric site as well as competitive antagonists. The orthosteric site is formed at the interface of subunits, usually an alpha and a beta (described elsewhere). The orthosteric site in homomeric receptors such as $\alpha7$ is formed at the interfaces of two alpha subunits. Recently, more binding interfaces have been found such as between $\alpha4/\alpha4$ in $(\alpha4)_3(\beta2)_2$ nAChRs (Wang & Lindstrom, 2018). Other unorthodox sites are found at $\alpha4/\alpha5$, $\alpha5/\alpha4$, and $\beta3/\alpha4$ (Wang & Lindstrom, 2018).

Function of nAChRs is often modulated and controlled by compounds that bind at allosteric sites. Allosteric sites can be located in multiple places including in the channel, at the lipid–receptor interface and the ECD. Agonists as traditionally defined don't bind at these allosteric sites, and allosteric modulators don't bind to the orthosteric sites and open channels. There are some exceptions, however. We will present an overview of a number of compounds that affect nAChR function. The list will include many of the most important compounds, but the list is not exhaustive.

3.4 Agonists

Agonists binding in the orthostatic site in general work by stabilizing the nAChR in the open state or increase the probability of the channel opening. Agonists for nAChRs can also stabilize the receptor into a desensitized state. Agonists for various nAChR subtypes include ACh, nicotine, epibatidine, cysteine, GTS-21, and others. Each of these has a different affinity for each nAChR subtype and can influence function in various ways.

3.4.1 Acetylcholine

ACh is the natural ligand and binds at the orthosteric site of all subtypes as well as mAChRs. It is present at high concentrations at the NMJ synapse and CNS synapses, but at much lower levels around neurons as part of volume transmission. While ACh binds at the orthosteric site, binding induces an allosteric change in the receptor that is coupled to channel opening. Not all AChRs are equally sensitive to ACh. For example, $(\alpha4)2(\beta2)3$ receptors have a higher sensitivity to ACh (EC_{50} 1 μM), while (α4)3(β2)2 are much less sensitive (EC_{50} 100 μM) (Terry & Callahan, 2019).

3.4.2 Nicotine

Although acetylcholine is the normal neurotransmitter acting at all nAChRs, nicotine is also widely used given the presence in cigarette smoke. Nicotine can activate multiple subtypes of nAChRs by binding at the orthosteric site, but is often referred to as a time-averaged antagonist since prolonged exposure leads to receptor desensitization. Nicotine has been shown to affect cognition, short-term memory, and attention among many other effects. It is the addictive component of tobacco covered elsewhere in this book. α4β2 nAChRs have the highest affinity for nicotine and desensitize slowly. Smokers can achieve a nicotine level in the brain that can saturate most of the α4β2* nAChRs after smoking a few cigarettes. Smokers achieve a brain level of at least 50 nM (Matta et al., 2007). α4β2 nAChRs are widely expressed, and thus, nicotine binding is rapid and widespread in the brain. Nicotine, however, has various levels of affinity for multiple nAChRs subtypes (Luetje & Patrick, 1991). For example, nicotine is much more potent for α4* and α2* nAChRs than α3* nAChRs (Luetje & Patrick, 1991). Even (α4)2(β2)3 and (α4)3(β2)2 differ greatly in nicotine sensitivity. Nicotine is a weak agonist for α3β2 nAChRs due to low affinity, but a potent agonist for α2β2 and α4β2. Nicotine acts as a time-averaged antagonist; with preincubation it potently inhibits α4β2 nAChR function (IC_{50} 180 nM (Xiao et al., 2006).

On the other hand, α7 nAChRs have a lower affinity for nicotine and desensitize quickly. Nicotine improves attention in humans with AD, schizophrenia, and ADHD (Rezvani et al., 2013). However, as described elsewhere, nicotine has numerous drawbacks to clinical use. A number of agonists described below have been used to study nAChR function and as possible treatments for improving cognition.

3.4.3 Epibatidine

Epibatidine is a high-affinity agonist for multiple nAChR subtypes. Its highest affinity is for the α4β2 nAChRs and has been used in binding studies to map nAChRs in the human brain. Epibatidine binds to two high-affinity sites in human cortex with K_{D}s of 0.3 and 28.4 pM (Paterson & Nordberg, 2000). Epibatidine binds to α-Bgt-binding sites (α7 nAChRs) with low affinity in mammals (Marks et al., 2010). Epibatidine has low nM or even pM affinity for multiple recombinant human nAChR subtypes including α3β2, α3β4, α3β4α5, α4β4, α2β2, and α2β4 (Paterson & Nordberg, 2000). Epibatidine is a potent analgesic, 100× more so than morphine. Epibatidine is also toxic even in small doses (Decker et al., 1998), and derivatives have been produced with less toxicity. However, due to the toxicity and nonspecificity, it is not used clinically.

3.4.4 Cytisine

Cytisine is a partial agonist often used to identify nAChRs containing β2 subunits. Cytisine competes with ACh and can reduce the activation of β2 nAChRs (Papke & Heinemann, 1994). Cytisine binds to α4β2 nAChRs with high affinity, but with low efficacy. This high-affinity binding ability has been used to map β2 subunit containing nAChRs in brain. Cytisine also interacts with α3β2 nAChRs with low efficacy and low affinity (Papke & Heinemann, 1994). Cytisine is almost a full agonist also for α6β2* nAChRs and stimulates dopamine release (Gotti & Clementi, 2021). Cytisine is a full agonist for α7 nAChRs, but binds with lower affinity. nAChRs containing a β4 (α2β4, α3β4, α4β4) subunit are more sensitive to cytisine than ACh compared with those containing β2 subunits (Luetje & Patrick, 1991), but cytisine binds poorly to β4 nAChRs (Gotti & Clementi, 2021). However, cytisine is a full agonist for α3β4 nAChRs, but has low efficacy for human nAChRs (Gotti & Clementi, 2021). Cytisine is a poor agonist for α2β2 and α3β2 nAChRs expressed in *Xenopus oocytes* (Luetje & Patrick, 1991). Cytisine may also act as an intracellular chaperone of nAChRs (Gotti & Clementi, 2021). Because of its ability to act as competitive partial agonist of α4β2 nAChRs, it was used as a structure from which to build varenicline.

3.4.5 Varenicline

Varenicline is used in smoking cessation therapy. The hypothesis is that during nicotine cessation, dopamine levels drop and lead to resumption of smoking. Varenicline would produce a lower release of dopamine, which would lessen the withdrawal effects and craving. Binding of varenicline would also reduce nicotine activation of nAChRs during smoking and lessen the reward (Coe et al., 2005).

Its effects come from its function as an α4β2 nAChR partial agonist and an α7 nAChR full agonist. It is also a partial agonist for α3β2* and α6* nAChRs (Hong et al., 2011). Partial agonists act in some ways as antagonists. Varenicline concentration near to that which can be achieved in human brain would provide some minimal activation and low level of desensitization of α7 nAChRs while activating and desensitizing α4β2 nAChRs (Rollema et al., 2009). It is generally thought that partial agonists are better at producing desensitization instead of activation (Rollema et al., 2009). Varenicline was derived from cytisine, another α4β2 nAChR partial agonist. Varenicline showed a binding affinity for rat α4β2 nAChRs higher than for α3β4, muscle, and α7 subtypes and also was more selective (Coe et al., 2005). Varenicline was shown to reduce mesolimbic dopamine turnover in Sprague Dawley rats to 32% of the level induced by nicotine and inhibited the nicotine–produced response when given together by 66% (Coe et al., 2005). In addition to smoking cessation, varenicline has been shown to improve performance in rats in several cognitive tasks. Work is being done to determine whether varenicline may be affective at treating other disorders. Varenicline shows some antidepressant activity in a mouse assay (Rollema et al., 2009). Varenicline treatment for 8 weeks reduced sensory gating deficits in schizophrenic patients (Hong et al., 2011). The effect was not seen in smokers (Hong et al., 2011), perhaps due to their already high level of nAChR activation/desensitization. Varenicline in some studies was also shown to improve cognition in schizophrenic patients (Terry et al., 2015). See more about how varenicline is used for smoking sensation in Chapter 4.

3.4.6 ABT-418

A number of compounds structurally related to nicotine were developed with the goals of improving cognition, memory, anxiety, or analgesia. ABT-418 (S)-methyl-5-(1-methyl-2-pyrrolidinyl) isoxazol) is an agonist that competes for cytisine binding and acts as a full agonist at α4β2 nAChRs. ABT-418 doesn't have activity at α7 nAChRs. ABT-418 in animal studies produced enhancements in associative learning, working memory, anxiolytic-like activity with less side effects than nicotine (Terry et al., 2015). ABT 418 has an EC_{50} of 6 μM for α4β2 nAChRs and high efficacy (Bertrand & Terry, 2018). ABT 418 has a much lower EC_{50} (188 μM) for α3β4 nAChRs. ABT 418 was used in phase II clinical trials in AD and schizophrenia patients, but was not efficacious enough or produced side effects (Bertrand & Terry, 2018). ABT-418 aided in smoking cessation in a phase II trial, but not further developed (Rollema & Hurst, 2018).

3.4.7 ABT-594

ABT-594 was developed from epibatidine and is much less potent and thus can be clinically used. ABT-594 retains high affinity for α4β2 nAChRs, but much less so for ganglionic and NMJ receptors. ABT-594 is an analgesic in rat models of pain, cutting thermal, chemical, and neuropathic pain (Decker et al., 1998). It is an α4β2 nAChR agonist and is 200× more potent than morphine in pain assays. However, phase II trials of ABT-594 were discontinued (Arneric et al., 2007), and it is no longer commercially available.

3.4.8 ABT-126

AB-126 is a selective partial agonist of α7 AChRs with an EC_{50} of 2.6 μM and 100% efficacy (Bertrand & Terry, 2018). In a study of schizophrenic patients, nonsmokers showed improvement with ABT-126 in working memory, attention, and verbal learning. The drug produced some side effects such as nausea and headache (Haig et al., 2016). However, the population size was small, and since most schizophrenics smoke, ABT-126 may not prove to be clinically useful in a broad population of schizophrenics.

3.4.9 GTS-21 (DXMBA)

GTS-21 is an anabaseine analog. It is a selective human α7 partial agonist (with lower efficacy and potency than on rat α7 nAChRs) and a weak α4β2 antagonist (Terry et al., 2015). The GTS-21 EC_{50} at α7 nAChRs is 100 μM with less than 20% efficacy (Bertrand & Terry, 2018). GTS-21 doesn't produce significant α7 nAChR upregulation (Kem, 2000). However, a metabolite of GTS-21 is as active at human α7 nAChRs as at rat receptors (Kem, 2000). GTS-21 is weak antagonist for the α3β4 subtype expressed in PC12 cells and for the NMJ nAChR (Kem, 2000). GTS-21 produced some improvement in tasks measuring cognition in rats such as the Morris water maze and radial maze (Kem, 2000). GTS-21 normalizes the P50 sensory gating deficit (hallmark of schizophrenia) in mice (Leonard et al., 2007).

GTS-21 was used in a human phase I trial and produced some improvement in attention and working memory in young adults (Kem, 2000). In a phase II trial of GTS-21 with schizophrenics, some improvement of negative symptoms was noted at the higher dose tested (Freedman et al., 2008). GTS-21 has proven to be a useful tool to study a number of functions of α7 nAChRs including in aggression (Lewis et al., 2018).

3.4.10 Sazetidine-A

Sazetidine-A is a highly selective agonist for α4β2 nAChRs with an EC_{50} of 4–8 nM at the human (α4)2(β2)3 nAChR (Carbone et al., 2009). It binds with high affinity at α4β2 nAChRs (K_i 0.5 nM) (Xiao et al., 2006).

Interestingly, sazetidine-A is a full agonist at (α4)2(β2)3 nAChRs (high sensitivity) but has little effect at (α4)3(β2)2 nAChRs (low sensitivity) (Carbone et al., 2009). Sazeidine-A is not efficacious at (α4)2(β2)2α5 nAChRs (Brown & Wonnacott, 2015). It does have high efficacy and potency with α4α6(β2)2β3 nAChRs (Brown & Wonnacott, 2015). Sazetidine-A is most likely a desensitizer of α6β2β3 nAChRs after long exposure (Kuryatov & Lindstrom, 2011). It doesn't block the action of nicotine if applied together, but if given before nicotine, it blocks nicotine-induced α4β2 nAChR function (Xiao et al., 2006). Sazetidine-A rapidly activates α4β2 nAChRs, but quickly leads to a long-lasting desensitization (Rezvani et al., 2013). Sazetidine-A is often referred to as a "silent desensitizer" due to it producing little, but rapid activation, followed by long-lasting desensitization. Sazetidine-A has low affinity for the resting state and high affinity for the desensitized state (Xiao et al., 2006). By putting α4β2 nAChRs in a desensitized state, sazetidine-A is 50 X more potent at inhibiting α4β2 function than DHβE (Xiao et al., 2006). Sazetidine-A has more recently been found to activate α7 nAChRs with an EC_{50} of 1.2 μM for human nAChRs (60 μM for rat) and also desensitize the human receptors (Brown & Wonnacott, 2015). This function is similar to that of nicotine, first activating nAChRs, then desensitizing. Chronic and acute sazetidine-A treatment were shown to improve attention in rats as well as reduce nicotine self-administration. Sazetidine-A at doses higher than those that produce behavioral effects doesn't upregulate α4β2 nAChRs (Rezvani et al., 2013). Sazetidine-A showed analgesic properties in rats using the formalin model of inflammatory pain (Cucchiaro et al., 2008). Chronic exposure of mice to sazetidine-A showed antidepressant effects and may indicate a potential use in humans (Turner et al., 2010).

3.4.11 Choline

Choline is formed from the breakdown of acetylcholine by acetylcholinesterase and is recycled back into neurons by a choline transporter. Choline is used in the synthesis of acetylcholine and is also present in foods. Choline was shown to act as a full, but weak agonist (10× less potent than ACh) for α7 nAChRs (Alkondon et al., 1997; Wonnacott & Barik, 2007). Choline has low activity as an antagonist at α3β4* and α4β2* nAChRs as well (Wonnacott & Barik, 2007). Choline is a partial agonist for α7β2 nAChRs (Wu et al., 2016). Choline is an agonist for α9α10 nAChRs, being a full agonist for chick α9α10 nAChRs, but only a partial agonist for rat α9α10 (Moglie et al., 2021). Choline is a full agonist for rat α9 homomeric nAChRs (Moglie et al., 2021).

Dietary choline also has effects on nAChR receptors and may decrease the number of functional nAChRs (Alkondon et al., 1997). Increased choline can lead to increased activity of the choline transporter and more ACh synthesis. One rationale for dietary choline supplements is to promote cognitive function (Maurer et al., 2021; Ylilauri et al., 2019). Volume transmission is important in cholinergic signaling, and choline obtained from diet or breakdown of acetylcholine may have widespread effects on signaling. Dietary choline can achieve levels that affect nAChR function. Choline deficiency during development can produce impairment in cognitive function (Derbyshire & Obeid, 2021). However, choline has other biochemical actions that are not dependent on nAChR function. Any role of choline in neural function must be examined as to the mechanism of action.

3.5 Antagonists

Antagonists usually act by blocking receptor function as a competitive agonist at the orthosteric site. Under various conditions, nAChRs require upregulation or activation by agonists as we have seen above, or reduced signaling for proper neuronal functioning. Nicotine is complex in that activates nAChRs, but also acts as a time-averaged antagonist. Use of antagonists such as the few described below has yielded a large body of information regarding nAChR function and location. Several are also being used as potential therapeutic compounds.

3.5.1 Conotoxins

Snails of the genus Conus produce a large number of small peptides that they use for defense. The mature peptides share a basic structure of 12–20 amino acids in length and having specifically spaced cysteine residues and disulfide bridges (Lebbe et al., 2014). Variations in the basic structure allow them to be placed in families (i.e., alpha, omega, and mu) and superfamilies such as B3, D, L M, S, T, and J (Abraham & Lewis, 2018).

Numerous conotoxins in the A superfamily, α-conotoxins, are antagonists of nAChRs. Some conotoxins in other superfamilies also can target nAChRs (Abraham & Lewis, 2018). Some conotoxins target muscle nAChRs, but we will focus on those that target neuronal nAChRs. The location of α-conotoxin binding to nAChRs was determined initially by co-crystalizing multiple α-conotoxins with the acetylcholine binding protein (AChBP). These studies demonstrated that the conotoxin was bound in the orthosteric site and that specific interactions between the α-conotoxin and the nAChR subunits provided a basis for the selectivity of different α-conotoxins for different nAChR subtypes (Lebbe et al., 2014). Detailed pharmacophores have been built to understand the binding of α-conotoxins to nAChRs. Site-specific mutagenesis has been used to create even more specific α-conotoxins for use against nAChRs. Some conotoxins may act as noncompetitive inhibitors that may bind outside of the orthosteric site (Abraham & Lewis, 2018).

Conotoxins are being tested for use as analgesics with Vc1.1, RgIA, and RgIA4 used to target α9α10 nAChRs in pain and inflammation (Grau et al., 2019). So far, translating the use of conotoxins in animal models of pain to human use has not had much success. Since these are peptides, they are generally administered in vivo by injection intraperitoneally or intracerebrally. Some conotoxins have been modified by the use of selenocysteines in place of cysteines to make the peptides more stable (Azam & McIntosh, 2009). However, α-conotoxins are important research tools that have been used to determine the function and location of numerous nAChR subtypes (Table 3.1). A number of nAChR subtypes are involved in controlling striatal dopamine release such as α6β2, α6β2β3, or α6α4β2*, and the α-conotoxins MII and PIA have been used to dissect the roles of these subtypes (Azam & McIntosh, 2009). α3β4 nAChRs are involved in hippocampal norepinephrine (NE) release, and AuIB has aided these studies. Hippocampal NE release is modulated by α6α4β2β3 and α6α4β2β4β3 nAChRs and MII, BuIA and PIA have been used in this work (Azam & McIntosh, 2009). While the list is not exhaustive, Table 3.1 shows some of the α-conotoxins used in nAChR research and the subtypes targeted by each. More conotoxins are being discovered and mutations are being made in existing ones to make them more selective. α-conotoxins have been important tools for studying nAChR function and may have many future therapeutic applications.

3.5.2 Mecamylamine

Mecamylamine is a broad spectrum, noncompetitive, voltage-dependent antagonist used in thousands of studies as general blocker of nicotinic signaling. It was first studied as a ganglionic locker, i.e., α3β4* nAChRs, and is sometimes used in humans for control of hypertension. Mecamylamine

Table 3.1 Conotoxins targeting nAChRs.

Conotoxin	Subtype
PnIA	$\alpha3\beta2 > \alpha7$
PnIB	$\alpha7 > \alpha3\beta2$
ImII	$\alpha7$
ArIB	$\alpha7 = \alpha6\beta2^* > \alpha3\beta2$
IMI	$\alpha3\beta2 > \alpha7 > \alpha9$
MII	$\alpha3\beta2 = \alpha6\beta2^* > \alpha6\beta4$
PIA	$\alpha6\beta2^* > \alpha6\beta4 = \alpha3\beta2$
OmIA	$\alpha3\beta2 > \alpha7 > \alpha6\beta2^*$
GIC	$\alpha3\beta2 = \alpha6\beta2^* > \alpha7$
GID	$\alpha7 = \alpha3\beta2 > \alpha4\beta2$
AnIA	$\alpha3\beta2, \alpha7$
AuIB	$\alpha3\beta4$
RgIA4	$\alpha9\alpha10 > > \alpha7$
ArIA	$\alpha7 = \alpha3\beta2$
BuIA	$\alpha6\beta2\beta3$
GID	$\alpha7 = \alpha3\beta2 > \alpha4\beta2$
Vc1.1	$\alpha9\alpha10$
[V11L;V16D]ArIB	$\alpha7$
VnIB	$\alpha6\beta4^*$

Adapted from Azam, L., & McIntosh, J. M. (2009). Alpha-conotoxins as pharmacological probes of nicotinic acetylcholine receptors. *Acta Pharmacologica Sinica, 30* (6), 771–783. https://doi.org/10.1038/aps.2009.47.

crosses the blood-brain barrier and has been shown to be a relatively non-selective antagonist of both central and peripheral nAChRs that can block many of the effects of nicotine in the brain (Damaj et al., 2005). While acting as an antagonist for all nAChRs, there are some differences in inhibitory efficacy for various subtypes. It is not generally used, however, to antagonize a specific subtype. Mecamylamine has some preference for $\beta4$-containing nAChRs such as $\alpha3\beta4$ vs. $\alpha4\beta2$ (Papke et al., 2008). Mecamylamine binds in the channel and blocks the flow of cations through the pore (Nickell et al., 2013). Mecamylamine doesn't prevent the channel from closing and can be trapped in the channel (Nickell et al., 2013). In animals, mecamylamine decreases nicotine-seeking behavior and reinforcement (Nickell et al., 2013). Mecamylamine pretreatment also blocks conditioned place preference (CPP) for nicotine (Nickell et al., 2013) and has antidepressant effects in animals in some assays (Nickell et al., 2013). In humans, mecamylamine doesn't work well as an adjunct to nicotine replacement therapy (NRT) and produces a number of side effects. These side effects are not unexpected given mecamylamine's wide range of nAChR targets. Mecamylamine is being tested in humans for possible use in treatment of alcohol addiction and depression with results suggesting that mecamylamine may not be useful for widespread treatment of these disorders. However, as a relatively non-selective antagonist, it has been and will continue to prove useful in many preclinical studies of nAChRs and nicotine actions.

3.5.3 Dihydro-β-erythroidine (DHβE)

DHβE is competitive antagonist of nAChRs with selectivity for α4β2 and α4β4 nAChRs. In general, DHβE is most effective on nAChRs with an α4 subunit (Papke et al., 2008) and is often used to block α4β2 nAChR activity. DHβE competes with ACh for ligand binding and shares some features with nicotine and acts at the orthosteric site (Harvey et al., 1996). DHβE nAChR subtype selectivity is influenced by both the alpha and beta subunits. α3β2 nAChRs are almost 60× more sensitive to block by DHβE than α3β4 nAChRs with important determinants of this selectivity between residues 54 and 63 of the beta subunit (Harvey & Luetje, 1996). α4β4 nAChRs are more than 100× more sensitive to DHβE than are α3β4 nAChRs with important determinants located in the first 100 residues of the N-terminus of the alpha subunit (Harvey et al., 1996). DHβE is a more potent inhibitor of α3β2 than α3β4 nAChRs (Harvey et al., 1996). DHβE has a high binding affinity for α4β2 nAChRs with a K_i for human α4β2 nAChRs of 410 nM and 600 nM for rat α4β2 nAChRs. This is180X lower than for rat α3β4 nAChRs (Xiao et al., 2006). DHβE is 100× more potent at human α4β2 nAChRs than for rat α3β4 nAChRs (Xiao et al., 2006). By measuring net charge for mouse α4β2 and α3β2 nAChRs expressed in oocytes, it was shown that DHβE has an IC_{50} 260× lower for α4β2 nAChRs, thus showing far greater potency at α4β2 nAChRs (Papke et al., 2010). DHβE can also act as a competitive antagonist for α7 nAChRs expressed in Xenopus oocytes (Harvey & Luetje, 1996). DHβE can also antagonize α2β2 and α2β4 nAChRs in addition to the ones described above (Davis & Gould, 2006).

While not in use clinically, DHβE has been widely used to study the role of α4β2 nAChRs. DHβE has been used to block nAChR-mediated signaling in many places in the brain that effect GABAergic, glutamatergic, dopaminergic, and cholinergic signaling. Many animal behavioral studies have used DHβE to dissect the role of α4β2 nAChRs. These include contextual fear conditioning (Davis & Gould, 2006), depression (Andreasen et al., 2009), attention and memory (Bertrand & Terry, 2018), and body weight control (Dezfuli et al., 2020).

3.5.4 Methyllycaconitine (MLA)

MLA is a competitive antagonist at the orthosteric site of nAChRs. MLA acts at α7 nAChRs with an IC_{50} of less than 1 nM (Bertrand & Terry, 2018). MLA binds to the same sites as α-Bgt (see below) but with different kinetics (Papke & Horenstein, 2021). However, in the presence of allosteric modulators of α7 nAChRs, MLA has a more complex mode of action as an "inverse agonist." By definition, an inverse agonist binds to the same site as an agonist, but reduces activity that is currently present, rather than stimulate. This is different from a competitive antagonist that competes for binding to the orthosteric site and blocks signaling.

MLA can also antagonize α4β2 nAChRs but with a much higher IC_{50} (1000× higher) than for α7 nAChRs (Absalom et al., 2013). MLA binds at the α4/α4 subunit interface of the (α4)3(β2)2 stoichiometry. MLA does not have this action at the (α4)2(β2)3 nAChRs that have only α4/β2 interfaces. The action at the α4β2 nAChRs may be different in that MLA inhibition of ACh activation seems insurmountable and requires MLA binding to reach equilibrium, not something usually seen for competitive antagonists. This α4/α4 interface may be a new site for targeting α4β2 nAChRs and has potential therapeutic implications (Absalom et al., 2013).

Due to its complex actions as an antagonist and inverse agonist, dosage and timing of application are important in determining the effects. MLA has been used to reveal a role of α7 nAChRs in neuroprotection (Telles-Longui et al., 2019). In rats, MLA was shown to improve attention (Levin et al., 2013). Care should be taken in interpreting the effects of MLA as simple antagonism since low-concentration MLA seems to potentiate human α7 function (van Goethem et al., 2019). Low levels also facilitated LTP in mice, increased hippocampal glutamate efflux and memory acquisition (van Goethem et al., 2019). MLA has been used in numerous preclinical studies to regulate the function of α7 nAChRs.

3.5.5 Alpha Bungarotoxin (α-Bgt)

α-Bgt was one of the first compounds used in the study of nAChRs. α-Bgt is highly potent and an almost irreversible antagonist of NMJ nAChRs (Papke & Horenstein, 2021). The use of the snake venom toxin α-Bgt, isolated from *Bungarus multicinctus*, to bind the nAChR was an important step to isolate nAChRs from the electric organ of *Torpedo*, since it was shown that α-Bgt blocked neuromuscular transmission and also blocked the flux of cations in the electroplaques (Changeux, 2012). Subsequently, purification methods were applied to the electroplaques of *Torpedo marmorata*, which also had a high concentration of nAChRs (Miledi et al., 1971). It was initially show to bind to α-Bgt "binding sites" or "α-Bgt receptors" in the brain, those distinct from high-affinity nicotine-binding sites (Clarke et al., 1985). It was later shown by molecular cloning that α-Bgt binding in the brain was to α7 nAChRs (Couturier et al., 1990; Schoepfer et al., 1990). α-Bgt antagonizes neuronal α7 nAChRs in addition to NMJ nAChR responses. It was not clear why a snake venom toxin would bind to human nAChRs, but comparison of the structure of the snake venom and "three finger proteins" showed structural homology (Nirthanan, 2020). Members of this family such as the Lynx family of proteins are endogenous regulators of nAChRs (see Chapter 9). α-Bgt is useful for mapping, quantifying, and localization of nAChRs in the brain and in cell lines and for blocking α7 nAChR function in many oocyte- or cell-based assays. However, due to the toxicity, it can't be used in vitro.

3.6 Allosteric modulators

Allosteric modulators regulate nAChR activity by interacting at sites other than the orthosteric site. Allosteric modulator function is observed in the presence of agonist by potentiating the effects of ACh as a positive allosteric modulator (PAM) or reducing activity as a negative allosteric modulator (NAM) (Chatzidaki & Millar, 2015). Allosteric modulators generally only have effects when the agonist is present; they don't alone open the channel. However, some compounds designated allosteric agonists can activate nAChRs without ACh present (Chatzidaki & Millar, 2015). In addition, silent allosteric modulators (SAMs) can bind at the allosteric site and competitively block activity of NAMs and PAMs (Chatzidaki & Millar, 2015). The structure of the orthosteric site is highly conserved among various nAChR subtypes (to accommodate ACh binding), and there are subtype-selective compounds available as are described above. In order to achieve the goal of designing compounds for treatment of various disorders with minimal site effects, subtype-specific drugs are needed. Since NAMs and PAMs don't bind at the orthosteric site, it is thought the allosteric sites may have more subtype specificity. Side effects might also be less because PAMs and NAMs work well only if ACh is present. A number of allosteric sites have been identified using various techniques including binding, molecular dynamic simulations, affinity labeling, and X ray crystallography (Chatzidaki & Millar, 2015; Pavlovicz et al., 2011). Allosteric sites on nAChRs have been located at intra-subunit cavities, transmembrane sites, α/α interfaces, β(+)/α(−) interfaces, and α(+)/β(−) interfaces distinct from the orthosteric site (Chatzidaki & Millar, 2015). PAMs were used to differentiate the ionotropic and metabotropic functions of α7 nAChRs. Just as agonists and antagonists have activity at multiple nAChR subtypes, this also applies to allosteric modulators. Some modulators have differential effects at nAChRs and are difficult to classify while some have multiple effects. Below, we summarize some of the major allosteric modulators of neuronal nAChRs and how they may be used in therapy. This list is by no means exhaustive but gives some context of how they function and might be used in research and clinically.

3.6.1 Positive allosteric modulators (PAM)

Allosteric modulators are often classified as Type I or Type II. In general, PAMs increase apparent affinity for the receptor and increase agonist potency. Type I PAMs generally act to increase channel activation producing increased amplitude but not duration. This occurs by stabilizing the open state and when occupancy is low, allow opening without going through a desensitized state. Type II PAMs increase signaling through the channels by destabilizing the desensitized state. Some can activate nAChRs directly to promote moving from the desensitized state to an open state (Papke & Horenstein, 2021).

Many of the PAMs identified have been focused on α7 nAChRs, but some are being used to study other subtypes such as α4β2 nAChRs as well. α7 nAChRs desensitize rapidly and a property of some of the α7 targeted PAMs is to reverse or prevent desensitization (i.e., Br-PBTC). Drugs with these properties may be more suitable to clinical applications. α7 nAChRs open at low levels of occupancy and low probability (Papke & Horenstein, 2021). More agonist induces a desensitized state without an increase in affinity for agonist as occurs for other subtypes (Papke & Horenstein, 2021). α7 nAChRs have high Ca^{2+} permeability, but it is seen that PAM-potentiated currents don't demonstrate this (Papke & Horenstein, 2021). As described elsewhere in this book, α7 AChRs are involved in AD and schizophrenia among other neurological disorders as well as in immune regulation. They present an important therapeutic target.

3.6.1.1 NS-1738

NS-1738 is a Type I PAM. NS-1738 produces increased current and peak amplitude at α7 nAChRs with an EC_{50} of 3.4 μM. (Timmermann et al., 2007). The drug also produced cognitive enhancement in rats (Timmermann et al., 2007). NS-1738 may act at the α7 M2-M3 extracellular loop (Williams, Wang, & Papke, 2011b).

3.6.1.2 PNU-120596

PNU-120596 acts as a Type II PAM (Terry et al., 2015) with selectivity for α7 nAChRs (Hurst et al., 2005). PNU-120596 also reactivates nAChRs from a desensitized state (Arias, 2010). PNU-120596 binds within the intra-subunit cavity in the transmembrane domain (Arias, 2010). PNU-120596 prevented current decay and reactivated agonist-bound desensitized receptors (Hurst et al., 2005). PNU-120596 was shown to improve auditory gating in rats after impairment with amphetamine and may be useful for in vivo testing (Hurst et al., 2005). The compound was also shown to enhance various cognitive functions in animals including recognition memory, working memory, and cognitive flexibility (Terry et al., 2015). PNU-120596 may also be useful to treat some aspects of pain (Bagdas et al., 2018). PNU-120596 applied with a silent agonist such as MrIC (a conotoxin) and NS6740 can produce large, long-lasting currents (Papke & Horenstein, 2021). PNU-120596 was used to demonstrate the existence of two distinct desensitized states (Williams et al., 2011a).

3.6.1.3 3a,4,5,9b-Tetrahydro-4-(1-naphthalenyl)-3H-cyclopentan[c] quinoline-8-sulfonamide (TQS)

TQS is another Type II PAM for α7 nAChRs. TQS was shown to reactivate desensitized receptors, moving from a D_s to an open, agonist-bound state (Williams et al., 2011b).

3.6.1.4 NS9283

PAMs also exist for nAChRs other than the α7 nAChRs. NS9283 is a PAM for nAChRs with the (α4)3(β2)2 stoichiometry, but not for nAChRs with two α4 subunits (Mazzaferro et al., 2019). It acts as a Type II PAM, reopening closed channels by affecting the rate and probability of moving from a closed to an open state (Mazzaferro et al., 2019) Fig. 3.3. The β2-α4 interface was shown to be needed for the PAM activity. NS9283 had no activity at α3-containing nAChRs (Timmermann et al., 2012). NS9283 didn't bind to the orthosteric site. NS9283 has been useful in some models of pain (Chatzidaki & Millar, 2015). NS9283 also improved episodic memory, reference memory, and attention in rats (Timmermann et al., 2012). NS9283 also potentiated α2β2 and α2β4 nAChRs with a similar potency to that on α4β2 nAChRs (Timmermann et al., 2012).

3.6.1.5 Desformylflustrabromine (dFBr)

dFBr is a PAM for α2β2 and α4β2 nAChRs, most likely acting in the channel (Pandya & Yakel, 2011). dFBr also blocked the interactions of Aβ1–42 on both of these subtypes, suggesting a possible role in treatment of AD (Pandya & Yakel, 2011).

3.6.2 Negative allosteric modulators (NAMs)

Several studies have identified NAMs for nAChRs. A number of NAMs with activity toward α3β4 nAChRs were shown to act as noncompetitive inhibitors at an allosteric site (González-Cestari et al., 2009). These compounds didn't compete well with epibatidine binding at the orthosteric site, and some increased the apparent affinity of epibatidine for the orthosteric suite. This may be due to the NAMs causing desensitization. The potential binding site of one of these compounds, COB-3, was mapped using a molecular dynamic simulation to the alpha/beta interface about 7 Angstroms from the orthosteric site (González-Cestari et al., 2009). These studies were extended to demonstrate sequence differences in this allosteric site between subtypes and show a basis for compounds with selectivity for human α4β2 nAChRs vs. α3β4 nAChRs (Henderson et al., 2010, 2012; Pavlovicz et al., 2011). The allosteric site on α4β2 and α3β4 nAChRs was defined by molecular dynamic simulations and free energy of binding calculations and then confirmed by mutagenesis of the target site. KAB-18 was developed with selectivity for α4β2 and no activity at α3β4 nAChRs (Henderson et al., 2010). Docking studies combined with mutagenesis showed that KAB-18 binds near (10 Angstroms) the orthosteric site at the alpha/beta interface with the interacting amino acids mostly on the beta subunit (Pavlovicz et al., 2011). The differences in selectivity between α4β2 and α3β4 nAChRs were accounted for by slight differences in the sequences of the allosteric sites between these subtypes (Henderson et al., 2010, 2012; Pavlovicz et al., 2011). Additional compounds such as DDR-13 with a greater potency at α4β2 nAChRs were developed. These may not have sufficient potency to be used clinically but provide good lead compounds for drugs that can specific inhibit α4β2 nAChRs without the potential side effects of conventional antagonists.

NAMs for α7 nAChRs have also been identified (Smelt et al., 2018). DB04763, DB08122, and pefloxacin behaved as NAMs for α7 nAChRs expressed in Xenopus oocytes. Docking studies and mutagenesis supported a location of action for these NAMs at a transmembrane site (Smelt et al.,

2018). Another α7 NAM, 1,2,3,3*a*,4,8*b*-hexahydro-2-benzyl-6-*N*, *N*-dimethylamino-1-methylindeno[1,2,-*b*]pyrrole (HDMP), inhibited α4β2, and α3β4 nAChRs at low μM levels, but α7 nAChRs at nM levels (Abdrakhmanova et al., 2010). HDMP didn't affect binding of epibatidine or MLA, suggesting a noncompetitive allosteric site of action. It also blocked nicotine-induced analgesic effects in a mouse tail-flick assay (Abdrakhmanova et al., 2010).

3.6.3 Allosteric agonists and Ago-PAMs

Allosteric agonists activate nAChRs without an orthosteric agonist being present (Papke & Horenstein, 2021). They act at the allosteric site that the PAM uses, not at the orthosteric site. Some of these compounds also increase activation as a normal PAM would do, thus they have been called "ago-PAMS" (Papke & Horenstein, 2021). GAT-107 (active isomer of 4BP-TQS) is an ago-PAM that alone produces activation of α7 nAChRs, but also augments the signal as a normal PAM. It has a transmembrane-binding site (Papke & Horenstein, 2021) and at allosteric sites in the ECD separate from the orthosteric site.

3.6.4 Silent agonists

Silent agonists don't activate the nAChR, but induce desensitized states (Papke & Horenstein, 2021). They bind at an extended orthosteric site. NS6740 is a silent agonist. The activity of silent agonists is revealed when a PAM is also applied. NS6740 has a low efficacy for channel opening, but application of PNU120596 can increase activity, and NS6740 with an ago-PAM GAT-107 can generate large currents (Papke & Horenstein, 2021).

3.6.5 Endogenous modulators

Several endogenous molecules can act to affect nAChR function without being agonists or antagonists. Some positive modulators include SLURP-1, Lypd6, choline, 17 *b*-estradiol (α4β2), and even Zn+ ions. Some compounds with negative modulatory effects include Lynx 1 and 2 and Zn+ (Arias, 2010). Some of the endogenous molecules have subtype specificity and others don't (Arias, 2010). These will be covered in Chapter 9.

3.7 Summary

The pharmacology of nAChRs summarized above presents only part of the vast array of modulators for nAChRs. These compounds have provided great insights into the structure of nAChRs. These have also been useful in localizing nAChRs and assigning subtype-specific functions. nAChR function is complex and can be modulated in multiple ways to activate, desensitize, reactivate from desensitization, and antagonize nAChRs. Control of the activity of specific nAChR subtypes is key to developing drugs targeted for treatment of schizophrenia, AD, depression, and for smoking cessation.

References

Abdrakhmanova, G. R., Blough, B. E., Nesloney, C., Navarro, H. A., Damaj, M. I., & Carroll, F. I. (2010). In vitro and in vivo characterization of a novel negative allosteric modulator of neuronal nAChRs. *Neuropharmacology*, *59*(6), 511–517. https://doi.org/10.1016/j.neuropharm.2010.07.006.

Abraham, N., & Lewis, R. J. (2018). Neuronal nicotinic acetylcholine receptor modulators from cone snails. *Marine Drugs*, *16*(6). https://doi.org/10.3390/md16060208.

Absalom, N. L., Quek, G., Lewis, T. M., Qudah, T., von Arenstorff, I., Ambrus, J. I., Harpsøe, K., Karim, N., Balle, T., Mcleod, M. D., & Chebib, M. (2013). Covalent trapping of methyllycaconitine at the α4–α4 interface of the α4β2 nicotinic acetylcholine receptor. *Journal of Biological Chemistry*, *288*(37), 26521–26532. https://doi.org/10.1074/jbc.m113.475053.

Alkondon, M., Pereira, E. F. R., Cortes, W. S., Maelicke, A., & Albuquerque, E. X. (1997). Choline is a selective agonist of α7 nicotinic acetylcholine receptors in the rat brain neurons. *European Journal of Neuroscience*, *9*(12), 2734–2742. https://doi.org/10.1111/j.1460-9568.1997.tb01702.x.

Andreasen, J., Olsen, G., Wiborg, O., & Redrobe, J. (2009). Antidepressant-like effects of nicotinic acetylcholine receptor antagonists, but not agonists, in the mouse forced swim and mouse tail suspension tests. *Journal of Psychopharmacology*, *23*(7), 797–804. https://doi.org/10.1177/0269881108091587.

Arias, H. R. (2010). Positive and negative modulation of nicotinic receptors. In *Vol. 80 (C). Advances in protein chemistry and structural biology* (pp. 153–203). Academic Press Inc. https://doi.org/10.1016/B978-0-12-381264-3.00005-9.

Arneric, S. P., Holladay, M., & Williams, M. (2007). Neuronal nicotinic receptors: A perspective on two decades of drug discovery research. *Biochemical Pharmacology*, *74*(8), 1092–1101. https://doi.org/10.1016/j.bcp.2007.06.033.

Azam, L., & McIntosh, J. M. (2009). Alpha-conotoxins as pharmacological probes of nicotinic acetylcholine receptors. *Acta Pharmacologica Sinica*, *30*(6), 771–783. https://doi.org/10.1038/aps.2009.47.

Bagdas, D., Meade, J. A., Alkhlaif, Y., Muldoon, P. P., Carroll, F. I., & Damaj, M. I. (2018). Effect of nicotine and alpha-7 nicotinic modulators on visceral pain–induced conditioned place aversion in mice. *European Journal of Pain (United Kingdom)*, *22*(8), 1419–1427. https://doi.org/10.1002/ejp.1231.

Bertrand, D, & Terry, AV. (2018). The wonderland of neuronal nicotinic acetylcholine receptors. *Biochemical Pharmacology*, *151*, 214–225.

Brown, J. L., & Wonnacott, S. (2015). Sazetidine-A activates and desensitizes native α7 nicotinic acetylcholine receptors. *Neurochemical Research*, *40*(10), 2047–2054. https://doi.org/10.1007/s11064-014-1302-6.

Campling, B. G., Kuryatov, A., Lindstrom, J., & Simon, S. A. (2013). Acute activation, desensitization and smoldering activation of human acetylcholine receptors. *PLoS ONE*, *8*(11), e79653. https://doi.org/10.1371/journal.pone.0079653.

Carbone, A., Moroni, M., Groot-Kormelink, P.-J., & Bermudez, I. (2009). Pentameric concatenated (α4)(2)(β2)(3) and (α4)(3)(β2)(2) nicotine acetylcholine receptors: Subunit arrangement determines functional expression. *British Journal of Pharmacology*, *156*, 970–981.

Cecchini, M., & Changeux, J.-P. (2022). Nicotinic receptors: From protein allostery to computational neuropharmacology. *Molecular Aspects of Medicine*, *84*, 101044. https://doi.org/10.1016/j.mam.2021.101044.

Changeux, J. P. (2012). The nicotinic acetylcholine receptor: The founding father of the pentameric ligand-gated ion channel superfamily. *Journal of Biological Chemistry*, *287* (48), 40207–40215. https://doi.org/10.1074/jbc.R112.407668.

Changeux, J.-P. (2018). The nicotinic acetylcholine receptor: a typical 'allosteric machine'. *Philosophical Transactions of the Royal Society B: Biological Sciences*, *373*(1749), 20170174. https://doi.org/10.1098/rstb.2017.0174.

Chatzidaki, A., & Millar, N. S. (2015). Allosteric modulation of nicotinic acetylcholine receptors. *Biochemical Pharmacology*, *97*(4), 408–417. https://doi.org/10.1016/j.bcp.2015.07.028.

Clarke, P. B. S., Schwartz, R. D., Paul, S. M., Pert, C. B., & Pert, A. (1985). Nicotinic binding in rat brain: Autoradiographic comparison of [3H]acetylcholine, [3H]nicotine, and [125I]-α-bungarotoxin. *Journal of Neuroscience*, *5*(5), 1307–1315. https://doi.org/10.1523/jneurosci.05-05-01307.1985.

Coe, J., Brooks, P., Vetelino, M., Wirtz, M. C., Arnold, E., Huang, J., Sands, S., Davis, T., Lebel, L., Fox, C., Shrikhande, A., Heym, J., Schaeffer, E., Rollema, H., Lu, Y., Mansbach, R., Chambers, L., Rovetti, C., Schulz, D., … Neill, B. T. (2005). Varenicline: An a4b2 nicotinc receptor partial agonist for smoking cessation. *Journal of Medicinal Chemistry*, *48*, 3473–3477.

Couturier, S., Bertrand, D., Matter, J. M., Hernandez, M. C., Bertrand, S., Millar, N., Valera, S., Barkas, T., & Ballivet, M. (1990). A neuronal nicotinic acetylcholine receptor subunit (α7) is developmentally regulated and forms a homo-oligomeric channel blocked by α-BTX. *Neuron*, *5*(6), 847–856. https://doi.org/10.1016/0896-6273(90)90344-F.

Cucchiaro, G., Xiao, Y., Gonzalez-Sulser, A., & Kellar, K. J. (2008). Analgesic effects of sazetidine-a, a new nicotinic cholinergic drug. *Anesthesiology*, *109*(3), 512–519. https://doi.org/10.1097/ALN.0b013e3181834490.

Damaj, M. I., Wiley, J. L., Martin, B. R., & Papke, R. L. (2005). In vivo characterization of a novel inhibitor of CNS nicotinic receptors. *European Journal of Pharmacology*, *521*(1–3), 43–48. https://doi.org/10.1016/j.ejphar.2005.06.056.

Davis, J. A., & Gould, T. J. (2006). The effects of DHBE and MLA on nicotine-induced enhancement of contextual fear conditioning in C57BL/6 mice. *Psychopharmacology*, *184*(3–4), 345–352. https://doi.org/10.1007/s00213-005-0047-y.

Decker, M. W., Curzon, P., Holladay, M. W., Nikkel, A. L., Bitner, R. S., Bannon, A. W., Donnelly-Roberts, D. L., Puttfarcken, P. S., Kuntzweiler, T. A., Briggs, C. A., Williams, M., & Arneric, S. P. (1998). The role of neuronal nicotinic acetylcholine receptors in antinociception: Effects of ABT-594. *Journal of Physiology Paris*, *92*(3–4), 221–224. https://doi.org/10.1016/S0928-4257(98)80014-4.

Derbyshire, E., & Obeid, R. (2021). Choline, neurological development and brain function: A systematic review focusing on the first 1000 days. *Nutrients, 12,* 1731. https://doi.org/10.3390/nu12061731.

Dezfuli, G., Olson, T., Martin, L., Keum, Y., Siegars, B., Desai, A., Uitz, M., Sahibzada, N., Gillis, R., & Kellar, K. J. (2020). a4b2 nicotinic acetylcholine receptors intrinsically influence body weight in mice. *Neuropharmacology, 166.*

Freedman, R., Olincy, A., Buchanan, R. W., Harris, J. G., Gold, J. M., Johnson, L., Allensworth, D., Guzman-Bonilla, A., Clement, B., Ball, M. P., Kutnick, J., Pender, V., Martin, L. F., Stevens, K. E., Wagner, B. D., Zerbe, G. O., Soti, F., & Kem, W. R. (2008). Initial phase 2 trial of a nicotinic agonist in schizophrenia. *American Journal of Psychiatry, 165*(8), 1040–1047. https://doi.org/10.1176/appi.ajp.2008.07071135.

Gharpure, A., Teng, J., Zhuang, Y., Noviello, C. M., Walsh, R. M., Cabuco, R., Howard, R. J., Zaveri, N. T., Lindahl, E., & Hibbs, R. E. (2019). Agonist selectivity and ion permeation in the α3β4 ganglionic nicotinic receptor. *Neuron, 104*(3), 501–511.e6. https://doi.org/10.1016/j.neuron.2019.07.030.

Gielen, M., & Corringer, P. J. (2018). The dual-gate model for pentameric ligand-gated ion channels activation and desensitization. *Journal of Physiology, 596*(10), 1873–1902. https://doi.org/10.1113/JP275100.

González-Cestari, T. F., Henderson, B. J., Pavlovicz, R. E., McKay, S. B., El-Hajj, R. A., Pulipaka, A. B., Orac, C. M., Reed, D. D., Boyd, R. T., Zhu, M. X., Li, C., Bergmeier, S. C., & McKay, D. B. (2009). Effect of novel negative allosteric modulators of neuronal nicotinic receptors on cells expressing native and recombinant nicotinic receptors: Implications for drug discovery. *Journal of Pharmacology and Experimental Therapeutics, 328*(2), 504–515. https://doi.org/10.1124/jpet.108.144576.

Gotti, C., & Clementi, F. (2021). Cytisine and cytisine derivatives. *Pharmacological Research, 170.*

Grau, V., Richter, K., Hone, A. J., & McIntosh, J. M. (2019). Conopeptides [V11L;V16D] ArIB and RgIA4: Powerful tools for the identification of novel nicotinic acetylcholine receptors in monocytes. *Frontiers in Pharmacology, 9.* https://doi.org/10.3389/fphar.2018.01499.

Haig, G. M., Bain, E. E., Robieson, W. Z., Baker, J. D., & Othman, A. A. (2016). A randomized trial to assess the efficacy and safety of ABT-126, a selective α7 nicotinic acetylcholine receptor agonist, in the treatment of cognitive impairment in schizophrenia. *American Journal of Psychiatry, 173*(8), 827–835. https://doi.org/10.1176/appi.ajp.2015.15010093.

Harvey, S. C., & Luetje, C. W. (1996). Determinants of competitive antagonist sensitivity on neuronal nicotinic receptor β subunits. *Journal of Neuroscience, 16*(12), 3798–3806. https://doi.org/10.1523/jneurosci.16-12-03798.1996.

Harvey, S. C., Maddox, F. N., & Luetje, C. W. (1996). Multiple determinants of dihydro-β-erythroidine sensitivity on rat neuronal nicotinic receptor α subunits. *Journal of Neurochemistry, 67*(5), 1953–1959. https://doi.org/10.1046/j.1471-4159.1996.67051953.x.

Henderson, B. J., González-Cestari, T. F., Yi, B., Pavlovicz, R. E., Boyd, R. T., Li, C., Bergmeier, S. C., & McKay, D. B. (2012). Defining the putative inhibitory site for a selective negative allosteric modulator of human α4β2 neuronal nicotinic receptors. *ACS Chemical Neuroscience, 3*(9), 682–692. https://doi.org/10.1021/cn300035f.

Henderson, B. J., Pavlovicz, R. E., Allen, J. D., González-Cestari, T. F., Orac, C. M., Bonnell, A. B., Zhu, M. X., Boyd, R. T., Li, C., Bergmeier, S. C., & McKay, D. B. (2010). Negative allosteric modulators that target human α4β2 neuronal nicotinic receptors. *Journal of Pharmacology and Experimental Therapeutics, 334*(3), 761–774. https://doi.org/10.1124/jpet.110.168211.

Hong, L. E., Thaker, G. K., McMahon, R. P., Summerfelt, A., RachBeisel, J., Fuller, R. L., Wonodi, I., Buchanan, R. W., Myers, C., Heishman, S. J., Yang, J., & Nye, A. (2011).

Effects of moderate-dose treatment with varenicline on neurobiological and cognitive biomarkers in smokers and nonsmokers with schizophrenia or schizoaffective disorder. *Archives of General Psychiatry, 68*(12), 1195–1206. https://doi.org/10.1001/archgenpsychiatry.2011.83.

Hurst, R. S., Hajós, M., Raggenbass, M., Wall, T. M., Higdon, N. R., Lawson, J. A., … Arneric, S. P. (2005). A novel positive allosteric modulator of the α7 neuronal nicotinic acetylcholine receptor: In vitro and in vivo characterization. *Journal of Neuroscience, 25*(17), 4396–4405. https://doi.org/10.1523/JNEUROSCI.5269-04.2005.

Kem, W. R. (2000). The brain α7 nicotinic receptor may be an important therapeutic target for the treatment of Alzheimer's disease: Studies with DMXBA (GTS-21). *Behavioural Brain Research, 113*(1–2), 169–181. https://doi.org/10.1016/S0166-4328(00)00211-4.

Kuryatov, A., & Lindstrom, J. (2011). Expression of functional human α6β2β3* acetylcholine receptors in *Xenopus laevis* oocytes achieved through subunit chimeras and concatamers. *Molecular Pharmacology, 79*(1), 126–140. https://doi.org/10.1124/mol.110.066159.

Lebbe, E. K. M., Peigneur, S., Wijesekara, I., & Tytgat, J. (2014). Conotoxins targeting nicotinic acetylcholine receptors: An overview. *Marine Drugs, 12*(5), 2970–3004. https://doi.org/10.3390/md12052970.

Leonard, S., Mexal, S., & Freedman, R. (2007). Genetics of smoking and schizophrenia. *Journal of Dual Diagnosis, 3*(3–4), 43–59. https://doi.org/10.1300/J374v03n03_05.

Levin, E. D., Cauley, M., & Rezvani, A. H. (2013). Improvement of attentional function with antagonism of nicotinic receptors in female rats. *European Journal of Pharmacology, 702*(1–3), 269–274. https://doi.org/10.1016/j.ejphar.2013.01.056.

Lewis, A. S., Pittenger, S. T., Mineur, Y. S., Stout, D., Smith, P. H., & Picciotto, M. R. (2018). Bidirectional regulation of aggression in mice by hippocampal alpha-7 nicotinic acetylcholine receptors. *Neuropsychopharmacology, 43*(6), 1267–1275. https://doi.org/10.1038/npp.2017.276.

Li, P., & Steinbach, J. H. (2010). The neuronal nicotinic α4β2 receptor has a high maximal probability of being open. *British Journal of Pharmacology, 160*(8), 1906–1915. https://doi.org/10.1111/j.1476-5381.2010.00761.x.

Luetje, C., & Patrick, J. (1991). Both alpha- and beta-subunits contribute to the agonist sensitivity of neuronal nicotinic acetylcholine receptors. *The Journal of Neuroscience, 11*(3), 837–845. https://doi.org/10.1523/jneurosci.11-03-00837.1991.

Marks, MJ, Laverty, DS, Whiteaker, P, Salminen, O, Grady, SR, McIntosh, JM, & Collins, AC. (2010). John Daly's compound, epibatidine, facilitates identification of nicotinic receptor subtypes. *Journal of Molecular Neuroscience, 40*(1–2), 96–104.

Matta, S. G., Balfour, D. J., Benowitz, N. L., Boyd, R. T., Buccafusco, J. J., Caggiula, A. R., Craig, C. R., Collins, A. C., Damaj, M. I., Donny, E. C., Gardiner, P. S., Grady, S. R., Heberlein, U., Leonard, S. S., Levin, E. D., Lukas, R. J., Markou, A., Marks, M. J., McCallum, S. E., … Zirger, J. M. (2007). Guidelines on nicotine dose selection for in vivo research. *Psychopharmacology, 190*(3), 269–319. https://doi.org/10.1007/s00213-006-0441-0.

Maurer, S. V., Kong, C., Terrando, N., & Williams, C. L. (2021). Dietary choline protects against cognitive decline after surgery in mice. *Frontiers in Cellular Neuroscience, 15*. https://doi.org/10.3389/fncel.2021.671506.

Mazzaferro, S., Bermudez, I., & Sine, S. M. (2019). Potentiation of a neuronal nicotinic receptor via pseudo-agonist site. *Cellular and Molecular Life Sciences, 76*(6), 1151–1167. https://doi.org/10.1007/s00018-018-2993-7.

Miledi, R., Molinoff, P., & Potter, L. T. (1971). Biological sciences: Isolation of the cholinergic receptor protein of Torpedo electric tissue. *Nature, 229*(5286), 554–557. https://doi.org/10.1038/229554a0.

Moglie, M. J., Marcovich, I., Corradi, J., Carpaneto Freixas, A. E., Gallino, S., Plazas, P. V., Bouzat, C., Lipovsek, M., & Elgoyhen, A. B. (2021). Loss of choline Agonism in the inner ear hair cell nicotinic acetylcholine receptor linked to the α10 subunit. *Frontiers in Molecular Neuroscience, 14.* https://doi.org/10.3389/fnmol.2021.639720.

Nickell, J. R., Grinevich, V. P., Siripurapu, K. B., Smith, A. M., & Dwoskin, L. P. (2013). Potential therapeutic uses of mecamylamine and its stereoisomers. *Pharmacology Biochemistry and Behavior, 108,* 28–43. https://doi.org/10.1016/j.pbb.2013.04.005.

Nirthanan, S. (2020). Snake three-finger a-neurotoxins and nicotinic acetylcholine receptors: Molecules, mechanisms and medicine. *Biochemical Pharmacology, 181.*

Norleans, J., Wang, J., Kuryatov, A., Leffler, A., Doebelin, C., Kamenecka, T. M., & Lindstrom, J. (2019). Discovery of an intrasubunit nicotinic acetylcholine receptor–binding site for the positive allosteric modulator Br-PBTC. *Journal of Biological Chemistry, 294*(32), 12132–12145. https://doi.org/10.1074/jbc.RA118.006253.

Pandya, A., & Yakel, J. L. (2011). Allosteric modulator desformylflustrabromine relieves the inhibition of α2β2 and α4β2 nicotinic acetylcholine receptors by β–amyloid 1-42 peptide. *Journal of Molecular Neuroscience, 45*(1), 42–47. https://doi.org/10.1007/s12031-011-9509-3.

Papke, R. L. (2014). Merging old and new perspectives on nicotinic acetylcholine receptors. *Biochemical Pharmacology, 89*(1), 1–11. https://doi.org/10.1016/j.bcp.2014.01.029.

Papke, R. L., Dwoskin, L. P., Crooks, P. A., Zheng, G., Zhang, Z., McIntosh, J. M., & Stokes, C. (2008). Extending the analysis of nicotinic receptor antagonists with the study of α6 nicotinic receptor subunit chimeras. *Neuropharmacology, 54*(8), 1189–1200. https://doi.org/10.1016/j.neuropharm.2008.03.010.

Papke, R. L., & Heinemann, S. F. (1994). Partial agonist properties of cytisine on neuronal nicotinic receptors containing the β2 subunit. *Molecular Pharmacology, 45*(1), 142–149.

Papke, R. L., & Horenstein, N. A. (2021). Therapeutic targeting of α7 nicotinic acetylcholine receptors. *Pharmacological Reviews, 73*(3), 1118–1149. https://doi.org/10.1124/pharmrev.120.000097.

Papke, R. L., & Lindstrom, J. M. (2020). Nicotinic acetylcholine receptors: Conventional and unconventional ligands and signaling. *Neuropharmacology, 168,* 108021. https://doi.org/10.1016/j.neuropharm.2020.108021.

Papke, R. L., Wecker, L., & Stitzel, J. A. (2010). Activation and inhibition of mouse muscle and neuronal nicotinic acetylcholine receptors expressed in xenopus oocytes. *Journal of Pharmacology and Experimental Therapeutics, 333*(2), 501–518. https://doi.org/10.1124/jpet.109.164566.

Paterson, D., & Nordberg, A. (2000). Neuronal nicotinic receptors in the human brain. *Progress in Neurobiology, 61*(1), 75–111. https://doi.org/10.1016/S0301-0082(99)00045-3.

Pavlovicz, R. E., Henderson, B. J., Bonnell, A. B., Boyd, R. T., McKay, D. B., & Li, C. (2011). Identification of a negative allosteric site on human α4β2 and α3β4 neuronal nicotinic acetylcholine receptors. *PLoS ONE, 6*(9). https://doi.org/10.1371/journal.pone.0024949.

Rezvani, A. H., Cauley, M., Xiao, Y., Kellar, K. J., & Levin, E. D. (2013). Effects of chronic sazetidine-A, a selective α4β2 neuronal nicotinic acetylcholine receptors desensitizing agent on pharmacologically- induced impaired attention in rats. *Psychopharmacology, 226*(1), 35–43. https://doi.org/10.1007/s00213-012-2895-6.

Rollema, H., Hajós, M., Seymour, P. A., Kozak, R., Majchrzak, M. J., Guanowsky, V., Horner, W. E., Chapin, D. S., Hoffmann, W. E., Johnson, D. E., Mclean, S., Freeman, J., & Williams, K. E. (2009). Preclinical pharmacology of the α4β2 nAChR partial agonist varenicline related to effects on reward, mood and cognition. *Biochemical Pharmacology, 78*(7), 813–824. https://doi.org/10.1016/j.bcp.2009.05.033.

Rollema, H., & Hurst, R. S. (2018). The contribution of agonist and antagonist activities of α4β2* nAChR ligands to smoking cessation efficacy: A quantitative analysis of literature data. *Psychopharmacology*, *235*(9), 2479–2505. https://doi.org/10.1007/s00213-018-4921-9.

Schoepfer, R., Conroy, W. G., Whiting, P., Gore, M., & Lindstrom, J. (1990). Brain α-bungarotoxin binding protein cDNAs and MAbs reveal subtypes of this branch of the ligand-gated ion channel gene superfamily. *Neuron*, *5*(1), 35–48. https://doi.org/10.1016/0896-6273(90)90031-A.

Smelt, C. L. C., Sanders, V. R., Newcombe, J., Burt, R. P., Sheppard, T. D., Topf, M., & Millar, N. S. (2018). Identification by virtual screening and functional characterisation of novel positive and negative allosteric modulators of the α7 nicotinic acetylcholine receptor. *Neuropharmacology*, *139*, 194–204. https://doi.org/10.1016/j.neuropharm.2018.07.009.

Telles-Longui, M., Mourelle, D., Schöwe, N. M., Cipolli, G. C., Malerba, H. N., Buck, H. S., & Viel, T. A. (2019). α7 nicotinic ACh receptors are necessary for memory recovery and neuroprotection promoted by attention training in amyloid-β-infused mice. *British Journal of Pharmacology*, *176*(17), 3193–3205. https://doi.org/10.1111/bph.14744.

Terry, A. V., & Callahan, P. M. (2019). Nicotinic acetylcholine receptor ligands, cognitive function, and preclinical approaches to drug discovery. *Nicotine and Tobacco Research*, *21*(3), 383–394. https://doi.org/10.1093/ntr/nty166.

Terry, A. V., Callahan, P. M., & Hernandez, C. M. (2015). Nicotinic ligands as multifunctional agents for the treatment of neuropsychiatric disorders. *Biochemical Pharmacology*, *97*(4), 388–398. https://doi.org/10.1016/j.bcp.2015.07.027.

Timmermann, D. B., Grønlien, J. H., Kohlhaas, K. L., Nielsen, E., Dam, E., Jørgensen, T. D., Ahring, P. K., Peters, D., Holst, D., Chrsitensen, J. K., Malysz, J., Briggs, C. A., Gopalakrishnan, M., & Olsen, G. M. (2007). An allosteric modulator of the α7 nicotinic acetylcholine receptor possessing cognition-enhancing properties in vivo. *Journal of Pharmacology and Experimental Therapeutics*, *323*(1), 294–307. https://doi.org/10.1124/jpet.107.120436.

Timmermann, D. B., Sandager-Nielsen, K., Dyhring, T., Smith, M., Jacobsen, A. M., Nielsen, E., Grunnet, M., Christensen, J. K., Peters, D., Kohlhaas, K., Olsen, G. M., & Ahring, P. K. (2012). Augmentation of cognitive function by NS9283, a stoichiometry-dependent positive allosteric modulator of α2- and α4-containing nicotinic acetylcholine receptors. *British Journal of Pharmacology*, *167*(1), 164–182. https://doi.org/10.1111/j.1476-5381.2012.01989.x.

Turner, J. R., Castellano, L. M., & Blendy, J. A. (2010). Nicotinic partial agonists varenicline and sazetidine-A have differential effects on affective behavior. *Journal of Pharmacology and Experimental Therapeutics*, *334*(2), 665–672. https://doi.org/10.1124/jpet.110.166280.

van Goethem, N. P., Paes, D., Puzzo, D., Fedele, E., Rebosiio, C., Gulisano, W., Palmeri, A., Wennogle, L., Peng, Y., Bertrand, D., & Prickaerts, J. (2019). Antagonizing a7 nicotinic receptors with methllyaconitine (MLA) potentiates receptors activity and memory acquisition. *Cellular Signalling*, *62*.

Wang, J., & Lindstrom, J. (2018). Orthosteric and allosteric potentiation of heteromeric neuronal nicotinic acetylcholine receptors. *British Journal of Pharmacology*, *175*(11), 1805–1821. https://doi.org/10.1111/bph.13745.

Williams, D. K., Stokes, C., Horenstein, N. A., & Papke, R. L. (2011). The effective opening of nicotinic acetylcholine receptors with single agonist binding sites. *Journal of General Physiology*, *137*(4), 369–384. https://doi.org/10.1085/jgp.201010587 (PubMed: 21444659).

Williams, D. K., Wang, J., & Papke, R. L. (2011a). Investigation of the molecular mechanism of the α7 nicotinic acetylcholine receptor positive allosteric modulator PNU-120596 provides evidence for two distinct desensitized states. *Molecular Pharmacology*, *80*(6), 1013–1032. https://doi.org/10.1124/mol.111.074302.

Williams, D. K., Wang, J., & Papke, R. L. (2011b). Positive allosteric modulators as an approach to nicotinic acetylcholine receptor-targeted therapeutics: Advantages and limitations. *Biochemical Pharmacology*, *82*(8), 915–930. https://doi.org/10.1016/j.bcp.2011.05.001.

Wonnacott, S., & Barik, J. (2007). Nicotinic ACh receptors. *Tocris Reviews*, *28*.

Wu, J., Liu, Q., Tang, P., Mikkelsen, J. D., Shen, J., Whiteaker, P., & Yakel, J. L. (2016). Heteromeric α7β2 nicotinic acetylcholine receptors in the brain. *Trends in Pharmacological Sciences*, *37*(7), 562–574. https://doi.org/10.1016/j.tips.2016.03.005.

Xiao, Y., Fan, H., Musachio, J. L., Wei, Z. L., Chellappan, S. K., Kozikowski, A. P., & Kellar, K. J. (2006). Sazetidine-A, a novel ligand that desensitizes α4β2 nicotinic acetylcholine receptors without activating them. *Molecular Pharmacology*, *70*(4), 1454–1460. https://doi.org/10.1124/mol.106.027318.

Ylilauri, M. P. T., Voutilainen, S., Lönnroos, E., Virtanen, H. E. K., Tuomainen, T. P., Salonen, J. T., & Virtanen, J. K. (2019). Associations of dietary choline intake with risk of incident dementia and with cognitive performance: The Kuopio Ischaemic Heart Disease Risk Factor Study. *American Journal of Clinical Nutrition*, *110*(6), 1416–1423. https://doi.org/10.1093/ajcn/nqz148.

Further reading

Brody, A. L., Mandelkern, M. A., London, E. D., Olmstead, R. E., Farahi, J., Scheibal, D., Jou, J., Allen, V., Tiongson, E., Chefer, S. I., Koren, A. O., & Mukhin, A. G. (2006). Cigarette smoking saturates brain α4β2 nicotinic acetylcholine receptors. *Archives of General Psychiatry*, *63*(8), 907–915. https://doi.org/10.1001/archpsyc.63.8.907.

van Hout, M., Valdes, A., Christensen, S. B., Tran, P. T., Watkins, M., Gajewiak, J., Jensen, A. A., Olivera, B. M., & McIntosh, J. M. (2019). α-Conotoxin VnIB from Conus ventricosus is a potent and selective antagonist of α6β4* nicotinic acetylcholine receptors. *Neuropharmacology*, *157*. https://doi.org/10.1016/j.neuropharm.2019.107691.

Wilkerson, J. L., Deba, F., Crowley, M. L., Hamouda, A. K., & McMahon, L. R. (2020). Advances in the in vitro and in vivo pharmacology of Alpha4beta2 nicotinic receptor positive allosteric modulators. *Neuropharmacology*, *168*. https://doi.org/10.1016/j.neuropharm.2020.108008.

CHAPTER FOUR

Nicotine addiction: The role of specific nAChR subtypes and treatments

4.1 Introduction

Nicotine is one of the most highly used and addictive drugs in the world. The WHO in 2021 estimated about 1.3 billion people smoke while many others chew or use vaping products with nicotine. About 34 million Americans smoke, 14% of the adult population (Rigotti et al., 2022). By comparison, there are approximately 1 million daily heroin users. The number of smokers is much lower than 50 years ago in the United States, but the remaining large number of smokers worldwide is a health crisis. There are about 7 million deaths a year directly from tobacco product use and another million from second-hand smoke. Most lung cancer deaths are directly attributed to smoking. Higher rates of smoking are seen among people with lower levels of educational achievement, mental health issues, and users of other drugs. Economic status may also be involved in smoking occurrence, and more men than woman smoke. Most smokers start at a young age, usually between ages 15 and 25. Prevention of smoking during this age results in a low risk of future smoking (LeFoll et al., 2022).

4.2 Nicotine metabolism

Nicotine is a plant alkaloid that exists as (S)-nicotine and (R)-nicotine with (S)-nicotine as the major pharmacologically active molecule (Matta et al., 2007). Smoke contains numerous other toxic components such as carbon monoxide, formaldehyde, and toluene in addition to other alkaloids such as anabasine and nornicotine. An average cigarette is about 1%–2% tobacco. The average smoker achieves a blood level of 60–300 nM nicotine (Matta et al., 2007). Levels 2–3 times this can be present in the breast milk of a human smoker. Higher levels can also be present in fetal tissue in a smoking

Nicotinic Acetylcholine Receptors in Health and Disease
https://doi.org/10.1016/B978-0-12-819958-9.00006-2

mother. Nicotine is absorbed rapidly in the lungs, with arterial nicotine levels reaching 600 nM (Matta et al., 2007). Nicotine reaches the brain in about 10 s. This rapid high level increases the reinforcing properties of nicotine. Nicotine is rapidly distributed to all body tissue over about 20 min after a cigarette (Matta et al., 2007). Nicotine crosses the blood-brain barrier and also enters the fetal circulation.

Nicotine is metabolized in the liver with multiple products produced. A small amount is excreted by the kidneys. The major metabolite is cotinine with about 75% of nicotine converted to cotinine (Matta et al., 2007). Chronic smokers can achieve 15 times higher cotinine levels than nicotine since it has a longer half-life and its presence can be used to determine that someone has recently smoked. Most species used to study nicotine and nAChRs such as primates and mice metabolize nicotine in this manner. However, rats and guinea pigs create as much nicotine-N'-oxide as cotinine (Matta et al., 2007). The half-life of nicotine is about 2 h in humans and non-human primates, in rats 45 min and mice 6–7 min (Matta et al., 2007). In chronic smokers, nicotine also has a slow terminal removal component reflecting nicotine stored in tissues being gradually released over time that helps to maintain plasma nicotine levels. Even after overnight, smokers still may have a plasma level of 30 nM (Matta et al., 2007). These differences in metabolism are considered when designing experiments examining nicotine addiction.

Cytochrome P450 Family 2 Subfamily A Member 6 (CYP2A6) metabolizes nicotine to cotinine in the liver in mice and humans. CYP1B1/2 metabolizes nicotine in rats, not CYP2A6 (Matta et al., 2007). A number of allelic variants of CYP2A6 exist and metabolize nicotine at various rates. These polymorphisms and other biological variations lead to a large variability in nicotine levels across human smokers (Matta et al., 2007). This is a consideration as to how easy or difficult it is to become addicted to nicotine. In human or animal studies, the route of administration affects the rate of metabolism and is carefully considered when using animals to study the effects of nicotine.

Nicotine metabolism also varies in humans depending on race, age, sex, and smoking history (smokers have a lowered metabolism) (Matta et al., 2007). Nicotine elimination half-life is quite variable among individuals (Shiffman et al., 1992). Elderly smokers as well as Asians and African Americans metabolize nicotine more slowly. African Americans achieve higher cotinine levels relative to nicotine (Matta et al., 2007). Since cotinine has some agonist activity on nAChRs (Tan et al., 2021), variations in nicotine and cotinine metabolism may have effects on addiction. Diseases of the liver and drug use alter metabolism of nicotine and can affect susceptibility to nicotine's effects.

4.3 Presentation of nicotine addiction in humans and in animal models

Nicotine use results in a pleasant feeling, a reward. Smokers typically smoke thousands of cigarettes a year. Several factors influence who starts smoking including genetics, peer pressure, and cost. A smoker is sometimes defined in epidemiological studies as someone who has smoked over 100 cigarettes. At this step, dependence is often seen. The first response to nicotine (i.e., how aversive the first cigarettes are) may also influence progression to dependence . Changes in brain signaling occur with chronic use and conditioning to cues occurs. Withdrawal is precipitated in many smokers after stopping smoking for a short time (Shiffman et al., 1992). This can lead to craving and relapse. Smokers develop dependence and tolerance rapidly. Tolerance to smoking is demonstrated by the use of more cigarettes to achieve the same effect, in some cases 20 or more cigarettes a day (Bierut, 2009). Smoking (nicotine addiction) is progressive. The reward increases the chances of future smoking. In addition to this positive reinforcement, nicotine also reduces the activity of brain circuits that produce aversive effects (Wills et al., 2022). This process occurs not only in humans but also in mice, dogs, rats, and nonhuman primates. These animals also self-administer nicotine under fixed-ratio regimes. This pattern of use produces an inverted U-shaped dose–response curve, similar to that seen for other drugs of abuse such as cocaine or opiates (Wills et al., 2022). The interpretation of an inverted U-shaped dose–response curve is that positive and negative effects are produced at different doses. Increasing self-administration of the upward curve demonstrates increased reward, while the decreased response on the downward portion represents increased aversive effects. Smokers who have quit often relapse due to exposure to environmental cues that have been associated with previous smoking and have achieved a degree of motivational salience or withdrawal symptoms. (LeFoll & Goldberg, 2009). Initially, people smoke and either feel good and continue to smoke to reinforce this behavior or don't like it due to adverse effects and stop. Chronic use leads to nicotine-induced changes in the brain and signaling and conditioned behavior. Abstinence induces withdrawal effects, craving, and relapse.

The smoker's brain circuitry, neurotransmitter, and receptors levels adapt smoking behavior to optimize the rewarding effects and minimize the aversive ones. Nicotine use reinforces the behaviors associated with smoking, such as the feel of the cigarette in the mouth or hand that enhance the salience of these other features of smoking and promote increased nicotine use (LeFoll et al., 2022). Abstinence produces withdrawal symptoms and signs such as stress, anxiety, irritability, headache, nausea, low heart rate and blood pressure, memory problems, or problems in focusing (LeFoll et al., 2022; LeFoll & Goldberg, 2009). These negative effects increase the drive of the smoker to use cigarettes so as to mitigate the withdrawal effects.

Smokers space out cigarettes to maintain a minimum level that leads to avoidance of withdrawal (Shiffman et al., 1992). Chronic nicotine exposure as seen in smokers results in numerous neuronal changes including nAChR upregulation, changes in synaptic activity such as increased long-term potentiation (LTP), reduced glutamatergic signaling, and changes in specific RNA populations (Tapper et al., 2004). Chronic nicotine leads to upregulation of nAChR function as well (Tapper et al., 2004).

Some people are able to smoke a few cigarettes a day for several days a week and not progress to nicotine dependence. These nondependent smokers are sometimes described as "chippers." It is not clear why "chippers" don't progress to full dependence, but it was shown that they don't differ in the half-life of nicotine elimination compared with regular smokers (Shiffman et al., 1992). "Chippers" also don't differ in the amount of nicotine received from each cigarette, so with the limited smoking can't maintain a nicotine level that would prevent withdrawal in regular smokers (Shiffman et al., 1992). While "chippers" may not be fully dependent and have lower symptoms, they do experience cravings and have a lower autonomy over smoking (Wellman et al., 2006). Thus, while the motivation for this limited smoking is not clear, they do experience at least some limited drive to smoke.

A number of tests are available to determine the degree of nicotine dependence. The Fagerstrom test for Nicotine Dependence is often used (Heatherton et al., 1991). This test asks a series of six questions (Fig. 4.1) (https://cde.drugabuse.gov/sites/nida_cde/files/FagerstromTest_2014Mar24. pdf) related with smoking behavior. Scores are assigned based on the answers. A score of 7 or higher (10 is the highest) indicates a high level of dependence, 4–6 moderately dependent (Heatherton et al., 1991). Another test, the Heaviness of Smoking Index (HSI), is also used and asks two questions; how many cigarettes do you smoke a day, and when is the

NIDA Clinical Trials Network

Fagerstrom Test for Nicotine Dependence (FND)

Segment: _ _

Visit Number: _ _

Date of Assessment: (mm/dd/yyyy)   _ _/_ _/_ _ _ _

Do you currently smoke cigarettes?

☐ No   ☐ Yes

If "yes," read each question below. For each question, enter the answer choice which best describes your response.

1. **How soon after you wake up do you smoke your first cigarette?**

   ☐ Within 5 minutes   ☐ 31 to 60 minutes
   ☐ 6 to 30 minutes   ☐ After 60 minutes

2. **Do you find it difficult to refrain from smoking in places where it is forbidden** (e.g., in church, at the library, in the cinema)?

   ☐ No   ☐ Yes

3. **Which cigarette would you hate most to give up?**

   ☐ The first one in the morning   ☐ Any other

4. **How many cigarettes per day do you smoke?**

   ☐ 10 or less   ☐ 21 to 30
   ☐ 11 to 20   ☐ 31 or more

5. **Do you smoke more frequently during the first hours after waking than during the rest of the day?**

   ☐ No   ☐ Yes

6. **Do you smoke when you are so ill that you are in bed most of the day?**

   ☐ No   ☐ Yes

Comments:

Heatherton TF, Kozlowski LT Frecker RC (1991). The Fagerström Test for Nicotine Dependence: A revision of the Fagerström Tolerance Questionnaire. British Journal of Addiction 86:1119-27.

Fig. 4.1 Fagerstrom test for nicotine dependence. This test asks a series of six questions related with smoking behavior. Scores are assigned based on the answers. A score of 7 or higher (10 is the highest) indicates a high level of dependence, 4–6 moderately dependent.

(Continued)

NIDA Clinical Trials Network

Fagerstrom Test for Nicotine Dependence (FND)

Instructions

Clinic personnel will follow standard scoring to calculate score based on responses.

Your score was: (your level of dependence on nicotine is):   _ _

Heatherton TF, Kozlowski LT Frecker RC (1991). The Fagerström Test for Nicotine Dependence: A revision of the Fagerström Tolerance Questionnaire. British Journal of Addiction 86:1119-27.

Fig. 4.1, cont'd

first cigarette after waking? (LeFoll et al., 2022). The DSM-5 criteria are also used, but this test is not unique for nicotine addiction. This includes questions regarding craving, time spent using or finding tobacco, amount of tobacco use, and whether tolerance has developed or withdrawal symptoms are present.

Rats, mice, and primates have been used to study nicotine addiction. Animal studies have been used to identify the pathways and methods of addiction in humans and to test various potential smoking cessation drugs. Animal models exhibit withdrawal, reward, and tolerance as do humans. Several behavioral tests have been used such as conditioned place preference (CPP). For CPP, a three-chambered (or sometimes two-chambered) apparatus is used. Two chambers are used for conditioning, and the central chamber is for acclimating (Bagdas et al., 2018). The two outer chambers are painted so that the animal can distinguish the different chambers. At the beginning of the experiment, the animal is free to roam the chambers. Later, nicotine is administered to the animals when present in a specific chamber during several days of conditioning sessions. After conditioning, the animals are free to roam, and time spent in each chamber is measured. Time spent in the drug-paired chamber is a measure of the rewarding effect of nicotine. CPP is used with knockout (KO) mice to determine the role of specific subunits in nicotine addiction or with treatment of compounds that act upon specific nAChRs. Other drugs such as varenicline can be given to test for efficacy as smoking cessation aids. For example, varenicline blocks nicotine-induced CPP (Bagdas et al., 2018).

Nicotine can be given to animals such as mice, rats, or nonhuman primates in multiple ways. Use of these animals has provided extensive data relevant to human nicotine addiction. Methods are chosen that mimic various aspects of smoking behavior, such as the intermittent dosing seen in smokers throughout the day, or continuous exposure paradigms. These include in water, intramuscular, or subcutaneous injections, osmotic minipumps, and intravenous injections (Matta et al., 2007). Injections are often used when measuring the acute effects of nicotine such as on learning or attention. However, human smokers are chronically exposed to nicotine daily for years. To test the chronic effects in rodents or nonhuman primates, animals are given nicotine in water, by time and dosage-controlled release from osmotic minipumps or by self-administration.

Rodents and nonhuman primates can self-administer nicotine as a positive reinforcer with proper training. KO mice for α4, α6, or β2 subunits don't self-administer nicotine (Brunzell et al., 2014). Drug administration reinforces lever pushing to obtain more drug (LeFoll & Goldberg, 2009). Under a progressive ratio schedule, the animal must push the lever an increasing number of times to get the reward. This is used as a measure of the reinforcing effect of nicotine (LeFoll & Goldberg, 2009). Nicotine is an effective reinforcer in humans as well (LeFoll & Goldberg, 2009). Tests for the aversive effects of nicotine in animals are also used.

In addition, animals undergoing withdrawal demonstrate chewing, scratching, head nods, and increased anxiety that can be measured. Many studies that have developed what we know about nicotine addiction come from studies with rodents.

4.4 Reward pathways for nicotine

Multiple neural changes in signaling, nAChR expression, transmitter release, and nAChR function occur as a result of tobacco use. Nicotine is an agonist of nAChRs and thus can stabilize nAChRs in an open state. Nicotine also can promote desensitization of nAChRs. nAChR activity influences the release of a number of neurotransmitters including dopamine and GABA. nAChRs are expressed on specific pathways that mediate the various properties of nicotine involved in drug addiction. Nicotine exposure over time upregulates nAChRs. More nicotine-binding sites are present in the brains of smokers and rodents after nicotine exposure (Feduccia et al., 2012). This is the opposite to that which is often seen after chronic use of other drugs of abuse; this usually promotes receptor downregulation. This is thought to normalize function of a network after drug use (Feduccia et al., 2012). However, nicotine can cause a rapid desensitization of nAChRs and thus loss of function. The nAChR upregulation would normalize the function of the network and produce a higher responsiveness to nicotine that is important for the addictive process (Feduccia et al., 2012).

The addictive reinforcing properties of multiple drugs act through the mesolimbic dopamine system (Koob, 1992). This consists of projections of ventral tegmental area (VTA) catecholaminergic neurons to the nucleus accumbans (NAc), amydgdala, frontal cortex, septal area, and olfactory tubercle (Koob, 1992). Lesions to the VTA or the NAc block nicotine self-administration and the locomotor effects of nicotine (Picciotto & Kenny, 2021). Rats self-administer nicotine, and this self-administration is blocked by nAChR antagonists in the VTA as well (Tapper et al., 2004).

Multiple nAChRs are expressed in the VTA, NAc (ventral striatum), and prefrontal cortex (Fig. 4.2). Dopaminergic cell bodies in the VTA express

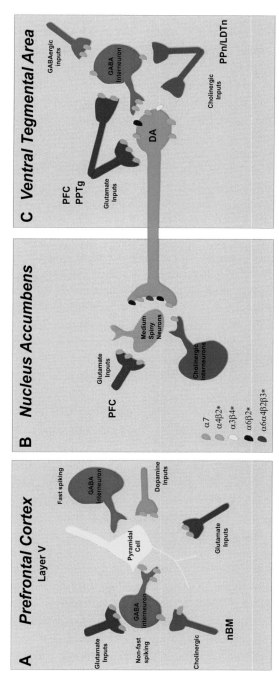

Fig. 4.2 nAChR distribution in the prefrontal cortex, nucleus accumbens, and ventral tegmental area. (A) Pyramidal cells in layer V of the PFC lack nAChRs, but their activity is modulated by excitatory and inhibitory neurons that do express them. There are two types of GABAergic interneurons, fast spiking and nonfast-spiking, with only the latter bearing nAChRs (α7 and α4β2*). Distinct populations of glutamatergic inputs express either α7 or α4β2* nAChRs while DA terminals projecting from the VTA contain α4β2*nAChRs. Cholinergic inputs into the PFC arise from the nucleus basalis of Meynert (nBM). (B) In the NAc, nAChRs (α4β2*, α6β2*, and α6α4β2*) expressed on DAergic terminals from the VTA mediate DA release based on the neuronal activity firing rate. A small population of tonically active cholinergic interneurons (∼2%) is synchronized with DA cell firing. Glutamatergic inputs from the PFC endow α7 nAChRs. (C) The VTA receives cholinergic innervation from the pedunculopontine (PPn) and laterodorsal tegmental nuclei (LDTn). In addition to the nAChRs localized on DA cell bodies, DAergic cell firing is modulated by α4β2* (and possibly α7) nAChRs expressed on GABAergic interneurons and excitatory glutamatergic afferents from the PFC and the PPn. *(From Feduccia, A. A., Chatterjee, S., & Bartlett, S. E. (2012). Neuronal nicotinic acetylcholine receptors: Neuroplastic changes underlying alcohol and nicotine addictions. Frontiers in Molecular Neuroscience, 2012. https://doi.org/10.3389/fnmol.2012.00083.)*

α4β2*, α6β2*, α7, and α3β4* nAChRs. In addition, GABAergic interneurons express both somatic and terminal α4β2* nAChRs. GABAeric neurons innervating the VTA express α4β2* nAChRs while glutamatergic neurons entering the VTA have presynaptic α7 nAChRs. Cholinergic input is provided by the pedunculopontine and laterodorsal tegmental nuclei (Feduccia et al., 2012). Dopaminergic projections from the VTA express terminal α4β2*, α6β2*, and α6α4β2β3* nAChRs. Glutamatergic inputs to the NAc express α7 nAChRs and cholinergic interneurons resident in the NAc expressed α4β2* (Feduccia et al., 2012). The firing of these neurons is synchronized with DA activity. Lastly, (Fig. 4.2) in the prefrontal cortex, dopaminergic afferents from VTA express α4β2* nAChRs, and cholinergic inputs from the nucleus basalis of Meynart (nBM) and glutamatergic inputs express terminal α7 and α4β2* nAChRs. GABAergic interneurons express both somatic and terminal α7 nAChRs and terminal α4β2* nAChRs. nAChRs are in position on cell bodies and dendrites to increase excitability of dopaminergic neurons in the VTA and stimulate dopamine release. Presynaptic nAChRs on GABAergic and glutamatergic neurons modulate release of these neurotransmitters in the VTA that can also influence the levels of dopamine released in the NAc. Multiple nAChR subtypes present on the dopaminergic terminals from the VTA in the NAc can also modulate dopamine release and thus, the rewarding effects of nicotine.

α6β2* nAChRs have a more restricted expression pattern than α4β2* nAChRs (Brunzell et al., 2014). α6β2* nAChRs may be responsible for the control of 80% of dopamine release in the NAc (Brunzell et al., 2014). A model proposed in Brunzell et al. (2014) depicts the potential roles of varied nAChR subtypes in nicotine reinforcement (Fig. 4.3). Note the locations of these nAChR subtypes in the VTA and NAc in (Fig. 4.2). At a level of nicotine found in smokers (300 nM), α7 nAChRs are not activated, while α4β2* nAChRs are desensitized. α4α6β2* nAChRs are activated at this level of nicotine and lead to higher dopamine release (Brunzell et al., 2014). At a higher concentration of nicotine, 1–3 μM, α4β2* nAChRs are now activated as well. Some are desensitized, but enough are active to reinforce the effects of nicotine. At higher levels around 10 μM, the reinforcing effects of nicotine are reduced (see also the role of nAChRs in the medial habenula (MHb) and interpeduncular nucleus (IPN) in aversion below). At this level, α7 nAChRs may be activated, and both the α4β2* and α6β2* nAChRs are now desensitized (Brunzell et al., 2014). Activation of α7 nAChRs inhibits the activities of β2* nAChRs.

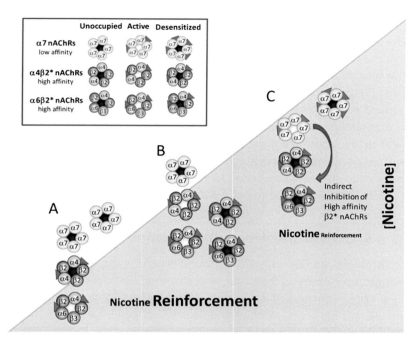

Fig. 4.3 Role of various nAChR subtypes in nicotine reinforcement. A model of nicotinic receptor subtype contributions to nicotine reinforcement and DA release, as might be expected with increasing concentrations of nicotine (depicted in *gray*). Nicotine and endogenous ACh are depicted as *blue* triangles. (A) Unbound α7 nAChRs are inactive and at rest at low concentrations of nicotine. A total of 300 nM nicotine that is sub-threshold to stimulate α7 nAChRs preferentially desensitizes α4β2* nAChRs. In the VTA, α4α6β2*nAChRs on DA neurons are persistently activated by this physiologically relevant concentration of nicotine believed to be achieved in the brains of smokers. (B) Higher concentrations of 1–3 μM nicotine activate α4β2* as well as α6β2*nAChRs. After activation, these receptors stabilize in a desensitized state, but fast application of nicotine activates sufficient subpopulations of these receptors to support nicotine reinforcement. Genetic and pharmacological manipulations that block activation of α6β2* or α4β2* nAChRs lead to reductions in nicotine self-administration and nicotine-stimulated DA release; hence, activation of β2*nAChRs appears necessary for nicotine reinforcement. (C) With higher concentrations of nicotine (10 μM), nicotine reinforcement would be satiated or reduced in smokers. The high-affinity α4β2* and α6β2* nAChRs would be shifted to the desensitized state and hence, block the rein-forcing efficacy of subsequent applications of the drug. Higher concentrations of nico-tine sufficient to stimulate populations of α7 nAChRs would reduce motivation to self-administer nicotine. Stimulation of α7 nAChRs by endogenous ACh would have a similar behavioral effect, and stimulation of α7 nAChRs would have an indirect effect in that it would inhibit nicotine reinforcement via stimulation of signaling pathways that inhibit β2* nAChR function on VTA DA neurons. Individuals who have blunted α7 nAChRs, such as those with schizophrenia, smoke more heavily than individuals with a full comple-ment of α7 nAChRs. α7 nAChR agonist ligand bound to more than β2-binding sites favors the desensitized state of these receptors. *Denotes possible assembly with other subunits. β3 is the typical accessory subunit that assembles with α6β2* nAChRs. The high-sensitivity α4β2* nAChRs, those with β2 rather than α4 or α5 at the accessory site, are depicted. (From Brunzell, D. H., Mcintosh, J. M., & Papke, R. L. (2014). Diverse strategies targeting α7 homomeric and α6β2* heteromeric nicotinic acetylcholine receptors for smoking cessation.* Annals of the New York Academy of Sciences, 1327 (1), 27–45. https://doi.org/10.1111/nyas.12421.)

As seen in (Fig. 4.2), multiple nAChRs are also present in the prefrontal cortex. Nicotine can alter the responsive of multiple neurons here as well. In summary, nAChRs are well placed to modulate neuronal excitability and as well modulate the release of dopamine, glutamate, and GABA in circuits critical to the addictive process. Activation of nAChRs by nicotine thus stimulates dopamine release and promotes the rewarding effects of smoking. The aversive effects are mediated in a major way by nAChRs expressed in the MHb–IPN pathway that will be described in more detail below.

4.5 Genetic involvement in nicotine addiction

Multiple studies support a strong genetic component of nicotine addiction, up to 50% as shown in twin studies (Lessov et al., 2004). Other twin studies also showed heritability of between 44% and 60% (McGue et al., 2000; True et al., 1999). Multiple nAChR genes have been linked to various aspects of smoking such as number of cigarettes smoked and difficulty in quitting. Genome-wide association studies (GWAS) have linked nAChR subunit genes on chromosome 15 in the CHRNA5-CHRNA3-CHRB4 cluster to nicotine dependence (Bierut, 2009; Oni et al., 2016). Multiple single-nucleotide polymorphisms (SNPs) in this region were identified in a candidate gene study (rs578776 in the 3′ untranslated region of α3, rs16969968 in the coding region of α5, and rs1051730 in α3) that correlated with nicotine dependence (Saccone et al., 2007). A strong association with nicotine dependence was also found for two SNPs (rs6474413, rs10958726) in the nAChR β3 gene (CHRNB3) (Saccone et al., 2007). These were found in the promoter region, one 17 kb upstream and another within 2 kb of the start site. Two other SNPs in CHRNB3 were also found. This region was also identified in another GWAS study (Bierut et al., 2007). Another SNP (rs1051730) in the CHRNA5-CHRNA3-CHRB4 cluster is correlated with difficulty in stopping smoking during pregnancy and heavier smoking before pregnancy, but not initiation (Freathy et al., 2009). The "at risk" variant at rs578776 in the α3 3′ untranslated region was associated with decreased intrinsic reward sensitivity (Robinson et al., 2013).

Some studies show a correlation with α4 nAChR gene SNPs and nicotine dependence, while others don't (Saccone et al., 2007). More recent work has demonstrated an association between an α4 SNP (rs2273500) at a splice acceptor site and nicotine dependence (Hancock et al., 2015). The minor allele leads to the production of an α4 nAChR subunit transcript that is removed by missense-mediated decay. This results in less α4* nAChR expression (Hancock et al., 2015). α4β2* nAChRs have a high affinity for nicotine and are important in circuits controlling dopamine release (Fig. 4.2) and below.

Since various studies using candidate genes or GWAS analyze different populations, and variations in results may be due to the study population used. Allele frequencies differ among different populations of humans, so that although a correlation with a specific marker with smoking is maintained, the frequency of an allele may vary in populations and thus the contribution to smoking behavior (Bierut, 2009). So far, the strongest associations are with SNPs in the CHRNA5-CHRNA3-CHRB4 cluster and CHRNB3.

One in particular is the α5 gene (rs16969968), which is linked with smoking and lung cancer. α5 subunits can assemble with α3 and β4 nAChR subunits and play a role in mediating the aversive effects of nicotine (Fowler et al., 2011). α5 subunits can also assemble with α4β2 subunits and increase Ca^{2+} permeability of the receptor compared with α4β2* nAChRs alone. In VTA neurons, the α5 nAChR subunit is needed for high α4 nAChR subunit expression (Sciaccaluga et al., 2015). Thus, mutations in the α5 gene may also effect the function of the major nicotine-sensitive nAChR population, α4β2* nAChRs in the VTA. A particular variant of α5 N398 (asparagine in 398 as the minor allele) has been associated with addiction and lung cancer. The major variant is D398 (aspartate).

α5-containing nAChRs were required for a Ca^{2+} increase in cultured ventral midbrain neurons in response to 100 μM nicotine in that KO mice for α5 didn't exhibit this increase while WT mice did (Sciaccaluga et al., 2015). Mice carrying the α5D398N variation showed a lower Ca2+ response to nicotine. In VTA slices, α5 subunits were required for high-amplitude currents, but α5D398N was less efficient than wild-type (Sciaccaluga et al., 2015). However, when α5(β2α4)2 and α5D398N (β2α4)2 nAChRs were expressed in a cell line, the levels of expression and channel Ca2+ permeability were the same (Sciaccaluga et al., 2015). The number of nicotine-evoked currents was similar between wild-type and D398N α5 nAChRs, but the latter desensitized more rapidly. The α5D398N mutation is present in the M3-M4 intracellular loop region. Thus, the effect on Ca^{2+} permeability of α5-containing nAChRs may be due to different interactions of the two α5 variants with Ca^{2+} in the loop region.

The N398 variant when expressed in induced pluripotent stem cells (iPSCs) differentiated into VTA dopaminergic neurons or glutamatergic neurons caused both to exhibit increased excitatory postsynaptic currents (EPSCs) after nicotine treatment (Oni et al., 2016). The glutamatergic neurons showed a declining response to nicotine consistent with nAChR desensitization. Derived dopaminergic neurons with the D398 variant showed increased expression of many genes including those involved in Ca^{2+} signaling (Oni et al., 2016).

To further examine the basis for the differential function of the N398 and D398 alleles, HEK293T cells were made that expressed either α4β2α5D398 or α4β2α5N398 nAChRs (Bierut et al., 2008). Function of the nAChRs was measured using Ca^{2+} assays. The variant N398 showed a lower maximum response to epibatidine (a potent nAChR agonist), with no changes in the EC50. Receptor expression levels were the same for both variants. The reduced function of the α4β2α5N398 nAChRs could indicate that people with this variant would require more nicotine to achieve the same reward, through activation of VTA DA neurons (Fig. 4.2). Expression on inhibitory GABAergic neurons could lead to reduced inhibition and more dopamine release in response to the same amount of nicotine in the VTA. Also, since α5* nAChRs were shown to be essential for the aversive effects of nicotine (Fowler et al., 2011), reduced function of α5α3β4* nAChRs in the MHb-IPN pathway could also reduce aversive affects and lead to higher nicotine consumption. In summary, these studies on the α5 mutation demonstrate a strong link between changes in the α5 gene associated with nicotine dependence and nAChR function.

4.6 nAChR subtypes involved in the effects of nicotine on reward and aversion

KO studies have shown that nAChRs containing β2 subunits are required for nicotine self-administration (Picciotto et al., 1998). Nicotine treatment doesn't increase the dopamine levels as detected by in vivo microdialysis in β2 KO mice (Picciotto et al., 1998). In slices of SN and VTA, nicotine increased firing frequency, which was blocked by DHβE (β2* nAChR antagonist). Using cytisine it was determined that the response was mediated by α4β2 nAChRs (Picciotto et al., 1998). α4 KO mice also don't increase dopamine release after nicotine treatment (Marubio et al., 2003). Thus, both α4 and β2 KO mice don't increase dopamine release, an important step in developing nicotine dependence. This makes sense given the presence of α4β2*nAChRs on the cell body and projections of dopaminergic neurons present in the VTA that project to the NAc

(Fig. 4.2). Nicotine increases the level of α4β2*nAChRs and β3 and β4-containing nAChRs as well (LeFoll et al., 2022).

To ascertain whether specific nAChR subtypes were sufficient for the effects of nicotine, knock-in mice with a hypersensitive α4 subunit were developed (Tapper et al., 2004). These α4 * nAChRs could be selectively activated by very low levels of nicotine and thus distinguish the effects of α4* nAChRs from other subtypes (Tapper et al., 2004). These nAChRs are expressed in the VTA neurons, can be activated by low doses of nicotine, and be upregulated by chronic low levels of nicotine (Tapper et al., 2004). Mice show CPP for a chamber paired with nicotine. This can be shown for other drugs of abuse as well. CPP is a measure of the rewarding effects of nicotine. The α4 nicotine-hypersensitive mice also showed CPP, and thus, α4* nAChRs were sufficient for the rewarding effects at nicotine levels that couldn't activate other nAChR subtypes (Tapper et al., 2004). Infusion of DHβE into the VTA of rats blocked the rewarding effects of nicotine (Laviolette & Kooy, 2003), also indicating the involvement of α4β2* nAChRs. β2 KO mice also don't increase dopamine release after nicotine exposure, since many α4 nAChRs (and α6*) contain the β2 subunit.

Nicotine can induce hypothermia. Tolerance to this can develop in mice with chronic nicotine treatment (Tapper et al., 2004). Hypothermia was induced after nicotine treatment in the α4* hypersensitive mice as well, but after 8 days of treatment, this effect was greatly reduced indicating that mice had become tolerant to this effect of nicotine (Tapper et al., 2004). Since at the dose given, the α4* hypersensitive mice could only act on α4* nAChRs, and this indicates that α4* nAChRs were sufficient for developing tolerance.

Sensitization to a drug develops when after some time of use at a specific dose the response increases rather than decreases (the opposite of tolerance). Rats develop sensitization to the effects of nicotine on locomotor activity (Tapper et al., 2004). Low levels of nicotine also produced this locomotor sensitization to nicotine in the α4* hypersensitive mice (Tapper et al., 2004). Mecamylamine blocked the effect, and cytisine (β2* nAChR partial agonist) produced this effect. This sensitization was shown to be caused by nAChR-controlled dopamine release in the NAc (Tapper et al., 2004). Overall, α4* nAChRs are sufficient for nicotine-induced sensitization, reward, and tolerance (Tapper et al., 2004). These are most likely part of nAChRs also containing α6 and β2 subunits and possibly others. α6 KO mice don't self-administer nicotine (Picciotto & Kenny, 2021). α6β2* nAChRs are involved in this in that selective blockade of α6β2 nAChRs in the VTA reduces nicotine self-administration (Picciotto & Kenny, 2021).

Since genetic studies showed a strong linkage with a SNP in the α5 gene, the role of nAChRs containing the α5 subunit in smoking was examined in α5 KO mice. α5 subunits are often expressed in combination with α3 and β4 subunits. The α5 KO mice also responded to self-administration in an inverted U dose-response curve as did wild-type mice; however, they had a higher level of response at high doses and achieved higher nicotine levels (Fowler et al., 2011). Normal mice titrate to a level comparable with human smokers, but the α5 KO mice continued to self-administer to higher levels. α5 KO mice exhibited nicotine reinforcement similar to wild-type mice, but demonstrated less negative or aversive effects at higher levels that normally tend to limit nicotine intake (Fowler et al., 2011).

α5* nAChRs are expressed in several brains areas including the hippocampus, VTA, SN, cortex, and importantly in the MHb-IPN pathway (Fowler et al., 2011). The MHb contains a high concentration of nicotine-binding sites and expresses α3, α4, α5, β2, and β4 subunits (Picciotto & Kenny, 2021). The MHb projects to the IPN and expresses presynaptic α5* nAChRs on the MHb neurons. To test the hypothesis that activation of α5* nAChRs in this pathway mediates aversion, Fowler et al. (2011) introduced the α5 nAChR subunit into the MHb of α5 KO mice. These mice administered nicotine in a pattern consistent with wild-type mice, not the α5 KO mice. α5 RNA was also detected in the treated mice in the IPN (Fowler et al., 2011). Thus, restored expression of α5 produced a normalized aversive response. Acetylcholine-mediated signaling was reduced in synaptosomes from IPN and habenula from α5 KO mice, but not reduced in the cortex and hippocampus (Fowler et al., 2011). This reduction was partially rescued in IPN, but not MHb after replacement of α5. α5* nAChRs may also be present on IPN neurons as well. α5 nAChRs are expressed presynaptically on MHb afferents. MHb signaling maybe also be affected by α5* nAChRs on inputs to the MHb from other brain areas (Fowler et al., 2011). The role of α5* nAChRs in habenula was confirmed when RNA interference was used to knock down α5 in the MHb of rats and showed a similar loss of aversion at higher levels of nicotine as in α5 KO mice (Fowler et al., 2011). α5 KO mice showed reduced IPN stimulation with nicotine, and this reduction increased nicotine self-administration. Nicotine stimulates release of glutamate from MHb neurons through α5* nAChRs. This produces an aversive or negative signal in the IPN neurons through activation of NMDA receptors. Loss of α5 nAChR function reduces this glutamate signal to the IPN and thus the aversive effects (Fowler et al., 2011). The MHb-IPN pathway is key to regulating nicotine intake by mediating the aversive effects of higher levels of nicotine. Humans with mutations producing reduced α5 * nAChR function have increased probabilities of smoking. This is consistent with genetic studies linking α5 to chronic obstructive pulmonary disease (COPD) and lung cancer (Fowler et al., 2011).

α5 nAChRs are often expressed with α3 and β4 nAChR subunits. Studies show that the α3 and β4 subunits are also important in mediating the aversive effects of nicotine and support the role of α5α3β4nAChRs in the MHb-IPN pathway as key to mediating aversion. Mice expressing a low level of α3* nAChRs were shown to self-administer larger amounts of nicotine compared with wild-type mice (Elayouby et al., 2021). Knocking down α3 expression in the MHb or IPN also specifically increased nicotine self-administration. Specifically blocking α3β4* nAChRs with α-conotoxin AuIB in the IPN did the same (Elayouby et al., 2021). α-Contoxin AuIB applied to MHb neurons in slice preparations blocked nicotine-induced currents (Elayouby et al., 2021). Overexpression of β4 * nAChRs reduced oral consumption of nicotine (Picciotto & Kenny, 2021), supporting β4 nAChR subunit contribution to aversion.

With the previous results with α5 KO mice, this supports the role of α3β4* nAChRs, many of which would also contain the α5 nAChR subunit, in regulating aversion to nicotine though activity in the MHb-IPN pathway. Other nAChR subtypes containing α5 such as α4β2α5* nAChRs also play a major role in nicotine aversion. The role of α4 in aversion is supported by work showing that DHβE (α4* nAChR antagonist) administered into the VTA also blocks some aversive effects of nicotine administration (Laviolette & Kooy, 2003).

In summary, a rich variety of nAChRs are involved in eliciting and controlling the rewarding, aversive, and withdrawal effects of nicotine. These subtypes provide a number of targets for the difficult task of developing better smoking cessation aids.

4.7 Pharmacotherapy for nicotine addiction

Most smokers try to quit at some point in their life; more than half attempted in 2018, but only a small percentage succeeded (Rigotti et al., 2022). Many make multiple attempts to quit (an average of 6) if they are successful at all (Rigotti et al., 2022). Even after quitting, few remain abstinent for more than 1 year. However, after repeated attempts and proper treatment, many smokers do eventually succeed in quitting. This is done using various nicotine replacement therapies (NRT) such as the nicotine patch, gum, lozenges, inhaler, and/or behavioral therapy. Drugs such as Chantix (varenicline) and bupropion are also used. Smoking cessation during pregnancy is very important given the various developmental effects that can occur to the fetus due to maternal smoking (see the chapter describing the developmental effects of nicotine exposure). Since many of the deleterious effects of smoking are mediated by nicotine alone, NRT is not recommended for pregnant women looking to stop smoking. NRT still provides nicotine that can mediate a number of negative effects throughout the body (see the chapter describing nonneuronal nAChRs). Achieving the best results for smoking cessation involves a combination of pharmacotherapy and counseling. Smokers who stop before age 40 can reverse damage caused by smoking and remove most of the mortality risk caused by smoking. Quitting before age 50 removes 2/3 of the lung cancer risk (LeFoll et al., 2022).

NRT involves providing a constant source of nicotine through a patch or periodic treatment with gum or an inhaler. There is a slower uptake of nicotine than occurs in a smoker (initial cigarettes in the morning deliver a high rapid dose). NRT only replaces nicotine, thus avoiding exposure to the numerous harmful compounds present in tobacco smoke. However, NRT can produce numerous negative effects on the body including the cardiovascular and immune systems through the actions of nicotine alone acting via nAChRs. Nicotine delivered in this manner may produce reduced reward and reduce withdrawal effects and aid in cessation.

Patches containing various amounts of nicotine up to 21 mg are used daily to administer a chronic sustained dose of nicotine. These are used for approximately 3 months, and the dosage may be reduced during this time. Side effects include vivid dreaming and skin irritation. This doesn't mimic the periodic smoking pattern and thus may not be as effective as other methods of cessation. Lozenges, gum, inhaler, or nasal sprays administer smaller periodic doses throughout the day (Rigotti et al., 2022). These sometimes cause mouth irritation, hiccups, cough, and heartburn (Rigotti et al., 2022). Data summarized from multiple studies demonstrated that use of any NRT product increased quit rates from 10% to 17%.

Varenicline has been the most successful aid in smoking cessation (Rigotti et al., 2022). Varenicline blocks reward and reduces withdrawal symptoms by acting as a partial agonist of α4β2* nAChRs and α6β2* nAChRs and a full agonist with low affinity for α7 nAChRs and α3β4* nAChRs (Brunzell et al., 2014). It is given as a 0.5 mg or 1 mg tablet daily.

This is normally used for 3–6 months. A warning was issued in 2009 about possible suicidal thoughts, aggression, and cardiac problems associated with varenicline use, but FDA concerns about potential behavioral changes produced by varenicline treatment were not confirmed, and the warning was removed (Anthenelli et al., 2016; Rigotti et al., 2022). A significant percentage of people taking varenicline report nausea, vivid dreams, headaches, and sleep problems (Rigotti et al., 2022). Data summarized from a number of trials showed that varenicline therapy resulted in smoking cessation after 6 months for 26% of the patients as opposed to 11% with placebo.

Bupropion (Wellbutrin) is a noradrenaline and dopamine reuptake inhibitor. Bupropion was shown to increase dopamine and norepinephrine levels in the NAc and prefrontal cortex in rodents and reduces craving and the effects of withdrawal in humans (Stahl et al., 2014). Bupropion may also mediate smoking cessation by actions at nAChRs. Bupropion is an antagonist of multiple nAChR subtypes including $\alpha3\beta2$, $\alpha4\beta2$, and $\alpha7$ with different levels of selectivity (Slemmer et al., 2000). The mechanism of action is noncompetitive (Slemmer et al., 2000). The blockage of nAChR function may also augment the effects on dopamine and norepinephrine signaling to aid smoking cessation. Bupropion (150 mg tablet) is given daily for about 3–6 months. Side effects include headaches, dry mouth, and insomnia (Rigotti et al., 2022). Data summarized from multiple studies showed that bupropion use increased the quit rate versus placebo from 11 to 19% after 6 months (Rigotti et al., 2022). Another study comparing the effectiveness of varenicline, bupropion, and nicotine patch after 12 weeks showed that varenicline was the most effective and the nicotine patch the least. All showed improvement over placebo (Anthenelli et al., 2016).

4.8 Other potential candidates for smoking cessation

Another strategy suggested by the model shown in Fig. 4.3 is the use of selective drugs to stimulate $\alpha7$ nAChRs and inhibit $\alpha6\beta2^*$ nAChRs. $\alpha7$ nAChRs may not be important for reinforcement but may be involved in the control of self-administration (Brunzell et al., 2014). $\alpha7$ nAChR agonists or positive allosteric modulators (PAMs) lead to indirect inhibition of $\alpha6\beta2^*$ nAChRs and thus reduced reinforcing effects of nicotine. Selective stimulation of $\alpha7$ nAChRs reduces nicotine self-administration. In addition, use of $\alpha6\beta2^*$ nAChR antagonists will also reduce nicotine reinforcement (Brunzell et al., 2014). Type I PAMs for $\alpha7$ lower the threshold for activation and thus cause $\alpha7$ nAChRs to be active at concentrations of ACh or nicotine that would normally not activate the receptors. Type II PAMs also could be used since they reduce $\alpha7$ desensitization ($\alpha7$ nAChRs normally desensitize rapidly) and continue to provide indirect inhibition of $\alpha6\beta2^*$ nAChRs (Brunzell et al., 2014). Since varenicline is a full agonist for $\alpha7$ nAChRs, some of it effectiveness in smoking cessation maybe due to effects on $\alpha7$ nAChRs that may indirectly inhibit $\alpha6\beta2^*$ nAChRs. These categories of compounds require much more study, but are potential candidates for smoking cessation therapy.

Cytisine has been used in Europe, but is not approved for use in the United States. Varenicline is a chemical derivative of cytisine, but further work with cytisine should be explored. Naltrexone, which is used for treatment of alcohol and opioid abuse, may have some efficacy when used with bupropion, but not with NRT or varenicline (LeFoll et al., 2022). Other preclinical studies indicate that more research should be done with GABAergic and glutamatergic compounds, 5HT-2C receptor agonists such as lorcaserin, dopamine D3 antagonists, and clonidine (LeFoll et al., 2022). Humans with damage to the insula can quit smoking easily, implicating another brain region in addiction. This may be a possible target for interventions. Deep insula/prefrontal cortex transcranial magnetic stimulation (TMS) is approved for smoking cessation therapy (LeFoll et al., 2022).

For smoking cessation, the best results are achieved with a combination of pharmacotherapy and counseling. The pharmacotherapy may consist of NRT, varenicline, bupropion, or some combination of these. More research needs to be done, and no therapy has been found to be to completely effective for everyone wishing to quit smoking. However, given the still large number of smokers in the United States and even more so in the world that desire to quit, more research is warranted.

References

Anthenelli, R. M., Benowitz, N. L., West, R., St Aubin, L., McRae, T., Lawrence, D., Ascher, J., Russ, C., Krishen, A., & Evins, A. E. (2016). Neuropsychiatric safety and efficacy of varenicline, bupropion, and nicotine patch in smokers with and without psychiatric disorders (EAGLES): A double-blind, randomised, placebo-controlled clinical trial. *The Lancet, 387*(10037), 2507–2520. https://doi.org/10.1016/S0140-6736(16)30272-0.

Bagdas, D., Alkhlaif, Y., Jackson, A., Carroll, F. I., Ditre, J. W., & Damaj, M. I. (2018). New insights on the effects of varenicline on nicotine reward, withdrawal and hyperalgesia in mice. *Neuropharmacology, 138*, 72–79. https://doi.org/10.1016/j.neuropharm.2018.05.025.

Bierut, L. J. (2009). Nicotine dependence and genetic variation in the nicotinic receptors. *Drug and Alcohol Dependence, 104*(1), S64–S69. https://doi.org/10.1016/j.drugalcdep.2009.06.003.

Bierut, L. J., Madden, P. A. F., Breslau, N., Johnson, E. O., Hatsukami, D., Pomerleau, O. F., Swan, G. E., Rutter, J., Bertelsen, S., Fox, L., Fugman, D., Goate, A. M., Hinrichs, A. L., Konvicka, K., Martin, N. G., Montgomery, G. W., Saccone, N. L.,

Saccone, S. F., Wang, J. C., … Ballinger, D. G. (2007). Novel genes identified in a high-density genome wide association study for nicotine dependence. *Human Molecular Genetics*, *16*(1), 24–35. https://doi.org/10.1093/hmg/ddl441.

Bierut, L. J., Stitzel, J. A., Wang, J. C., Hinrichs, A. L., Grucza, R. A., Xuei, X., Saccone, N. L., Saccone, S. F., Bertelsen, S., Fox, L., Horton, W. J., Breslau, N., Budde, J., Cloninger, C. R., Dick, D. M., Foroud, T., Hatsukami, D., Hesselbrock, V., Johnson, E. O., … Goate, A. M. (2008). Variants in nicotinic receptors and risk for nicotine dependence. *American Journal of Psychiatry*, *165*(9), 1163–1171. https://doi.org/10.1176/appi.ajp.2008.07111711.

Brunzell, D. H., Mcintosh, J. M., & Papke, R. L. (2014). Diverse strategies targeting α7 homomeric and α6β2* heteromeric nicotinic acetylcholine receptors for smoking cessation. *Annals of the New York Academy of Sciences*, *1327*(1), 27–45. https://doi.org/10.1111/nyas.12421.

Elayouby, K., Ishikawa, M., Dukes, A., Smith, A., Lu, Q., Fowler, C., & Kenny, P. J. (2021). 3* nicotinic acetylcholine receptors in the habenula-interpeduncular nucleus circuit regulate nicotine uptake. *The Journal of Neuroscience*, *41*(8), 1779–1787.

Feduccia, A. A., Chatterjee, S., & Bartlett, S. E. (2012). Neuronal nicotinic acetylcholine receptors: Neuroplastic changes underlying alcohol and nicotine addictions. *Frontiers in Molecular Neuroscience*, *2012*. https://doi.org/10.3389/fnmol.2012.00083.

Fowler, C. D., Lu, Q., Johnson, P. M., Marks, M. J., & Kenny, P. J. (2011). Habenular α5 nicotinic receptor subunit signalling controls nicotine intake. *Nature*, *471*(7340), 597–601. https://doi.org/10.1038/nature09797.

Freathy, R. M., Ring, S. M., Shields, B., Galobardes, B., Knight, B., Weedon, M. N., Smith, G. D., Frayling, T. M., & Hattersley, A. T. (2009). A common genetic variant in the 15q24 nicotinic acetylcholine receptor gene cluster (CHRNA5-CHRNA3-CHRNB4) is associated with a reduced ability of women to quit smoking in pregnancy. *Human Molecular Genetics*, *18*(15), 2922–2927. https://doi.org/10.1093/hmg/ddp216.

Hancock, D. B., Reginsson, G. W., Gaddis, N. C., Chen, X., Saccone, N. L., Lutz, S. M., Qaiser, B., Sherva, R., Steinberg, S., Zink, F., Stacey, S. N., Glasheen, C., Chen, J., Gu, F., Frederiksen, B. N., Loukola, A., Gudbjartsson, D. F., Brüske, I., Landi, M. T., … Stefansson, K. (2015). Genome-wide meta-analysis reveals common splice site acceptor variant in CHRNA4 associated with nicotine dependence. *Translational Psychiatry*, *5*. https://doi.org/10.1038/tp.2015.149.

Heatherton, T. F., Kozlowski, L. T., Frecker, R. C., & Fagerstrom, K.-O. (1991). The Fagerstrom test for nicotine dependence: A revision of the Fagerstrom tolerance questionnaire. *Addiction*, *86*(9), 1119–1127. https://doi.org/10.1111/j.1360-0443.1991.tb01879.x.

Koob, G. F. (1992). Drugs of abuse: Anatomy, pharmacology and function of reward pathways. *Trends in Pharmacological Sciences*, *13*(C), 177–184. https://doi.org/10.1016/0165-6147(92)90060-J.

Laviolette, S. R., & Kooy, D. (2003). The motivational valence of nicotine in the rat ventral tegmental area is switched from rewarding to aversive following blockade of the α7-subunit-containing nicotinic acetylcholine receptor. *Psychopharmacology*, *166*(3), 306–313. https://doi.org/10.1007/s00213-002-1317-6.

LeFoll, B., & Goldberg, S. R. (2009). Effects of nicotine in experimental animals and humans: An update on addictive properties. *Handbook of Experimental Pharmacology*, *192*, 335–367. https://doi.org/10.1007/978-3-540-69248-5_12.

LeFoll, B., Piper, M., Fowler, C., Tonstad, S., Bierut, L., Lu, L., Jha, P., & Hall, W. D. (2022). Tobacco and nicotine use. *Nature Reviews Disease Primers*, *8*.

Lessov, C. N., Martin, N. G., Statham, D. J., Todorov, A. A., Slutske, W. S., Bucholz, K. K., Heath, A. C., & Madden, P. A. F. (2004). Defining nicotine dependence for genetic research: Evidence from Australian twins. *Psychological Medicine*, *34*(5), 865–879. https://doi.org/10.1017/S0033291703001582.

Marubio, L. M., Gardier, A. M., Durier, S., David, D., Klink, R., Arroyo-Jimenez, M. M., McIntosh, J. M., Rossi, F., Champtiaux, N., Zoli, M., & Changeux, J. P. (2003). Effects of nicotine in the dopaminergic system of mice lacking the alpha4 subunit of neuronal nicotinic acetylcholine receptors. *European Journal of Neuroscience*, *17*(7), 1329–1337. https://doi.org/10.1046/j.1460-9568.2003.02564.x.

Matta, S. G., Balfour, D. J., Benowitz, N. L., Boyd, R. T., Buccafusco, J. J., Caggiula, A. R., Craig, C. R., Collins, A. C., Damaj, M. I., Donny, E. C., Gardiner, P. S., Grady, S. R., Heberlein, U., Leonard, S. S., Levin, E. D., Lukas, R. J., Markou, A., Marks, M. J., McCallum, S. E., … Zirger, J. M. (2007). Guidelines on nicotine dose selection for in vivo research. *Psychopharmacology*, *190*(3), 269–319. https://doi.org/10.1007/s00213-006-0441-0.

McGue, M., Elkins, I., & Iacono, W. G. (2000). Genetic and environmental influences on adolescent substance use and abuse. *American Journal of Medical Genetics—Neuropsychiatric Genetics*, *96*(5), 671–677. https://doi.org/10.1002/1096-8628(20001009)96:5<671::AID-AJMG14>3.0.CO;2-W.

Oni, E. N., Halikere, A., Li, G., Toro-Ramos, A. J., Swerdel, M. R., Verpeut, J. L., Moore, J. C., Bello, N. T., Bierut, L. J., Goate, A., Tischfield, J. A., Pang, Z. P., & Hart, R. P. (2016). Increased nicotine response in iPSC-derived human neurons carrying the CHRNA5 N398 allele. *Scientific Reports*, *6*. https://doi.org/10.1038/srep34341.

Picciotto, M. R., & Kenny, P. J. (2021). Mechanisms of nicotine addiction. *Cold Spring Harbor Perspectives in Medicine*, *11*(5). https://doi.org/10.1101/cshperspect.a039610.

Picciotto, M. R., Zoli, M., Rimondini, R., Léna, C., Marubio, L. M., Pich, E. M., Fuxe, K., & Changeux, J. P. (1998). Acetylcholine receptors containing the β2 subunit are involved in the reinforcing properties of nicotine. *Nature*, *391*(6663), 173–177. https://doi.org/10.1038/34413.

Rigotti, N. A., Kruse, G. R., Livingstone-Banks, J., & Hartmann-Boyce, J. (2022). Treatment of tobacco smoking: A review. *JAMA—Journal of the American Medical Association*, *327*(6), 566–577. https://doi.org/10.1001/jama.2022.0395.

Robinson, J. D., Versace, F., Lam, C. Y., Minnix, J. A., Engelmann, J. M., Cui, Y., Karam-Hage, M., Shete, S. S., Tomlinson, G. E., Chen, T. T. L., Wetter, D. W., Green, C. E., & Cinciripini, P. M. (2013). The CHRNA3 rs578776 variant is associated with an intrinsic reward sensitivity deficit in smokers. *Frontiers in Psychiatry*, *4*. https://doi.org/10.3389/fpsyt.2013.00114.

Saccone, S. F., Hinrichs, A. L., Saccone, N. L., Chase, G. A., Konvicka, K., Madden, P. A. F., Breslau, N., Johnson, E. O., Hatsukami, D., Pomerleau, O., Swan, G. E., Goate, A. M., Rutter, J., Bertelsen, S., Fox, L., Fugman, D., Martin, N. G., Montgomery, G. W., Wang, J. C., … Bierut, L. J. (2007). Cholinergic nicotinic receptor genes implicated in a nicotine dependence association study targeting 348 candidate genes with 3713 SNPs. *Human Molecular Genetics*, *16*(1), 36–49. https://doi.org/10.1093/hmg/ddl438.

Sciaccaluga, M., Moriconi, C., Martinello, K., Catalano, M., Bermudez, I., Stitzel, J. A., Maskos, U., & Fucile, S. (2015). Crucial role of nicotinic α5 subunit variants for Ca^{2+} fluxes in ventral midbrain neurons. *The FASEB Journal*, *29*(8), 3389–3398. https://doi.org/10.1096/fj.14-268102.

Shiffman, S., Zettler-Segal, M., Kassel, J., Paty, J., Benowitz, N. L., & O'Brien, G. (1992). Nicotine elimination and tolerance in non-dependent cigarette smokers. *Psychopharmacology*, *109*(4), 449–456. https://doi.org/10.1007/BF02247722.

Slemmer, J. E., Martin, B. R., & Damaj, M. I. (2000). Bupropion is a nicotinic antagonist. *Journal of Pharmacology and Experimental Therapeutics*, *295*(1), 321–327.

Stahl, S., Pradko, J., Haight, B., Modell, J., Rockett, C., & Learned-Coughlin, S. (2014). A review of the neuropharmacology of bupropion, a dual norepinephrine and dopamine reuptake inhibitor. *Primary Care Companion to the Journal of Clinical Psychiatry*, *6*(4), 159–166.

Tan, X., Vrana, K., & Ding, Z. M. (2021). Cotinine: Pharmacologically active metabolite of nicotine and neural mechanisms for its actions. *Frontiers in Behavioral Neuroscience*, *15*. https://doi.org/10.3389/fnbeh.2021.758252.

Tapper, A. R., McKinney, S. L., Nashmi, R., Schwarz, J., Deshpande, P., Labarca, C., Whiteaker, P., Marks, M. J., Collins, A. C., & Lester, H. A. (2004). Nicotine activation of α4* receptors: Sufficient for reward, tolerance, and sensitization. *Science*, *306*(5698), 1029–1032. https://doi.org/10.1126/science.1099420.

True, W. R., Xian, H., Scherrer, J. F., Madden, P. A. F., Bucholz, K. K., Heath, A. C., Eisen, S. A., Lyons, M. J., Goldberg, J., & Tsuang, M. (1999). Common genetic vulnerability for nicotine and alcohol dependence in men. *Archives of General Psychiatry*, *56*(7), 655–661. https://doi.org/10.1001/archpsyc.56.7.655.

Wellman, R. J., DiFranza, J. R., & Wood, C. (2006). Tobacco chippers report diminished autonomy over smoking. *Addictive Behaviors*, *31*(4), 717–721. https://doi.org/10.1016/j.addbeh.2005.05.043.

Wills, L., Ables, J. L., Braunscheidel, K. M., Caligiuri, S. P. B., Elayouby, K. S., Fillinger, C., Ishikawa, M., Moen, J. K., & Kenny, P. J. (2022). Neurobiological mechanisms of nicotine reward and aversion. *Pharmacological Reviews*, *74*(1), 271–310. https://doi.org/10.1124/PHARMREV.121.000299.

CHAPTER FIVE

Disease associations—Alzheimer's disease, schizophrenia, Parkinson's disease, autism, and cancer

nAChRs and cholinergic signaling are involved in numerous diseases. We will focus on what we know about nAChR function in Alzheimer's, Parkinson's, schizophrenia, autism, and some cancers. Cholinergic signaling through nAChRs is certainly involved in other diseases such as epilepsy, Tourette's, and depression as well as in skin disorders and inflammatory conditions. However, the first five are major disorders affecting millions of people, and we will summarize the basic information regarding the role of nAChRs in each of these in this chapter.

5.1 Alzheimers' disease

5.1.1 Background

Alzheimer's disease (AD) is a leading cause of death in the United States. Although there are multiple dementias that occur such as Lewy Body, vascular and frontotemporal, Alzheimer's is the most common dementia and the most prevalent dementia occurring in people over 60. About 10% of people over 65 have Alzheimer's. The incidence increases with age with about 32% having it at age 85 (Alzheimer's Disease Facts and Figures, 2020). The total cost of treatment was estimated to be over $300 billion in 2020. The disease was first described by Alois Alzheimer in 1906. His patient had memory loss, language and behavioral problems. Upon autopsy, she was found to have what are now known as amyloid plaques and neurofibrillary tangles (tau protein). Today, AD is still characterized by beta amyloid and tau protein deposits.

Currently known behavioral symptoms include apathy, depression, problems with memory such as recalling names or events (Thies & Bleiler, 2013). Problem-solving abilities are eroded. Completing tasks becomes more difficult, and patients lose interest in work and social activities. As the disease progresses, patients have trouble with judgment, mental confusion and show motor difficulties in speaking, walking, and eating (Thies & Bleiler, 2013). Diagnosis before death is difficult, but levels of β-amyloid and tau in cerebral spinal fluid (CSF) may be diagnostic. Positron emission tomography (PET) imaging can also be used to detect β-amyloid in the brain (Alzheimer's Disease Facts and Figures, 2020 and references therein). Amyloid beta and tau accumulate, with β-amyloid increasing before tau and perhaps promoting tau deposition (Alzheimer's Disease Facts and Figures, 2020 and references therein). Inflammation and atrophy of the brain also occur, and microglia become activated. Multiple factors are involved, and brain changes may occur up to 20 years before onset of symptoms (Thies & Bleiler, 2013).

Most cases are sporadic and later onset (65 and older), but some cases have a genetic component and can occur earlier (Alzheimer's Disease Facts and Figures, 2020 and references therein). As in any complex disease, multiple genes are involved. A specific APOE allele (e4) is associated with increased risk of late-onset Alzheimer's. The e4 allele has been associated with increased β-amyloid levels (Jansen et al., 2015). Someone with a first-degree relative with Alzheimer's is more likely to develop it (Alzheimer's Disease Facts and Figures, 2020). One percent of cases can be due to specific mutations. Amyloid precursor protein (APP) is encoded on chromosome 21, and people with Downs Syndrome (trisomy 21) have a greatly increased chance of getting Alzheimer's. Mutations in the presenilin 1 and 2 genes also lead to increased incidence of Alzheimer's. These changes lead to Alzheimer's, sometimes as early as age 30 (Thies & Bleiler, 2013).

5.1.2 Cholinergic role in Alzheimer's disease

Cholinergic deficits in signaling are key to the development of Alzheimer's disease. Symptoms of early AD are memory loss, attention and spatial memory deficits. The cortex, hippocampus, and basal forebrain regions are involved. Basal forebrain neurons are the major cholinergic output to the hippocampus, cortex, and amygdala. Basal forebrain connections play a vital role in cognitive processing (Ballinger et al., 2016). AD patients suffer a major loss of white matter and loss of basal forebrain cholinergic neurons (Lombardo & Maskos, 2015). Basal forebrain neurons degenerate, and there is a reduction in ACh, vesicular acetylcholine transporter (VAChT), and choline acetyltransferase (ChAT) expression. Binding sites for nicotine and ACh are reduced in the brains of AD patients, and ChAT enzyme activity is decreased (Lombardo & Maskos, 2015). Severe loss of half of the α4β2 nAChRs present in the hippocampus and neocortex occurs in patients with advanced AD (Dineley et al., 2015). α7 nAChRs remain fairly stable as do levels of mAChRs. Alzheimer's disease is thought to progress as amyloid beta (Aβ) and tau accumulate leading to death of cholinergic neurons. Aβ can interact with α7 nAChRs, α7β2 nAChRs as well as α4β2 nAChRs. Since α4β2 nAChRs are lost early in AD, radio-labeled A85380 (targets α4β2 nAChRs) is being used in PET to scan for early loss of nAChRs.

α4β2 nAChRs account for 90% of the high-affinity nicotine binding in the brain and are expressed in many brains areas suffering neuronal loss in AD (Dineley et al., 2015). Homomeric α7 nAChRs are also widely detected in brain areas affected by AD. As described previously, α7 nAChRs can act as homomers or form heteromers with β2 subunits. α7 nAChRs are widely expressed, have a relatively high Ca^{2+} permeability, and desensitize rapidly (Dani & Bertrand, 2007).

α7β2 heteromeric receptors also are present in basal forebrain neurons, and these are also sensitive to amyloid β as are α7 nAChRs (Dineley et al., 2015). α7β2 nAChRs are also expressed in GABAergic interneurons of the hippocampus (Liu et al., 2012) and in basal forebrain (Liu et al., 2009). α7β2 nChRs have a similar pharmacological profile to cholinergic compounds as do α7 nChR homomers (Murray et al., 2012). Agonist-induced current amplitude of α7β2 nAChRs was half that seen for α7 homomers (Murray et al., 2012). The function of α7β2 nAChRs expressed in hippocampal CA1 interneurons in response to choline was inhibited by 1 nM Aβ1–42 (Liu et al., 2012). α7β2 nAChRs expressed in basal forebrain neurons were more sensitive to oligomeric Aβ-1-42 than homomeric α7 nAChRs expressed in ventral tegmental area (VTA) neurons. (Liu et al., 2009).

α4β2, α7β2, and α7-containing nAChRs are present in basal forebrain and within their wide projection areas. As described elsewhere in this book, nAChRs can be present presynaptically, postsynaptically, or even extrasynaptically. α7 and α4β2 nAChRs (and other subtypes as well) have now been shown to produce postsynaptic responses. nAChRs can also act presynaptically to regulate the release of other neurotransmitters such as glutamate or GABA. nAChRs are also expressed perisynaptically and thus are involved in nonsynaptic signaling (Dineley et al., 2015). Many cholinergic terminals don't form synapses, but release ACh by volume transmission that then can signal to high-affinity receptors in the vicinity. ACh can thus act widely as a diffuse modulatory neurotransmitter. Given the many roles nAChRs play in signaling, dissecting their function and involvement in Alzheimer's disease is challenging. However, data about the function of specific subtypes, especially α7-containing nAChRs, in animal models of AD and how treatment with nicotinic cholinergic compounds affects symptoms, provide clues to possible mechanisms of cell death and malfunction in human Alzheimer's disease due to changes in cholinergic signaling.

5.1.3 Amyloid beta

We will focus on APP and Aβ, although other molecules are clearly involved in the genesis and progression of Alzheimer's disease. This is due to the clear role of nAChRs in cholinergic dysfunction in Alzheimer's and the identified interactions between Aβ and nAChRs. APP is cleaved by either α-secretase or β-secretase (Fig. 5.1). Cleavage by β-secretase produces sAPPβ and C99, which is further processed by γ-secretase to Aβ. APP can also be processed by α-secretase to sAPPα and then further by γ-secretase to P3 (Dineley et al., 2015). Aβ monomers can associate to form more complex and larger forms such as oligomers, protofibrils, and fibrils (Li & Selkoe, 2020). Aggregation of these latter forms leads to insoluble plaques (Li & Selkoe, 2020). Monomers and dimers of Aβ1–42, Aβ1–40, low n oligomers, and nonfibrillar forms are all found in the brain, blood, and CSF of Alzheimer's brain samples

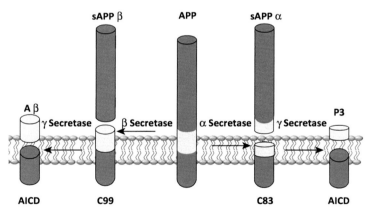

Fig. 5.1 Amyloid precursor protein (APP) cleavage produces amyloid beta. APP is cleaved by either β-secretase or α-secretase to produce fragments of the APP. Cleavage by β and γ-secretase produces amyloid-β and AICD. AICD (APP intracellular domain); C99, C83, C-terminal fragments of 99 and 83 amino acids; sAPP, soluble forms of APP. *(From Dineley, K. T., Pandya, A. A., & Yakel, J. L. (2015). Nicotinic ACh receptors as therapeutic targets in CNS disorders.* Trends in Pharmacological Sciences, 36 *(2), 96–108. https://doi.org/10.1016/j.tips.2014.12.002.)*

(Li & Selkoe, 2020). Besides any interactions with nAChRs, mechanisms of action by Aβ1–42 include effects on choline uptake and release, effects on ChAT activity, and decreases in glucose uptake (Wang, Lee, D'Andrea, et al., 2000).

Even in early AD, loss of synaptic density in the hippocampus is detected (Li & Selkoe, 2020). As is well known, β-amyloid plaques are a hallmark of AD. However, effects on synaptic plasticity have been observed in animal models of AD before insoluble plaques are detected. Soluble Aβ oligomers block long-term potentiation (LTP) in in vitro and animal studies (Li & Selkoe, 2020). Aβ extracts from AD brains containing small monomers (4 kDa) don't inhibit LTP. Large insoluble plaques don't either (Li & Selkoe, 2020). Small oligomers (7–16 kDa) and some larger soluble forms (16–60 kDa) impair LTP (Li & Selkoe, 2020).

Monomers don't appear to block synaptic function, but small oligomers and larger Aβ aggregates do (Li & Selkoe, 2020). Importantly, Aβ extracts isolated from human brains produce a much higher level of toxicity and synaptic dysfunction that synthetic aggregates (Li & Selkoe, 2020). As we discuss the interactions between nAChRs and Aβ, these different preparations and forms must be kept in mind. The presence of various forms (Aβ1-40, Aβ1-42, oligomers, and fibrils) may also explain the observations that Aβ in some forms activates nAChRs of specific subtypes and in other forms inhibits. The interplay between nAChR signaling and Aβ is complex, but understanding these will be the key to finding treatments for AD by targeting these interactions.

5.1.4 Interactions between amyloid beta and nAChRs

The focus has been on the Aβ1–42 protein. Aβ is a major factor in degeneration and is present in neuritic plaques (Wang, Lee, D'Andrea, et al., 2000). Immunochemistry was used to co-localize α7 nAChRs with Aβ1–42 in plaques and neurons of human Alzheimer's disease postmortem sections from the hippocampus and cortex (Wang, Lee, D'Andrea, et al., 2000). α4*nAChRs and NMDA R1 glutamate receptor proteins were not localized. Fluorescent labeling demonstrated that α7 and Aβ1–42 were expressed in the same cortical neurons from Alzheimer's disease postmortem samples. α7 nAChR protein was co-immunoprecipitated with Aβ1–42 in membrane extracts from Alzheimer's disease hippocampal samples. This co-precipitation occurred when either anti-Aβ1–42 or anti-α7 nAChR was used for the pull-down step. No co-precipitation occurred with antibodies to α1, α3, α4, α5, α8, or β2 subunits (Wang, Lee, D'Andrea, et al., 2000). Some co-immunoprecipitation occurred between α7 nAChRs and Aβ1–42 in membrane extracts from age-matched control hippocampus, but at a 20 X lower level (Wang, Lee, D'Andrea, et al., 2000). The interaction between Aβ1–42 and α7 nAChRs was also demonstrated in a cell line overexpressing α7 and incubated with Aβ1–42. An Aβ 12–28 peptide almost completely blocked interactions between α7 nAChRs and Aβ 1–42, confirming specificity and indicating that the epitope required for interaction was between amino acids 12–28 (Wang, Lee, D'Andrea, et al., 2000). Overexpression of human APP in PC12 cells further confirmed a tight association between Aβ 1–42 derived from human APP and α7 nAChR. α-Bgt (α7 antagonist) was shown to block binding of Aβ1–42 to α7 nAChRs on cells in culture overexpressing α7 and reciprocally Aβ 1–42 blocked binding of α-Bgt to α7 nAChRs as did controls MLA and epibatidine (Wang, Lee, D'Andrea, et al., 2000). These results were also reproduced with hippocampal membranes. In the α7 overexpressing cells, the IC_{50} for Aβ 1–42 inhibition of α-Bgt binding (competitive) indicated two sites with varying affinity, 0.2pM and 5 nM (Wang, Lee, D'Andrea, et al., 2000). Aβ1–42 stable oligomers were not detected.

Binding studies in vitro showed that Aβ1–42 displaced ^{3}H MLA (α7 antagonist) in rodent membrane preps from the cortex and hippocampus. Binding to two sites was indicated, one with a K_i of 4–5 pM (depending on the species) and the other site with about a $100\times$ lower affinity (Wang, Lee, Davis, & Shank, 2000). ^{3}H cytisine binding (β2 containing nAChRs) was displaced with a K_i of between 23 and 30 nM (depending on rodent species). A second site was also detected with a higher affinity of between 140 and 360 pM (Wang, Lee, Davis, et al., 2000). They also showed that Aβ1–42 had little or no affinity for NMDA, 5-HT-3, or M1 and M2 muscarinic receptors. Thus, Aβ1–42 binds at least two nAChR subtypes and to the α7-containing nAChRs with very high affinity.

Given that there are interactions between α7 nAChRs and Aβ1–42, the effects of Aβ1–42 on cell survival were tested using a cell line overexpressing α7 nAChRs. In total, 95% of the overexpessing cells were killed after a 24-h exposure to 100 nM Aβ 1–42, while over 85% of the normal cells survived (Wang, Lee, D'Andrea et al., 2000). This effect was specific in that Aβ 40–1 caused only a small amount of death at 10 μM. Nicotine or epibatidine pretreatment protected the α7 expressing cells from death (Wang, Lee, D'Andrea, et al., 2000). Thus, while the mechanisms are not all defined as yet, Aβ interacts with nAChRs, especially the α7 subtype and can activate or block α4β2 and/or α7 nAChRs under specific condition of Aβ structure, concentration, solubility, and conformation (Dineley et al., 2015).

Aβ 1–42 also blocks function of β2* heteromeric nAChRs (Lamb et al., 2005). The functions of α2β2, α4β2, and α4α5β2 (but to lesser degree) expressed in oocytes were all inhibited by 1 μM Aβ 1–42. In this study (Lamb et al., 2005), α7 nAChR function was not affected by Aβ1–42, in contrast to previous studies. Desformylflustrabromine (dFBr) is a positive allosteric modulator (PAM) for α2β2 and α4β2 nAChRs (Pandya & Yakel, 2011). Aβ 1–42 blocked function of α2β2 and α4β2 nAChRs expressed in oocytes, but this blockade was eliminated by 1 μM dFBr for both subtypes (Pandya & Yakel, 2011).

Oligomeric Aβ 1–42 activates α7 nAChRs as well as α7β2 heteromeric nAChRs. Basal forebrain cholinergic neurons exposed to oligomeric Aβ 1–42 increased action potential firing and increased intrinsic levels of excitability (George et al., 2021). This demonstrates that α7β2 receptors are also a target for Aβ, and these interactions may influence the Alzheimer's disease phenotype.

The effects of amyloid β on various subtype functions are varied and sometimes conflicting. Previous work showed physical interactions between Aβ 1–42 and nAChRs, but the effects varied. These differences stem from using electrophysiology to study channels in oocytes, or cell lines, as well as primary neurons or slices (Lombardo & Maskos, 2015). Variations depending on the composition of the Aβ 1–42 used have also been seen. Concentration-dependent effects were also present.

Application of Aβ 1–42 (form was not clear, but should be soluble oligomeric form based on the timing and method of preparation) inhibited whole cell and single-channel ACh currents in stratum radiatum interneurons (Pettit et al., 2001). Even at only 100 nM, Aβ 1–42 produced a 32% inhibition of currents. The effects of the inhibition were shown to be reversible (Pettit et al., 2001). Aβ 40–1 had no effect on nAChRs, and Aβ 1–42 was specific for nAChRs as it had no effect on glutamate signaling in this system, and atropine did not block the effects of Aβ 1–42. These result support the effect of Aβ 1–42 as being on postsynaptic nAChRs (Pettit et al., 2001). The effects of Aβ 1–42 on stratum radiatum interneuron nAChRs were to decrease the open probability of the channels. Aβ1–42 inhibited both α7nAChRs and non-α7 nAChRs but to different degrees (Pettit et al., 2001).

Inhibition of nAChR channel function was also demonstrated for α7 and α4β2 nAChRs expressed in Xenopus oocytes (Tozaki et al., 2002). Aβ 1–40 and Aβ 1–42 (100 nM) both reduced whole cell currents produced from α4β2 nAChRs. There was a slower smaller and more gradual reduction of α7 nAChR whole cell currents (Tozaki et al., 2002).

Over a concentration range from 1 pM to 100 nM, neither Aβ 1–40 or Aβ1–42 activated human α7, α4β2, or α3β4 nAChRs expressed in Xenopus oocytes (Pym et al., 2005). This is similar to another study that showed Aβ1–42 didn't activate human α7 nAChRs over this range of concentration (Grassi et al., 2003). Grassi et al. (2003), also showed that 100 nM Aβ 1–42 reduced the current induced by ACh in oocytes expressing human α7 in a noncompetitive manner. In total, 100 nM Aβ 40–1 didn't produce this effect, but there was some inhibition by exposure to 100 nM Aβ 42–1, but that this effect was more reversible (Grassi et al., 2003). Aβ 1–42 acted as agonist at a mutated α7 nAChR (L248T). This is consistent with other α7 nAChR antagonists and is consistent with the functional inhibition being due to interaction at the receptor and not other effectors (Grassi et al., 2003). Pym et al. (2005) also showed that Aβ 1–42 and 1–40 at 10 nM reduced current through α7 nAChRs in oocytes, in a noncompetitive manner, but that 10 nM of Aβ 1–42 or Aβ 1–40 enhanced ACh current through α4β2 nAChRs. Analysis showed that few fibrils were present in this study, soluble Aβ 1–42 and 1–40 were used. Wu et al. (2004) reported a reduction in α4β2 function, but they dissolved Aβ in water, and the preparation may have contained more aggregates. Aβ 1–42 and Aβ 1–40 produced no changes in current amplitude of α3β4 human nAChRs expressed in oocytes (Pym et al., 2005). However, another study showed that rat α7 nAChRs were activated by Aβ 1–42 (prepared to minimize aggregation) at low concentrations (1–100 pM) (Dineley et al., 2002). Aβ 1–40 produced a reduced response, and Aβ 40–1 didn't work. Less activation was seen at nM levels, and the actions of Aβ 1–42 were desensitizing. This is consistent with the effects seen in Pettit et al. (2001), higher amounts may have desensitized the response of human α7 nAChRs in that study. Interestingly, low concentrations didn't desensitize the response to nicotine, but higher amounts did (Dineley et al., 2002). They speculate that perhaps the α7 nAChRs have two binding sites for Aβ. Inward currents were carried by Ca^{2+} that could lead to ERK/MAPK activation (Dineley et al., 2002).

Dougherty et al. (2003) observed in brain synaptosome (presynaptic) preparations from rat hippocampus and cortex that low levels of Aβ 1–42 (1–100 pM) activated α7 and produced sustained increases in presynaptic Ca^{2+}. They first observed that Aβ 1–42 (and Aβ 12–28) blocked the response to nicotine. However, further work showed this was because

Aβ 1–42 and Aβ 1–28 activated the presynaptic receptors and prevented further nicotine action. Aβ affected both α7 and non-α7 nAChRs, but the non α7-nAChRs were affected at a higher concentration (Dougherty et al., 2003). The nonaggregated form was involved.

Whole-cell patch clamp using dissociated rat hippocampal neurons showed that soluble Aβ1–42 inhibited α7 nAChRs with an IC_{50} of 7.5 nM and in a noncompetitive manner (Liu et al., 2001) and didn't appear to increase desensitization. This is a higher concentration used in the brain synaptosome studies from rat hippocampus that activated α7 nAChRs described above. Rat Aβ 1–42, Aβ 1–40, and human Aβ 1–40 all produced the block. The effect was specific to α7 nAChRs (Liu et al., 2001). The block observed in the hippocampal neurons was not use or voltage-dependent (Liu et al., 2001) and depended on the extracellular domain of α7 being present. The β-amyloid effect was on presynaptic receptors as a nicotine-induced increase in mEPSCs was blocked by 100 nM Aβ 1–42 (Liu et al., 2001).

As stated before, different systems have provided different results. For example, α4β2 nAChRs, expressed in a rat neural cell line, were exposed to soluble oligomeric 100 nM Aβ 1–42 (Arora et al., 2013). Increases in Ca^{2+} in axonal-like projections were observed, supporting that Aβ 1–42 acted on α4β2 nAChRs located presynaptically. Three-day exposure to 100 nM Aβ 1–42 augmented the Ca^{2+} response to acute Aβ 1–42 as well as nicotine (Arora et al., 2013). DHβE blocked these effects. Chronic Aβ 1–42 exposure caused increased surface expression of α4β2 nAChRs, possible due to increased receptor recycling. Mitochondrial dysfunction was also observed after Aβ 1–42 exposure at these low levels. Apoptosis, nuclear fragmentation, and increased ROS occurred in the α4β2 expressing cells in the presence of Aβ 1–42, and the reverse peptide Aβ 42–1 was used as a negative control. Thus, Aβ 1–42 working through α4β2 nAChRs may cause death after sustained exposure to Aβ 1–42 (Arora et al., 2013). Fig. 5.2 summarizes some interactions that take place between various forms of Aβ and nAChRs,

Since both fibrillar and oligomeric forms of Aβ 1–42 and Aβ 1–40 were used in many studies and are present in the AD brain, it will be important to dissect the effects of each form. Fibrillar Aβ 1–40 (100 nM) produced cell death in PC12 cells, but this was mitigated by α7 nChR agonists varenicline and JN403 (Lilja et al., 2011). Oligomeric Aβ 1–40 didn't have much effect at this concentration, and both oligomeric and fibrillar Aβ 1–42 produced cell loss only at μM concentrations (Lilja et al., 2011). Varenicline increased

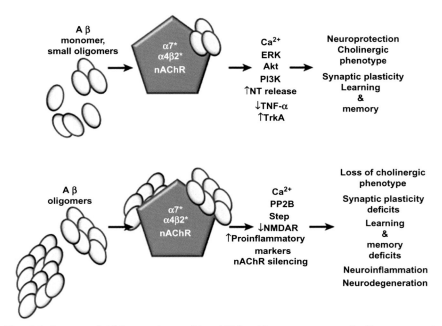

Fig. 5.2 Beta-amyloid interactions with nAChRs. Aβ monomer, small oligomers, and oligomers interact with either α7* or α4β2* nAChR and alter multiple singling pathways that mediate plasticity, neuroprotection, and degeneration and inflammation. ERK, extracellular signal-regulated kinase; NMDAR, NMDA receptor; NT, neurotrophin; PI3K, phosphatidylinositide 3-kinase; PP2B, protein phosphatase 2B; TNF-α, tumor necrosis factor-α; TrkA, tropomyosin receptor kinase A. *(From Dineley, K. T., Pandya, A. A., & Yakel, J. L. (2015). Nicotinic ACh receptors as therapeutic targets in CNS disorders.* Trends in Pharmacological Sciences, 36 *(2), 96–108. https://doi.org/10.1016/j.tips.2014.12.002.)*

an amyloid ligand binding to fibrils in human AD brain samples, showing that nAChR agonists may block Aβ/nAChR interactions. Oligomeric Aβ 1–40 displaced ³H-epibatidine from AD brain homogenates, indicating specific binding to nAChRs, perhaps acting allosterically (Lilja et al., 2011). Oligomeric (but not fibrillar) Aβ 1–40 increased intracellular Ca^{2+} in SH-SY5Y cells acting through varenicline-sensitive α7 nAChRs. The oligomeric form may act as an agonist to stimulate nAChRs signaling, but the fibrillar form may at least partly block nAChR function and produce toxic effects (Lilja et al., 2011).

Monomeric 100 nM Aβ 1–40 inhibits the nicotine-induced increase in dopamine (DA) release from rat brain synaptosomes. α4β2* (α-conotoxin MII resistant) nAChRs are implicated in the DA release that is blocked by

Aβ 1–40 (Olivera et al., 2014). A noncompetitive mechanism is supported. Positive allosteric modulators (i.e., dFBr) mitigated the inhibitory effect of Aβ 1–40 on DA release (Olivera et al., 2014). Aβ 1–40 doesn't affect noradrenaline release mediated by α3β4 nAChRs (Olivera et al., 2014).

Knockout mice lacking α7 were used to define further the role of α7 nAChRs in AD pathology. This seemed like a direct approach, but two studies using this approach produced conflicting results. Briefly, Dziewczapolski et al. (2009) crossed an APP mouse line containing the Swedish and Indiana mutations (two seen in human AD). The APP line had memory and spatial deficits at 13–16 months, but an APP line with the α7 nAChRs knocked out had improved memory over the APP mice. Even though the APP-α7 KO mice expressed high levels of amyloid β, the mice had good memory function in the Morris water maze, reduced gliosis, LTP remained stable, and synaptic markers such as MAP2 and synaptophysin were preserved (Dziewczapolski et al., 2009). This study supported the concept that without α7 nAChRs Aβ could not enter cells and accumulate, causing neuropathology. Previous studies had indicated that binding of Aβ1–42 to α7 nAChR produced internalization of the complex causing neurotoxicity (Nagele et al., 2002). Signaling through α7 to intracellular pathways was also prevented due to the absence of α7 nAChRs. Another study by Hernandez et al. (2010) using the Swedish mutation in mice to overproduce APP and crossed with α7 KO showed that loss of α7 produced more severe learning and memory deficits in 5-month-old mice. Neurodegeneration and loss of function in the hippocampus and basal forebrain (Hernandez et al., 2010) were increased in the APP-α7KO mice. They proposed that increased levels of soluble oligomeric Aβ lead to the loss in function. This is in contrast to (Dziewczapolski et al., 2009) that showed protection, although their mice were tested at an older age. Hernandez et al. (2010) proposed that α7 protects by Aβ-α7 interactions initiating protective signals to the neuron and that α7 may sequester Aβ and lower the increasing amount of oligomeric forms that produce toxicity and death. Ultimately, α7 nAChRs maybe more protective at early stages but as Aβ builds up, interaction with α7 nAChRs leads to internalization and death (Hernandez et al., 2010). α7 maybe upregulated in early stages to compensate for desensitization and functional downregulation for a while as Aβ accumulates. As was shown in Dziewczapolski et al. (2009), in older animals (and potentially humans), loss of α7 nAChRs protects from toxic effects of higher levels of Aβ accumulating internally. This might suggest that in later stages, blocking interactions of α7 and Aβ might be beneficial to treat later-stage disease.

α7 nAChRs are also present on astrocytes. Astrocytes release gliotransmitters and can regulate neuronal excitability as part of the "tripartite synapse." Soluble oligomeric Aβ 1–42 levels seen in the brain can activate astrocytes and increase intracellular Ca^{2+} leading to release of glutamate mediating synaptic damage through extrasynaptic NMDA receptors on neurons (Talantova et al., 2013).

5.1.5 Potential therapies

Both agonists and antagonists regulate Aβ and nAChR interactions (Lombardo & Maskos, 2015). Nicotine and cotinine have been shown to improve memory performance. Cotinine is a metabolite of nicotine but also an α7 PAM. MLA (α7 antagonist) and an α7 partial agonist have produced protection in synaptosomes and cell cultures. Nicotine produced improvements in memory in old APP mice and reduced oligomeric Aβ. α7 levels and synapse numbers increased after nicotine treatment (Lombardo & Maskos, 2015). However, the addictive nature of nicotine makes it a poor candidate for therapy. AChE inhibitors such as Donezipil (Aricept) and galantamine (Razadyne) were used to treat AD and have produced some improvements in memory (Dineley et al., 2015). The actions of these drugs may be complex in that galantamine is also a PAM of α4β2 and α7nAChRs.

Numerous α7 and α4β2 nAChR agonists and partial agonists have been tested, but none so far have proved very effective (Hurst et al., 2013), and others produced undesirable side effects. These include varenicline, AZD-3480, EVP-6124, ABT-418, ABT-126, and RG3487 (Hoskin et al., 2019). Other agonists have been or are being tested. Most compounds tested act as partial or full agonists of α7 or α4β2 nAChRs or as PAMs of α7 nAChRs (Hoskin et al., 2019). This makes sense given what is known about the interactions of Aβ with specific nAChR subtypes. While studies are ongoing, so far no "magic cholinergic bullet" for AD treatment has been developed yet. It is possible that by the time many of these drugs being tested were given, the disease was too far advanced. Trials of new compounds commencing earlier with patients experiencing mild cognitive impairment or early-onset Alzheimer's disease may give better results.

5.1.6 Summary

Multiple forms of Aβ have been tested in different species, cell lines, various preparations of primary neuronal cultures and on different AChR subtypes. Many of these were synthetic Aβ peptides in various forms (Li & Selkoe,

2020). Activation/blockade of nAChRs clearly influences signaling and death. Differences in results may be due to other molecules, which may or may not be present in a specific preparation (i.e., slice vs oocytes vs human brain vs cell lines) and that may be needed for functional effect. More studies using Aβ complexes from human brain should be used in studies evaluating the interactions between Aβ and nAChRs since these may be more relevant (Li & Selkoe, 2020). Since different levels of various forms of β-amyloid occur during different stages of AD, changes in cell viability and function mediated by β-amyloid and nAChR interactions are complex and most likely vary over the time course of the disease. Amyloid signaling through nAChRs may have a normal role in the brain to promote function, but higher levels of Aβ may promote receptor malfunction and cell death (Dineley et al., 2001). The sum of the data supports an important role for nAChRs in Alzheimer's disease. Therapies targeting nAChRs at various disease time points could provide an important tool to mitigate some of the symptoms of AD.

5.2 Schizophrenia

5.2.1 Basics of schizophrenia

The term "schizophrenia" was coined by Paul Bleuler to describe what appeared to be a group of diseases. Psychiatrist Emil Kraeplin described it as adolescent insanity due to its time of onset (Wallace & Bertrand, 2015). Schizophrenia affects more than 2.6 million adults (more than 1% of the adult population) in the United States. Schizophrenia is a chronic neurological disorder characterized by multiple positive and negative symptoms. Positive symptoms refer to traits not seen in normal people. These include hallucinations, delusions (paranoia), and movement problems. Negative symptoms (lack of functions seen in normal people) include a flat affect (difficulty showing emotions), reduced joy gained from normal life experiences, social withdrawal, and reduced ability to speak. Cognitive problems develop due to poor executive function, lack of focus, and poor working memory (https://www.treatmentadvocacycenter.org/evidence-and-research/learn-more-about/25-schizophrenia-fact-sheet). Due to these neurological problems, people have trouble interacting with others, interpreting others' facial expression, living in an altered world, and thus are isolated, often without being able to hold jobs and living on the street.

Schizophrenia is a developmental neurological disorder that manifests in late adolescence or early adulthood. It is rare before puberty with the onset in men generally between ages 21 and 25 and woman between 25 and 35 or even later. The disease occurs equally in men and women (https://www.psychiatry.org/patients-families/schizophrenia/what-is-schizophrenia).

There is a genetic component to schizophrenia, but multiple genes are certainly involved, and no one gene can be used to predict the onset of disease. Environmental components may contribute to development of the disease including stress and virus infection. Brain structural changes have been noted such as larger ventricles and lower gray matter volume (Wallace & Bertrand, 2015). Connectivity changes and differences in neurotransmitter levels such as dopamine, glutamate, and acetylcholine have also been observed. Increased dopaminergic signaling has been a hallmark of schizophrenia and the basis for development of antipsychotic drugs that act as dopamine receptor antagonists. Many antipsychotics used to treat schizophrenia target the dopamine D2 receptor. These help reduce positive symptoms, but don't work well at alleviating negative ones (Featherstone & Siegel, 2015) (https://www.nimh.nih.gov/health/topics/schizophrenia/index.shtml).

Some drugs used to treat schizophrenia such as chlorpromazine have actions on multiple neurotransmitter systems, and the exact contributions of specific neurotransmitters to schizophrenia have been difficult to decipher.

5.2.2 Cholinergic role in schizophrenia

A cholinergic role in schizophrenia was suggested by the observation that most (at least 80%) of schizophrenics smoke (Leonard et al., 2007). The hypothesis was developed that schizophrenics smoke or use nicotine, as a form of self-medication or to remedy a nicotinic signaling imbalance. The cholinergic system has a major role in executive function and attention; functions that are affected in schizophrenia. Stimulation of nAChRs (or perhaps desensitization) by nicotine ameliorates or corrects some signaling malfunctions. nAChRs are present, pre-, post-, and perisynaptically. $\alpha 4\beta 2^*$ nAChRs are often found presynaptically while $\alpha 7$ nAChRs can be located in all of the above locations. Transmitter releases including glutamate, dopamine, acetylcholine, norepinephrine, and GABA are modified by nAChR function. The expression of several neurotransmitters is altered in schizophrenia including glutamate and dopamine. The multiple-neurotransmitter hypothesis includes a role for α7 nAChRs in increasing cholinergic, glutamatergic, and GABA signaling, which results in normalized DA levels (Bencherif et al., 2012). α7 nAChRs gate Ca^{2+} and can affect the release of other neurotransmitters as well and lead to activation of second messenger systems. α7 nAChRs are present presynaptically on glutamatergic neurons.

Smoking does in fact improve some cognitive and sensory problems in smokers (Dineley et al., 2015). Specifically schizophrenics have a deficit in inhibiting an auditory stimulus, P50. This deficit is common among schizophrenics and is genetic. The mutation linked to the P50 deficit has been localized to a region on chromosome 15 containing the α7 gene. Smoking normalizes this response (Leonard et al., 2007). Sequence variants in the human α7 gene promoter were shown to produce reduced transcription, and these variants were more prevalent in schizophrenic patients (Leonard et al., 2002). A large genetic study linked a deletion on chromosome 15 containing the human α7 gene to schizophrenia (Stefansson et al., 2008). This region contained multiple genes, but further studies demonstrated that patients with deletions containing only α7 have intellectual disability (Featherstone & Siegel, 2015). Schizophrenics also have a deficit in prepulse inhibition and eye tracking, which are also improved by smoking (Leonard et al., 2007). Working memory and attention deficits are also ameliorated by smoking (Leonard et al., 2007). These and many other behavioral observations have linked the nicotinic cholinergic system to schizophrenia (Featherstone & Siegel, 2015).

5.2.3 nAChRs and schizophrenia

Deficits in specific nAChRs have also been observed in the schizophrenic brain. α7 gene deletions have been detected in schizophrenics (Schaaf, 2014). α7 nAChR levels are reduced in several brain regions including the hippocampus and the cingulate and frontal cortex (Schaaf, 2014). Human postmortem hippocampus from schizophrenics also has reduced α-Bgt binding (Stevens et al., 1996), and α7 expression was also low in cortex and reticular nucleus of the thalamus (Leonard et al., 2007). The reticular nucleus projects to the dorsal thalamic nuclei, which are important in mediating attention and sensory gating (Wallace & Bertrand, 2015) Some studies show no change in α-Bgt binding in the hippocampus however (Gibbons & Dean, 2016). α4β2* nAChRs comprise most of the high-affinity nicotine-binding sites in the brain. $^{3}H-$ nicotine binding is also reduced by 50% in the hippocampus, caudate, and cortex in the schizophrenic brain (Breese et al., 2000), indicating that in addition to α7 nAChRs, the α4β2* subtypes may also have a role. Smoking leads to significantly increased nicotine binding (50%) in the hippocampus, but only a slight increase is seen in schizophrenic hippocampus (Leonard et al., 2007). This observation also supports some change in the regulation of α4β2 nAChRs. A genetic study using a family-based association study indicated that neither α4 or β2 alone was sufficient to influence development of schizophrenia, and the two genes appeared to interact to affect the schizophrenic phenotype (De Luca et al., 2006). This was not replicated in another study however (Featherstone & Siegel, 2015). A study by Radek et al. (2006) indicated that α4β2* receptors blocked by DHβE may also contribute to the effects of α7 nAChRs in sensory gating in DBA/2 mice in response to nicotine.

What is the role of reduced levels of nAChRs or altered nAChR signaling in schizophrenia? Lower receptor levels may lead to changes in neurotransmitter release or response of neurons to endogenous acetylcholine. Changes in cholinergic signaling in schizophrenic brains could lead to altered signaling and gene expression with widespread effects that schizophrenic smokers may be unconsciously attempting to ameliorate by smoking. Support for this hypothesis came from a study whereby expression of a large number of genes assessed by a microarray showed changes in expression levels of 277 genes when compared between the brains of control smokers and nonsmokers (Mexal et al., 2005). This came from a panel of 4246 unique genes tested. In total, 77 genes were differentially expressed when comparing schizophrenic smokers to schizophrenic nonsmokers. Many of the changes observed were in genes expressed at postsynaptic densities containing NMDA glutamate receptors, including NCAM1, α-catenin, and calcineurin Ag (Mexal et al., 2005). Interestingly, KO mice for calcineurin Ag, NCAM, and αN-catenin exhibit deficits similar to those seen in human schizophrenics (Mexal et al., 2005). Smoking seemed to normalize expression of these genes and others that were altered in schizophrenics (Leonard et al., 2007; Mexal et al., 2005). Since changes in glutamate signaling are seen in schizophrenia, smoking (nicotinic signaling) may alter glutamatergic signaling and normalize the schizophrenic phenotype to some degree. α7 and NMDA receptors are sometimes colocalized and may have functional interactions (Wallace & Bertrand, 2015).

5.2.4 Cholinergic drugs for possible schizophrenia treatment

nAChRs have clearly been linked to schizophrenia and have been the target of drugs designed to treat schizophrenia. Both α7 and α4β2* nAChRs may be involved. Mice strains that vary in levels of α7 nAChRs have been used to study the role of these in the schizophrenic phenotype. Mice that express lower levels of hippocampal CA3 α7 nAChRs (DBA/2j) than other strains such as C3H demonstrate a sensory gating deficit (Stevens et al., 1996). Antagonists of α7 nAChRs can reproduce the P50 sensory gating deficits in animals (Leonard et al., 2007). Treatment with an α7 agonist DMXB-A (GTS-21) normalizes the P50 sensory gating deficit in mice (Leonard et al., 2007).

Varenicline (Chantix) is an α4β2 partial agonist and an α7 nAChR full agonist. Treatment with varenicline improved cognitive function in some human patients (Dineley et al., 2015). α7 partial agonists (DMXB-A, EVP 6124) have improved the defects in human schizophrenics. DMXB-A treatment in human patients showed an improvement in default network function (overactivity may lead to disconnections from reality). Pharmacogenetic studies indicated this was mediated by α7 nAChRs (Tregellas et al., 2011). EVP 6124 used in a proof-of-concept randomized trial showed improvement in cognitive tests involving learning, memory, and executive function (Preskorn et al., 2014). Positive allosteric modulators of α7 nAChRs such as PNU-120596 and JNJ-1930942 improved auditory gating deficits in mice (Dineley et al., 2015). Galantamine has produced improvements in cognition and negative symptoms (Gibbons & Dean, 2016). Cholinesterase inhibitors including galantamine (also a nAChR PAM) also had some success in treating schizophrenia (Gibbons & Dean, 2016).

5.2.5 Summary

Cholinergic signaling through multiple nAChR types is clearly involved in schizophrenia. Much of the clinical treatment relating nAChRs to schizophrenic has focused on the α7 nAChRs. This is reasonable given that genetic linkage studies implicate α7 nAChRs as well as the reduction of α7 receptor numbers observed in the hippocampus and other brain areas and reduced nAChR function in mice and humans. Evidence for a cholinergic role includes that many schizophrenics smoke and that nicotine ameliorates some sensory and cognitive deficits. There is evidence for a role for α7β2 nAChRs as well as α4β2* nAChRs. Unfortunately, longer clinical trials have not shown the compounds tested so far to be effective, and more effective treatments targeting the cholinergic system are needed.

5.3 Autism

5.3.1 Basics of autism

Autism is a spectrum of related developmental neurological disorders defined as autism spectrum disorder (ASD). Cognitive impairment, repetitive behaviors, and lack of verbal and social skills are often seen (Dineley et al., 2015). Children have trouble making eye contact, often prefer not to be touched, and may not understand other's feelings. People with autism can be very high functioning and intelligent. Children with autism can be very focused on one interest and can be very good at it. The number of children affected by autism is generally thought to be about 1/54 children (austismspeaks.org) to 1/68 children (Jung et al., 2017) to 1/100 (Ebert & Greenberg, 2013). There are genetic and environmental components that are not well understood. Many genes are involved, and many may be important in regulating signaling, neuronal activity, synapse development, and plasticity (Ebert & Greenberg, 2013).

Several genetic studies have shown associations between genomic copy number variations and single-nucleotide polymorphisms and ASD (Jung et al., 2017; Shishido et al., 2014). Twin studies have shown that a strong genetic component contributes to the development of autism (Gaugler et al., 2014). Hundreds of ASD risk genes have been proposed with around 50 as high-risk (De Rubeis & Buxbaum, 2015). ASD is sometimes associated with other neurological disorders such as fragile X syndrome.

Autism signs usually manifest at ages 2–3 (https://www.cdc.gov/ncbddd/autism/facts.html). Autism occurs at a time when significant synaptic development is being modified by experience. Abnormalities in several neurotransmitters may be involved including GABA, serotonin, norepinephrine, glutamate, and acetylcholine (Martin-Ruiz et al., 2004). Changes in the size of neurons and cell packing are altered in several regions in autistic brains including the hippocampus, amygdala, septum, and cingulate gyrus (Deutsch et al., 2010). Cerebellar Purkinje cell loss is also often observed. Many neurological disorders use animal models for studies of basic mechanisms and drug testing. ASD is complicated because good animal models for specific behaviors are problematic, but some do exist and are shedding light on possible mechanisms.

5.3.2 Evidence for a cholinergic role in ASD

Nicotinic receptors play many roles in neurodevelopment, and autism is a disorder characterized by deficits that occur early in development. As such, changes in cholinergic markers, receptors, and anatomy of cholinergic brain regions might be expected. Cholinergic neurons from the basal forebrain such as the medial septal nuclei project to the hippocampus and neurons from the nucleus basalis of Meynert provide input to wide areas of the cortex. The pontine tegmental projection innervates the thalamus among other regions. These communicate through both nicotinic and muscarinic receptors (Deutsch & Burket, 2020). $\alpha7$ nAChRs are involved with circuits regulating attention and cognition, deficits of which are seen in ASD (Deutsch & Burket, 2020). One model suggests that in ASD excitation is increased and inhibition decreased. Since $\alpha7$ nAChRs are expressed on GABAergic (inhibitory) neurons, deficits or changes in $\alpha7$ nAChR function or expression may cause some aspects of the ASD phenotype due to loss of inhibitory tone.

Changes in cholinergic neurons and projections have been noted in ASD postmortem samples (Deutsch & Burket, 2020). Cholinergic pathology in basal forebrain neurons is observed in autism (Martin-Ruiz et al., 2004). Neurons in the cholinergic diagonal band of Broca were found to be smaller in children with autism and larger in adults than normal (Deutsch et al., 2010). In the BTBR mouse model of autism, nicotine decreased spontaneous grooming, a hallmark of ASD in some animal models (Deutsch & Burket, 2020). Since nicotine upregulates nAChR expression, this supports a role of nAChRs in ASD.

Evidence for a cholinergic role in ASD comes from human and animal studies that suggest a possible relationship between maternal smoking and ASD or ASD-like behaviors in offspring (Jung et al., 2017). Fetal nicotine exposure is associated with a number of neurological changes in animals and increased attention deficit disorder (ADD) and antisocial behavior in humans (Jung et al., 2017). A metaanalysis of studies focusing on maternal smoking during pregnancy found no significant association with ASD. However, when population-level metrics were included, an association was shown, possibly due to second-hand exposure of the fetus and mother due to paternal smoking (Jung et al., 2017). Changes in ratios of excitatory/inhibitory (E/I) signaling have also been implicated in ASD (John & Berg, 2015). Cholinergic modulation of circuits through control of glutamate and GABA release is important in the E/I balance of circuits and brain regions. Changes in cholinergic signaling (release of ACh or nAChR expression) may be involved in ASD.

Children with ASD exhibit increased pain and touch sensitivity and decreased thermal sensitivity (Wang et al., 2015). The BTBR mouse, used as a model of autism, shows increased tolerance to heat and cold as occurs in humans with ASD as well as hypersensitivity to pain (Wang et al., 2016). Cholinergic activation lowers social deficits and repetitive behavior in the BTBR mouse (Wang et al., 2015). Nicotine at lower doses increased the hyporesponse to heat, but mitigated the response to cold. At higher doses, nicotine normalized the response to heat, but increased the hyporesponse to cold (Wang et al., 2015). These responses to nicotine differed from those seen in B6 mice and implicated differences in the cholinergic system between ASD mice and normal.

5.3.3 nAChRs and autism

These changes in behavior correlate with the expression of nAChR RNAs. In BTBR mice $\alpha 3$, $\alpha 4$, $\alpha 7$, $\beta 2$, and $\beta 4$ RNAs were reduced in embryonic day 13 mice compared with B6 mice (Wang et al., 2016). In adult BTBR mice, expression of $\alpha 5$ was lower in the prefrontal cortex, $\alpha 4$, $\alpha 5, \beta 2$ lower in the hippocampus, and $\alpha 3$, $\alpha 4$, $\alpha 5$, $\alpha 7$, $\beta 2$, and $\beta 3$ lower in the cerebellum than in B6 mice (Wang et al., 2016).

No differences in the levels of ChAT and acetycholinesterase activity were observed in the basal forebrain and frontal and parietal cortex in brain samples from normal or autistic brains (Perry et al., 2001). However, epibatidine binding was reduced by 2/3 in the frontal and parietal cortex and levels of $\alpha 4$ and $\beta 2$ proteins detected by Western blotting were reduced in the parietal cortex of autism brain samples. No differences in α-Bgt binding in the cortex were noted (Perry et al., 2001). Nicotine binding was the same in the basal forebrain of normal and autistic brains. Epibatidine binding was reduced in the cerebellum of autism brains while $\alpha 7$ nAChRs detected by α-Bgt binding were increased. The $\alpha 4$ nAChR subunit protein levels were reduced while $\alpha 7$ nAChR protein increased (Lee et al., 2002). Further studies showed that $\alpha 4$ nAChR RNA and protein as well as epibatidine binding were decreased in the parietal cortex of autistic brain (Martin-Ruiz et al., 2004). Parietal cortex $\alpha 7$ nAChR RNA, protein, and α-Bgt binding were conserved in autistic brains compared with normal. $\alpha 4$ nAChR protein and epibatidine binding were reduced in the cerebellum of autistic brain while paradoxically $\alpha 4$ nAChR RNA levels increased. α-Bgt binding was increased in the cerebellum, but without RNA or protein increases in autistic brain (Martin-Ruiz et al., 2004). The thalamus is an important relay center in the brain and expresses multiple nAChR subtypes. The $\alpha 7$ and $\beta 2$ subunit proteins were reduced in some thalamic nuclei (paraventricular nucleus and nucleus reuniens) in autistic samples while α4 nAChR protein levels were unchanged (Deutsch et al., 2010; Ray et al., 2005).

α7 nAChR protein was increased in astrocytes in the paraventricular nucleus and nucleus reunions (Deutsch & Burket, 2020). Changes in nAChR expression in an area of the brain that communicates with several brain areas implicated in autism (i.e., hippocampus, striatum, prefrontal) could support a cholinergic role. α7 nAChRs are expressed in parvalbumin positive GABAergic neurons. These control timing of neuronal firing in assemblies involved with working memory and other higher-order processing (Deutsch & Burket, 2020). Defects in α7 signaling in parvalbumin positive GABAergic interneurons may disrupt timing of pyramidal neuron firing leading to some of the deficits seen in ASD. α7 nAChR and NMDA glutamate receptors are expressed in some GABAergic neurons, which are involved in ASD animal models (Deutsch & Burket, 2020).

Genetic studies have focused on the mutations/deletion/duplication of the α7 gene in neurological disorders. There has been an association with schizophrenia (see above); however, links to ASD have been more tenuous. Reduced α7 nAChR expression due to deletion has been linked to ASD, but seems to be a minor contributor (Deutsch & Burket, 2020). A closely linked gene OTUD7A may be involved in ASD and complicates understanding of the role of the α7 gene (Deutsch & Burket, 2020).

α7 nAChR heterozygous KO mice with reduced α7 expression in the hippocampus demonstrate altered auditory sensory processing and increased pyramidal neuron firing in the CA3 region (Adams et al., 2012). GAD-65 (enzyme involved in GABA synthesis from glutamate) expression was reduced in these mice, and GABA-A receptors reduced in male mice only (Adams et al., 2012). GABA levels were normal, however, perhaps due to synthesis of GABA by another enzyme GAD 67. That perturbations in α7 nAChR expression can alter hippocampal circuit function and produce phenotypes suggests a role for α7 nAChRs in sensory processing deficits seen in ASD.

5.3.4 Cholinergic approaches to treating autism

In contrast to schizophrenics, fewer people with autism smoke than in the average population (Bejerot & Nylander, 2003). Some studies show loss of α4β2 (high-affinity nAChRs) in autism, and thus, the rewarding effects of nicotine would be lower. Studies have been in conflict as to the best approaches to treatment. Work suggests that antagonism and lowering of cholinergic tone may be a useful approach (Lippiello, 2006). A hypercholinergic tone in ASD has been proposed due to neuronal size changes in basal ganglion (Lippiello, 2006). A decrease in α4β2 nAChRs then might be expected as a compensation. Smoking is low as well because this would be expected to exacerbate symptoms. Autistic patients also have a high pain tolerance. Hypercholinergic activity through descending pathways to the spine could explain this and supports a role of excessive cholinergic tone in ASD (Lippiello, 2006).

Other mechanisms may be possible based on studies using drugs that increase cholinergic tone. Antagonists such as mecamylamine weren't successful in treating autism symptoms, but cholinesterase inhibitors such as donepezil and galantamine as well as agonists and PAMs may provide symptomatic relief (Dineley et al., 2015). A small trial using galantamine improved ASD patient function by reducing anger, social withdrawal, and irritability over a 12-week trial (Nicolson et al., 2006). Yet another small trial using galantamine in combination with risperidone showed improvement in irritability and lethargy/social withdrawal (Ghaleiha et al., 2014). An α7 PAM, AVL-3288, improved ASD-associated behaviors in the BTBR mouse model in social approach and self-grooming assays (Yoshimura et al., 2017). A limited test using two doses of DMXB-A (α7 nAChR partial agonist) with two ASD patients produced improvement in attention (Olincy et al., 2016). In a small trial, the nicotine patch was shown to reduce aggression and improve sleep in ASD patients (Lewis et al., 2018).

5.3.5 Summary

Cholinergic signaling is involved in some of the pathways exhibiting deficits in ASD. Cholinergic receptors such as the α7 and α4β2 subtypes are altered as well in ASD brain. Given that nicotine and some cholinergic compounds produce symptomatic relief in a mouse model of ASD and in some human studies, more studies of modulators of cholinergic signaling in humans may be worthwhile.

5.4 Parkinson's disease
5.4.1 Background

Approximately 1% of people over 60 years old develop Parkinson's disease (Quik et al., 2019). Parkinson's disease (PD) results from the loss of a number of neuronal populations, but prominent is the death of nigrostriatal dopaminergic (basal ganglia) neurons. This results in less dopamine release in the striatum and lack of signaling through medium spiny neurons expressing dopamine D1 and D2 receptors. This leads to a decrease in stimulation and increase in inhibition and production of multiple motor deficits (Quik et al., 2019). PD manifests with movement difficulties including tremor, loss of facial expression, muscle stiffness, freezing of gait, posture and balance problems, and bradykinesia or akinesia. Nonmotor deficits are also present such as sensory problems, cognitive issues, autonomic nervous system dysfunction, and GI disturbances. Symptoms often develop slowly and become progressively worse. PD usually begins in middle age and is more frequent in males (Mayo Clinic). Writing and speech become difficult, coordination and unconscious movements are impaired. Depression, sleep disturbance, and cognitive decline are sometimes present (Quik, Bordia, et al., 2015). The etiology is complex and results from death of neurons of the substantia nigra (SN) that project to the striatum, producing impaired dopaminergic transmission. Lewy bodies containing α-synuclein are often present.

Other neurotransmitter systems degenerate including cholinergic. Neurons in the central and peripheral nervous systems degenerate in addition to those in the nigrostriatal pathway (Quik & Wonnacott, 2011). Neuroinflammation mediated by microglia and astrocytes may also produce some pathogenesis observed in PD (Jurado-Coronel et al., 2016). Specific genes (such as multiple parkin genes and α-synuclein) predispose to the development of Parkinson's and environmental factors such as pesticides and other chemical or metal exposure may slightly increase risk. Drugs that increase dopamine transmission have been used as treatment such as L-DOPA. L-DOPA treatment usually is effective for only a few years, and prolonged treatment may produce motor problems called L-DOPA-induced dyskinesias (LIDS) (Getachew et al., 2019; Quik et al., 2019), LIDS are produced in about 30% of patients on L-DOPA (Lieberman et al., 2019). Dopamine agonists and monoamine oxidase inhibitors are also used for PD treatment (Quik, Bordia, et al., 2015).

Smoking or smokeless tobacco use is correlated with reduced incidence of PD (Tizabi & Getachew, 2017). In twin studies, the smoking twin had a reduced occurrence of PD (Quik, Zhang, et al., 2015). Other lines of evidence supporting the link between smoking and reduced incidence of PD are the length of smoking and number of cigarettes, and that smoking cessation reestablished the expected risk (Quik et al., 2019). Nicotine enhances dopaminergic signaling, so may be expected to improve symptoms.

Interactions between the cholinergic and dopaminergic systems in the striatum and substantia nigra are key to normal motor function and are altered in PD (Tizabi & Getachew, 2017). Nicotine and modulation of cholinergic signaling may be beneficial for treatment of various symptoms of PD. nAChRs are widespread in the striatum and SN, and we will focus on how nAChRs and cholinergic signaling are involved in PD.

5.4.2 Cholinergic role in Parkinson's disease

Motor function is controlled by a number of cortical areas in addition to the basal ganglia. The basal ganglia receive cortical input and ultimately communicate back to the cortex through the thalamus (Fig. 5.3).

Components of the basal ganglia include the SN, caudate nucleus, putamen (together with the caudate nucleus make up the striatum), globus pallidus (GP), and the subthalamic nucleus. Two pathways or circuits through the basal ganglia are important for coordinated motor function. The direct pathway produces enhancements in initiation of motor movements. Cortical stimuli excite striatal putamen neurons, which inhibit internal GP (GPi, EPN) neurons (which are inhibitory) and lead to relief of ventral lateral nucleus thalamic neurons and thus increased activity to the cortex. The indirect pathway antagonizes the functions of the direct pathway and inhibits the thalamus. Lower DA levels after SN neuron death in PD reduce activation of the direct pathway, but fail to inhibit striatal neurons, which inhibit the GPe through the indirect pathway. Death of SN neurons leads to an imbalance in the functions of these pathways producing motor deficits including rigidity, freezing, bradykinesia, gait problems, and tremor.

Since PD involves death of SN neurons, we will focus on cholinergic signaling in the striatum (target of SN) and the SN. SN input modulates activity of the striatum and regulates both the direct and indirect pathways. The main dopaminergic projections relative to PD are from the SN to the dorsal striatum. Dopaminergic neurons also project from the VTA to the ventral striatum or nucleus accumbens core, but this circuit is generally not affected in PD (Quik & Wonnacott, 2011). The degeneration described here mostly occurs in the dorsal striatum (Quik & Wonnacott, 2011). Large amounts of acetylcholine are present in the striatum. Cholinergic interneurons in the striatum release ACh tonically under basal conditions (Quik et al., 2019). ACh release from these cholinergic neurons is modulated by GABA, glutamate, and dopamine. Dopamine D1 receptor agonists increase

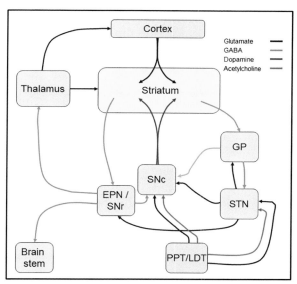

Fig. 5.3 Basal ganglia direct and indirect pathways. Dopaminergic projections from the substantia nigra pars compacta (SNc) and cortical glutamatergic afferents synapse onto the medium spiny neurons (MSNs) of the striatum. These neurons are classically subdivided into the "direct" or "indirect" pathways based on their expression of D1 or D2 dopamine receptors, respectively. Direct pathway D1 MSNs project directly to the enteropeduncular nucleus (EPN; internal segment of the globus pallidus in primates) or the substantia nigra pars reticulata (SNr), and thence to the brain stem or thalamus/cortex, respectively. Indirect pathway D2 MSNs project to the globus pallidus (GP; external segment of the globus pallidus in primates) en route to the EPN and SNr via the SNc or the subthalamic nucleus (STN). Depicted are also the cholinergic projections from the pedunculopontine tegmental (PPT) and laterodorsal tegmental (LDT) nuclei to the striatum, STN, and SNc, which in addition to cholinergic interneurons regulate basal ganglia function. *(From Quik, M., Boyd, J. T., Bordia, T., & Perez, X. (2019). Potential therapeutic application for nicotinic receptor drugs in movement disorders.* Nicotine and Tobacco Research, 21 *(3), 357–369. https://doi.org/10.1093/ntr/nty063.)*

ACh release, and dopamine D2 receptor agonists (and D1 antagonists) lower ACh release (Quik et al., 2019).

The dopaminergic neurons from the SN synapse on GABAergic medium spiny neurons from nigrostriatal pathway (Quik & Wonnacott, 2011). Medium spiny neurons are part of both the direct (D1 expressing) and indirect pathways (D2 expressing) involved in motor function. The direct pathway communicates to basal ganglia structures SN pars reticulata and the entopeduncular nucleus (rodents) or internal segment of the globus pallidus (GPi) in primates (Quik & Wonnacott, 2011). These communicate to the brain stem or thalamus, which communicates to cortical motor areas. The indirect pathway communicates to the external segment of the globus pallidus GPe (primates) and on to the subthalamic nucleus and then the SN pars reticulata and the entopeduncular nucleus (GPi) (Quik & Wonnacott, 2011).

The nigrostriatal DA input and glutamatergic inputs reciprocally regulate medium spinal neurons, which are part of the direct and indirect pathways (Quik & Wonnacott, 2011). In total, 1%–3% of striatal interneurons are cholinergic with varicosities that extend throughout the striatum. ACh is released through volume transmission over a wide area from tonically active neurons. SN dopaminergic neurons also fire rhythmically and can burst when stimulated by cholinergic PPT inputs. Nicotine increases SN neuron firing. The higher level of firing silences ACh release (Quik & Wonnacott, 2011). This provides for interactions between dopaminergic and cholinergic neurons (Quik & Wonnacott, 2011), in addition to acetylcholine release, multiple populations of neurons express nAChRs. In the striatum (Quik & Wonnacott, 2011), nAChRs of multiple subtypes are present presynaptically on GABAergic interneurons, the dopaminergic nigrostriatal afferents, the glutamatergic corticostriatal afferents as well as on raphe nucleus afferents and the cholinergic interneurons themselves (Fig. 5.4). nAChRs are also present postsynaptically on the medium spiny neurons. Subtypes detected in the striatum include $\alpha4\beta2$, $\alpha4\alpha5\beta2$, $\alpha4\alpha6\beta3$, $\alpha6\beta2\beta3$, and $\alpha6\beta2$ on the nigrostriatal afferents (presynaptic), $\alpha7$ on the glutamatergic corticostriatal afferents, and $\alpha4\beta2^*$ on the medium spiny neurons and GABAergic interneurons (Quik & Wonnacott, 2011). These subtypes are potential targets for pharmacological intervention.

The SN also contains multiple populations of neurons expressing nAChRs (Quik & Wonnacott, 2011) (Fig. 5.5). The dopaminergic neurons of the nigrostriatal projection express $\alpha7$, $\alpha6\beta2^*$, and $\alpha4\beta2^*$ nAChRs on the soma, the GABAergic afferents (presynaptic) and interneurons (soma) express $\alpha4\beta2^*$, and the glutamatergic afferents express at least presynaptic $\alpha7$ nAChRs and possibly $\beta2^*$ nAChRs as well. Dopamine is released in the SN, and this release is regulated by nAChRs (Quik & Wonnacott, 2011). The SN nAChRs are stimulated by PPT input and thus release more dopamine in the striatum.

Multiple targets for nicotine and other cholinergic drugs are present in the striatum and SN. A variety of subtypes allow for a multiple sites of regulation by ACh and nicotine of signaling in the striatum and thus an important role in motor function and PD. Postsynaptic nAChRs can influence the

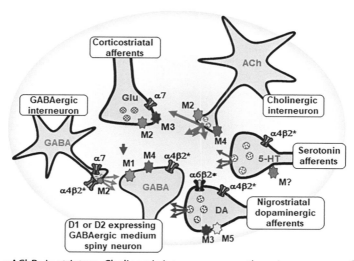

Fig. 5.4 nAChRs in striatum. Cholinergic interneurons are the primary source of striatal acetylcholine (ACh) and regulate its function via pre- and postsynaptic nAChRs and muscarinic receptors. Acetylcholine regulates the activity of direct and indirect GABAergic medium spiny neurons (MSNs) by acting at $\alpha4\beta2^*$ nAChRs, as well as M1 and/or M4 muscarinic receptors. In addition, acetylcholine modulates striatal dopamine (DA) release via an interactionat $\alpha6\beta2^*$ and $\alpha4\beta2^*$ nAChRs along with M2 and/or M4 muscarinic receptors on nigrostriatal dopaminergic and serotonergic (5-HT) terminals, which further regulates the output of direct and indirect pathway MSNs. Likewise, acetylcholine can modulate GABAergic interneuron activity via $\alpha7$ and $\alpha4\beta2^*$ nAChRs, as well as M2 muscarinic receptors. Acetylcholine can further control striatal function via $\alpha7$ nAChRs and M2 and M3 muscarinic receptors located on the excitatory glutamatergic (GLU) inputs arising from the cortex. *(From Quik, M., Boyd, J. T., Bordia, T., & Perez, X. (2019). Potential therapeutic application for nicotinic receptor drugs in movement disorders.* Nicotine and Tobacco Research, 21 *(3), 357–369. https://doi.org/10.1093/ntr/nty063.)*

excitability of the neurons, while presynaptic nAChRs can have effects on the level of transmitter release from neurons releasing dopamine, glutamate, or GABA. As shown below, some positive effects of nicotine on PD symptoms come from activation of nAChRs while others may be due to desensitization of nAChR function after prolonged exposure. Dopamine and acetylcholine levels in the striatum are balanced and can affect each other's levels (Quik & Wonnacott, 2011) Reduced ACh levels were shown to lower DA release in the mouse striatum (Zhou et al., 2001). ACh action through nAChRs was shown to be key as inhibition of nAChRs lowered DA release, but inhibition of muscarinic nAChRs did not (Zhou et al., 2001). Excess ACh after AChE inhibition was shown to lower DA release,

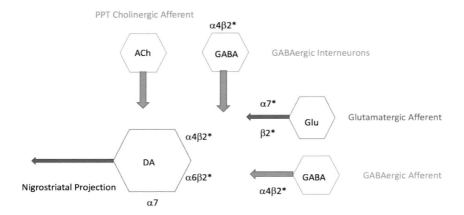

Adapted from Quik and Wonnacott, 2011

Fig. 5.5 nAChR localization in the substantia nigra. GABAergic interneurons express α4β2* nAChRs on the cell body, while GABAergic afferents to the SN express α4b2* nAChRs on their terminals. Glutamatergic afferents also express terminal α7* and β2* nAChRs. Dopaminergic neurons that are part of the nigrostriatal projection express α7, α4β2*, and α6β2* nAChRs on the cell. PPT, pedunculopontine tegmental nucleus.

indicating that desensitization of nAChRs by excess ACh was the cause. β2* nAChRs were shown to be important in nAChR control of release by using β2 KO mice and the β2* antagonist DHβE. Treatment with α7 nAChRs antagonists didn't lower dopamine release (Zhou et al., 2001). Burst firing of DA neurons can silence cholinergic interneurons in the striatum. The lowered ACh could lead to decreased DA release as described above. In the SN then, nicotine through β2* nAChRs increases firing while desensitization of β2 * nAChRs on GABAergic interneurons would increase dopamine neurons activity through reduced inhibition (Quik & Wonnacott, 2011).

The balance between DA and cholinergic signaling in the striatum is delicate, and death of SN neurons in PD upsets this balance. Nicotine or nAChR subtype selective drug treatments may be expected to help restore the balance in the face of chronically lower DA due to SN death in PD. Nicotine or nAChR activity may also work through other mechanisms that inhibit or stimulate DA or ACh release. α7 nAChRs have high permeability to Ca^{2+}. Ca^{2+} entering through nAChRs may activate pathways mediated by calcium calmodulin kinase and PKA to stimulate extracellular signal-regulated kinase (ERK). Ca^{2+}-independent pathways downstream from nAChR activation can stimulate Janus kinase-signal transducer and activator of transcription (JAK2-STAT3) signaling (Quik et al., 2019).

5.4.3 Treatment using cholinergic modulation

Smoking has been negatively correlated with prevalence of PD (see above). In addition, nicotine improves dopamine release in lesioned animals (Quik, Cox, et al., 2007). Consistent with this are several lines of evidence showing that nAChR activation or modulation has positive effects against some of the PD symptoms. The multiple nAChR subtypes expressed in the striatum and SN are potential targets for nicotine and other compounds.

5.4.4 Effects on neurodegeneration

Neuronal death is the hallmark of PD and is progressive. While numerous genes may be involved, exposure to metals, fungicides, and pesticides may also play a role. A number of neuroprotective compounds unrelated to cholinergic effects are being tested (Quik et al., 2012). Whether death is caused by action of specific genes or environmental insults, is there a role for modulation of cholinergic signaling in the reduction or slowing of neuronal death?

Nicotine was protective against various toxins and other compounds such as rotenone and salsolinol in dopaminergic cells such as SH-SY5Y cells (Getachew et al., 2019). In SH-SY5Y cells, nicotine protected the cells from death after salsolinol treatment. This protective effect was blocked by mecamylamine and a conotoxin that blocked $\alpha3$-nAChRs (Copeland et al., 2005). Small numbers of $\alpha3\beta2^*$ nAChRs are also detected in the striatum (Quik & Wonnacott, 2011). Blockade of $\alpha4\beta2$ nAChRs with DHβE or $\alpha7$ with MLA didn't reduce the protective effect of nicotine (Copeland et al., 2005). Nicotine was also shown to provide protection from other compounds associated with the death of dopaminergic neurons such as 1–methyl-4–phenyl-1,2,3,6-tetrahydropyridine (MPTP), 6–hydroxydopamine (6-OHDA), and methamphetamine (Tizabi & Getachew, 2017). Nicotine acting through non-$\alpha7$ nAChRs protected ventral mesencephalic neurons from 1–methyl-4-phenylpyridinium (MPP+) induced cell death (Jeyarasasingam et al., 2002).

Nicotine also protects against exposure to amounts of manganese and iron that produce toxicity in SH-SY5Y cells. Interestingly, both MLA and DHβE blocked the protective effects of nicotine (Getachew et al., 2019). This could indicate that different nAChR subtypes mediate protection from salsalinol compared with iron or manganese.

Studies with MPTP-treated monkeys or mice and 6-OHDA-treated rats or mice showed that nicotine or nAChR agonists protected against damage to nigrostriatal dopaminergic neurons (Quik et al., 2012, 2019). Multiple nAChR subtypes are involved including those containing α4, α7, α6, and β2* subunits (Quik et al., 2019). Nicotine was protective when delivered by multiple methods including in water (Huang et al., 2009), injection or osmotic minipumps.

Animals pretreated with nicotine have reduced damage resulting from subsequent lesioning of dopaminergic pathways (Quik & Wonnacott, 2011). Chronic nicotine protected striatal dopaminergic terminals in rats after lesion with 6-OHDA, while acute nicotine treatment was shown to protect neurons from methamphetamine-induced degeneration (Ryan et al., 2001). Intact α4-containing nAChRs were required for protection from methamphetamine. α7 nAChRs may also be involved in the neuroprotective effects of nAChR activation. PNU-282987, an α7 agonist, protected mice against MPTP-induced SN dopaminergic cell loss (Stuckenholz et al., 2013). Another α7 nAChR agonist as well as nicotine also protected dopaminergic neurons from loss in 6-OHDA lesioned rats as measured by dopamine transporter levels (DAT) (Bordia et al., 2015). DAT levels were not increased in intact striatum. Nicotine acting through α6β2* and α4β2* nAChRs as well as ABT-107 (α7 nAChR agonist) reduced the loss of dopamine release in striatum due to lesioning (Bordia et al., 2015). Nicotine pretreatment, but not post treatment, improved amphetamine-induced rotation in rats. Pretreatment preserved dopaminergic neurons after lesioning in rats, but posttreatment did not (Huang et al., 2009). Nicotine treatment after lesioning in monkeys did not prevent the decrease in dopamine levels and nAChRs, measures of dopaminergic neuron survival, but pretreatment did (Huang et al., 2009). The effects, at least in rat striatum, were mediated by α6α4β2* nAChRs, even though α6 (nonα4)β2* nAChRs were also present (Huang et al., 2009).

Nicotine reduces neuron loss in the SN in animal models of PD (Meshul et al., 2002). Pretreatment of mice with nicotine 5× a day at 2-h intervals for 4 weeks, but not cotinine, showed a small degree of protection against MPTP-induced toxicity of dopaminergic neurons (Parain et al., 2001). Several studies give mixed or conflicting results, however, using the MPTP model. Fung et al. (1991) showed that nicotine treatment didn't improve the locomotor response or maintain DA levels. Neuron loss was not measured, and they used a lower dose of nicotine and 2× a day treatment. However, Gao et al. (1998) showed that nicotine treatment did protect from

MPTP-induced reduction in locomotor behavior and striatal dopamine loss in mice.

In summary, multiple studies using various treatment paradigms indicate a possible role of dosage and timing in the effectiveness of cholinergic modulation in the protection of dopaminergic neurons. In monkeys and rats and in some mice studies, nicotine administered in various methods (i.e., pump, injection, and water), and with multiple timing and dosages protected nigrostriatal dopaminergic neurons from death induced by chemical lesioning or hemisectioning (Quik et al., 2007). Neuroprotection seems to rely more on activation than desensitization. These studies do support the idea that modulation of nAChR signaling should be explored in reducing death of dopaminergic neurons in PD. These treatments may be needed early in disease progression before significant cell loss has occurred.

5.4.5 Treatment of LIDs

A many of PD patients treated with L-DOPA develop dyskinesias (LIDs) or abnormal involuntary movements. There are few good treatment options other than lowering the dose of L-DOPA.

The nicotinic cholinergic system is important in motor signaling and also for dyskinesias induced by prolonged L-DOPA treatment. The importance of striatal cholinergic interneurons to this process was shown when specific lesions to these neurons before L-DOPA treatment prevented LIDs development and didn't prevent the palliative effects of L-DOPA on PD symptoms (Won et al., 2014). Selective optogenetic activation of cholinergic striatal interneurons in mice after 6-OHDA lesioning reduced LIDs produced by L-DOPA treatment. Long pulses were required supporting a role of nAChR desensitization in the process (Bordia et al., 2016). LIDs were reduced by more than 50% in parkinsonian rats, mice, and monkeys after treatment with nicotine (Quik et al., 2019), also consistent with the desensitizing effects of nicotine after prolonged treatment. No tolerance developed to the therapeutic effects of nicotine Quik et al., 2019).

Further studies implicated the role of nAChR subtypes containing α4, α6, α7, and β2 subunits in mice (Quik et al., 2019). β2* and α7 nAChRs were specifically shown to be involved in reduction of LIDs in monkey as well as rodents. After dyskinesia was induced in MPTP lesioned monkeys with L-DOPA treatment, two β2* nAChR agonists ABT-894 (full) and ABT-089 (partial) reduced LIDs (Zhang, Bordia, et al., 2014). Effects of ABT-894 lasted 6 weeks after washout (Zhang, Bordia, et al., 2014).

ABT 894 was show to have a similar potency to nicotine and was much more potent than ABT-089 for both α4β2* and α6β2* nAChRs in striatum (Zhang, Bordia, et al., 2014). Mice lesioned with 6-OHDA and treated with nicotine had reduced contralateral rotations induced by apomorphine suggesting that nicotinic agonists could be used to treat dyskinesias that result from Parkinson's treatment with long-term dopamine (Meshul et al., 2002).

α7 nAChR activation also reduces LIDs in a monkey model of parkinsonism (Zhang, McGregor, et al., 2014). The α7 agonist ABT-107 reduced LIDs by 50%–60% without developing tolerance. Combined treatment with ABT-894 and ABT-107 didn't further reduce LIDs, indicating that they may act on a common pathway (Zhang, McGregor, et al., 2014). Other drugs shown to reduce LIDs in rats include varenicline, sazetidine, and TC8831, all β2* nAChR agonists. Varenicline and TC8831 also worked in nonhuman primates (Zhang, McGregor, et al., 2014). Human clinical trials with oral nicotine reduced LIDS (Quik et al., 2019).

In summary, these and other studies (Quik, Cox, et al., 2007) support the use of nAChR agonists in the treatment of LIDs, to be administered alongside L-DOPA. More human clinical trials are needed to develop and test the best cholinergic drugs for treatment of LIDs in humans.

5.4.6 Improvement in motor symptoms

Some work with animals suggests that treatment with drugs targeting nAChRs may improve motor symptoms of parkinsonism. A nAChR agonist SIB-1508Y was shown to mildly improve multiple measures of motor function in MPTP lesioned primates. It also improved the effect of L-DOPA when co-administered by IV (Schneider et al., 1998). SIB-1508Y was later shown to be an α4β2 agonist, but was not shown to be effective in human phase II trials (Parkinson Study Group, 2006). 6-OHDA lesioned rats showed some improvement in motor function after treatment with nicotine or ABT-107 delivered via osmotic minipump (Bordia et al., 2015). This was thought to occur by sparing dopaminergic neurons.

However, other studies in parkinsonian animal models were not supportive. Parkinsonian rats treated with either constant or intermittent nicotine showed improvement in LIDs, but not in other motor symptoms (Bordia et al., 2008). Nicotine treatment of lesioned monkeys also showed no motor improvement either with or without L-DOPA treatment (Quik, Cox, et al., 2007). Drugs that improve LIDs such as ABT 107 and ABT-894 didn't improve parkinsonism in lesioned monkeys (Zhang, McGregor, et al., 2014).

As far as treatment with nAChR-targeted drugs improving human motor symptoms, the results are generally negative. Methods used have included IV nicotine, patch, and gum. Several studies showed no improvement in randomized, or blinded, or double-blinded or placebo-controlled studies for nicotine or SIB-1508Y ($\alpha4\beta2$ agonist) (Clemens et al., 1995; Shoulson, 2006; Vieregge et al., 2001). Some studies showed improvement, but were not double-blinded (Kelton et al., 2000) or not randomized and placebo-controlled (Villafane et al., 2007). A recent study using NC001 (nicotine bitartrate) was shown to reduce LIDs in primates and also showed some reduction of falls and freezing of gait in humans (Lieberman et al., 2019) They hypothesize this is due to activation of nAChRs in the peduncular pontine nucleus (PPN), since (PPN) neuron loss has been linked to human PD symptoms. NC001 may also reduce LIDs (Lieberman et al., 2019).

However, for the most part, there is no clear evidence of improvement in human double-blinded studies or animal models of parkinsonism, suggesting little effectiveness treating acute motor symptoms (Quik et al., 2019) with nicotine or cholinergic drugs. Some primate work (Schneider et al., 1998) however demonstrated the principle that cholinergic drugs may be used to ameliorate symptoms in Parkinson's disease patients. Studies with human are less clear. Clearly more work needs to be done.

5.4.7 Summary

Nicotine and agonists protect against degeneration ($\alpha4\beta2^*$, $\alpha6\beta2^*$, and $\alpha7$ involved in this), but don't improve motor symptoms in human trials. Some motor symptoms are improved in animals. They do reduce LIDs, even for 6 weeks after discontinuation of therapy (Quik, Bordia, et al., 2015) Varenicline and mecamylamine reduce LIDs supporting a role of nAChR in treatment. Animal models show that nicotine and agonists protect against degeneration, but may need to be administered early in the progression of the disease. nAChR modulation may prevent death, but not restore damaged neurons (Huang et al., 2009). Nicotine seems to act to reduce LIDs via nAChRs desensitization while protecting from degeneration by activation. Given the diverse expression of nAChR subtypes in the striatum, SN, and PPT, nicotine and nAChR-directed drugs may have effects on PD symptoms by actions of specific nAChR populations modulating DA release, glutamate and GABergic signaling and by other Ca2+ dependent pathways. While the diversity of nAChR subtypes and signaling locations are complex and thus studies may be difficult, finding better treatments for a disease that affects a significant number of people each year is needed.

5.5 Cancer
5.5.1 Introduction

Smoking is a major risk factor in several cancers including the pancreatic, lung, skin, and breast (Gaudet et al., 2013; Hecht, 2012; Kolodecik et al., 2014; Leonardi–Bee et al., 2012; WHO, n.d.). In total, 45 million Americans currently smoke. Cigarette smoke contains numerous chemicals that can promote tumor growth such as nicotine, and nitrosamines found in tobacco, NNN (N'-nitrosonornicotine) and NNK ([4-(methylnitrosoamino)-1-(3-pyridyl0-1-butanone]}. NNN and NNK signal through nAChRs of various subtypes. These molecules are not thought to initiate cancer, but can act as mitogens and promote cancer cell growth. Nicotine also enhances antiapoptotic functions, and in this way also promotes tumor growth. Smoking has a particularly deadly role in lung and pancreatic cancers. We will focus here on lung and pancreatic cancer and the role of nAChRs in these. In total, 90% of lung cancer in the United States stems from smoking. Tobacco use is the most significant risk factor for 70% of worldwide lung cancer deaths and 20% of all global cancer deaths (WHO, n.d.). The lung cancer 5-year survival rate is 15% and hasn't changed much in 30 years. Most lung cancers are classified as nonsmall cell lung carcinoma (NSCLC) and small cell lung carcinoma (SCLC). Most of the tumors are NSCLC (80%), but the less common SCLC is very aggressive. Smoking plays a role in the development of either class.

Pancreatic cancer is the fourth leading cause of cancer-related death in the United States (NIH). In total, 44,000 new cases occurred in 2011, with approximately 37,000 deaths. The 5-year survival rate is less than 5% (Hidalgo, 2010). Smoking is the most established risk factor for pancreatic cancer. Risk is 2.5–3.6× higher in smokers than in nonsmokers and even greater with long-term smoking. Smokeless tobacco use has also been associated with increased pancreatic cancer incidence in some studies (Duell, 2012), although not as strongly as cigarette smoking. Childhood exposure to maternal smoking was also associated with pancreatic cancer (Hidalgo, 2010). Some studies are consistent with smoking exposure contributing to both early and late-stage effects on pancreatic cancer (Duell, 2012).

The basic mechanisms by which tobacco leads to the development of pancreatic cancer are unclear. However, the actions of nicotine and nitrosamines found in tobacco, NNN and NNK, are clearly involved in any action of tobacco, which causes or promotes pancreatic cancer. NNN and NNK are involved in signaling in other cancers as well (Schuller, 2009). NNK has been demonstrated to cause pancreatic cancer in animals (Duell, 2012). NNK can be distributed to a fetus, and it has been shown that the offspring of pregnant hamsters given NNK one day before delivery developed pancreatic adenocarcinoma at 1 year of age (Duell, 2012). Although acetylcholine is the endogenous agonist, nicotine, NNN and NNK act by signaling through nAChRs. In conjunction with evidence for a role of maternal smoking in the development of pancreatic cancer in offspring, this work supports a role for nAChR signaling in cancer development.

5.5.2 Nicotinic cholinergic systems and nAChRs are present in normal and cancerous lung and pancreas

Acetylcholine (ACh) is present in normal lung, with ACh being released from the parasympathetic nervous system into conducting airways and being produced by human airway epithelial cells (Spindel, 2009; Wessler et al., 1998). nAChRs are also present in normal lung tissue (Spindel, 2009; Wessler et al., 1998). A diversity of nicotinic subunits are expressed by human and mouse bronchial epithelial cells including $\alpha3$, $\alpha5$, $\alpha7$, $\beta2$, and $\beta4$ (Maus et al., 1998; Sekhon et al., 1999). nAChR subunits are also expressed in monkey submucosal glands, airway wall fibroblasts, and alveolar epithelial cells (Sekhon et al., 1999). $\alpha1$, $\alpha3$, $\alpha5$, and $\alpha6$ subunits were also detected in alveolar macrophages and in some cells lining the developing alveolar airspaces (Sekhon et al., 1999). Homomeric and heteromeric nAChRs are present in normal lung tissue. Nicotine-induced currents were detected in cultured human bronchial epithelial cells (HBECs) as well as epibatidine binding, indicating the presence of functional, assembled nAChRs in normal lung tissue (Maus et al., 1998). Two major types of lung cancer cells, NSCLC and SCLC also express a diversity of nAChR subunit genes. NSCLC express $\alpha3$-$\alpha7$, $\alpha9$, $\beta2$, and $\beta4$ subunits (Improgo et al., 2013) while SCLC contains $\alpha3$, $\alpha5$, $\alpha7$, $\alpha9$, $\beta2$, and $\beta4$ subunits (Improgo et al., 2013; Song et al., 2003; Tarroni et al., 1992). Most squamous lung cell carcinomas (SCCs) examined expressed $\alpha3$, $\alpha4$, $\alpha5$, $\alpha7$, $\beta2$, $\beta3$, and $\beta4$ nAChR subunit proteins (Song et al., 2008). Other components of the cholinergic system are also present in lung cancer. SCLCs express choline acetyltransferase

(ChAT), choline transporter (ChT), vesicular acetylcholine transporter (VAChT), and mAChRs (Song et al., 2003). ChAT is also expressed in biopsied SCLC (Song et al., 2003). Squamous cell carcinoma (SCC) cells also have increased $\alpha5$, $\beta3$ nAChR subunits and ChAT and ACh levels compared with normal tissue (Song et al., 2008). Levels of $\alpha5$, $\alpha7$, $\alpha9$, and $\beta2$ nAChR subunits are elevated in various NSCLC lines (Lam et al., 2007).

Smoking history and nicotine exposure influence cholinergic receptor expression and function. Genetics may play a role as a specific $\alpha5$ nAChR allele is linked to lung cancer (Olfson et al., 2016). This may be due to increased risk of addiction and thus increased smoking.

NSCLC tumors from smokers showed higher levels of $\alpha6$ and $\beta3$ nAChR RNA expression than from nonsmokers (Lam et al., 2007). $\alpha4$ RNA expression was lower in NSCLC tumors and $\beta4$ expression higher than in normal lung (Lam et al., 2007). $\alpha1$, $\alpha5$, and $\alpha7$ RNA levels were increased in HBEC lines exposed to nicotine (Lam et al., 2007). The activities of $\alpha7$ and $\alpha3$-containing nAChRs were upregulated by nicotine in monkey lung (Wang et al., 2001), while the function of $\alpha7$ nAChRs was upregulated in HBEC by nicotine (Sekhon et al., 1999). SCC-L cells have increased $\alpha7$ nAChR mRNA levels that correlate with the smoking history of the patients (Brown et al., 2013). In SCC, all nAChR RNA levels trended higher than in normal tissue (Song et al., 2008). The H520 SCC line showed increased ACh levels and AChR expression and activity after nicotine exposure (Song et al., 2008). Even after smoking cessation, nAChR signaling may be increased.

Several lung cancer cell lines express various combinations of nAChR subunits. NSCLC A549 cells (pulmonary type II epithelial cells, adenocarcinoma) express several nAChR subunit proteins/RNAs including $\alpha3$, $\alpha4$, $\alpha5$, $\alpha7$, $\beta2$, and $\beta4$ (Brown et al., 2012; Dasgupta, Kinkade, et al., 2006; Improgo et al., 2013; Ma et al., 2014). Nicotine has been shown to reduce etoposide-induced apoptosis in A549 lung cancer cells (Zhang et al., 2009). A549 cells may express nAChR subtypes including $\alpha7$ homomers, and $\alpha3\alpha5\beta4$, $\alpha3\alpha5\beta2$, $\alpha3\beta4$, $\alpha3\beta2$, and $\alpha4\beta2$ heteromers.

NSCLC H520 cells (a squamous cell carcinoma line) express the same nAChR subunits as A549 (Song et al., 2008) but also express the $\beta3$ subunit. Thus, the same nAChR subtypes present in A549 cells can be expressed in H520 cells, but $\beta3$-containing subtypes may also be expressed. Clearly the $\alpha7$–nAChR subtype is functional since $\alpha7$ nAChR-mediated currents are present in H520 cells that are upregulated by nicotine (Song et al., 2008). SCLC DMS 53 cells express $\alpha3$, $\alpha5$, and $\beta4$ nAChR subunits. It is not known if DMS-53 cells express α7 nAChRs, although α7 RNA expression is increased in SCLC cell lines compared with normal lung cell lines (Improgo et al., 2010). DMS-53 cells express at least α3β4α5 heteromers and possibly α7 nAChRs as well. SCLC H82 cells express α3, β4, and α5 nAChR proteins and RNAs (Osborne et al., 2014; Zhang et al., 2010). They also express α7 nAChR RNA (Osborne et al., 2014). Thus, H82 cells express at least two types of nAChRs, α3β4α5 heteromers, and α7 homomers.

There may be an important role for nicotinic cholinergic signaling in pancreatic cancer as well, especially given the linkage to smoking as the major preventable cause. Acetylcholine is present in the exocrine regions of the pancreas, due to autonomic nervous system innervation as well as localized production of acetylcholine by pancreatic alpha cells (Rodriguez-Diaz et al., 2011). Indirect evidence indicates that some pancreatic cancer cell lines also produce acetylcholine (Al-Wadei et al., 2012). nAChR subunit RNAs have also been detected in the pancreas and pancreatic cancer cell lines including α6 and β3 nAChR subunits (Al-Wadei et al., 2012; Boyd, 2012; Wessler & Kirkpatrick, 2008). Some signaling may be through adrenergic pathways activated by nAChR-stimulated epinephrine/norepinephrine release, but others may not be. Not much is known about nAChR-mediated pathways in pancreatic cancer. Thus, autocrine and/or paracrine signaling by ACh may be part of normal pancreatic function, but perturbation of these pathways may produce or promote pancreatic cancer as it does for some other types.

Nicotine accumulates in the pancreas of smoke-exposed animals. Nicotine can cause pancreatitis in rats exposed to nicotine, with features similar to that seen in human pancreatitis (Bose et al., 2005). This inflammation of the pancreas is often a precursor to development of pancreatic cancer. Animal exposure to nicotine also produces other morphological changes in the pancreas and affects pancreatic function (Chowdhury et al., 2002). Importantly the pancreas responds to nicotine exposure, not just to cigarette smoke. Nicotine itself causes proliferation of pancreatic and lung cancer cells in vitro (Bose et al., 2005; Egleton et al., 2008). Nicotine is not thought to initiate cancer, but can act as a mitogen and thus promote pancreatic cancer cell growth. Nicotine signals though nAChRs containing subunit proteins α3, α4, α7, and α5 in pancreatic cancer cells and pancreatic duct epithelial cells (PDAC) (Al-Wadei et al., 2012). Nicotine exposure increases α7 expression in PDAC (Sullivan et al., 2011). Long-term pretreatment of pancreatic cancer cells with nicotine in vitro increases subsequent proliferation and migration (metastatic potential). Thus, a smoker may have upregulated cholinergic signaling, which would promote tumor growth once initiated (Momi et al., 2012). In addition to effects on pancreatic cancer cells in vitro, nicotine can also act as a mitogen to xenografted pancreatic tumors in mice and also increase tumor migration (Al-Wadei et al., 2009; Momi et al., 2012). The metastasis is associated with increased MUCIN 4 expression, which is in turn dependent on α7 nAChR expression (Momi et al., 2012). In summary, components required for cholinergic signaling are present in normal lung and pancreas as well as in cancerous tissues and cell lines. A multitude of various nAChR subtypes are also available for normal signaling in response to ACh and abnormal signaling in the presence of nicotine, NNN, or NNK.

5.5.3 Mechanisms of nicotine action in lung and pancreatic cancer

In addition to signaling in normal pathways in several cell types, nAChR signaling is being recognized as important in the pathogenesis of cancer (Schuller, 2009; Schuller & Al-Wadei, 2010). Nicotine present in the blood of a smoker can also stimulate proliferation, angiogenesis, migration and reduce or confer resistance to apoptosis acting through nAChRs. The proliferative effects of nicotine were first identified in lung cancer, but also occur in pancreatic and colon cancer cell lines (Egleton et al., 2008). There is some evidence that nicotine can't only stimulate growth, but perhaps initiate pancreatic cancer as well (Schaal et al., 2015). The α7 nAChR subtype is involved in this, given the ability of α7 nAChR specific antagonists to block this action of nicotine.

Other subtypes may be involved as well. Nicotine can also protect several types of cancer cells including breast, lung, and oral from apoptosis normally induced by anticancer drugs (Dasgupta, Kinkade, et al., 2006; Xu et al., 2007; Zeidler et al., 2007) Stimulation of different nAChR subtypes activates specific pathways in normal and cancer cells. Nicotine can increase the level of several growth factors (i.e., VEGF-C, TGF-b, and BDNF) and growth factor receptors (i.e., PDGFR, EGFR, and VEGFR-2). Nicotine can promote growth of cancer cells by activating ERK and PI3K/mTOR signaling pathways (Egleton et al., 2008). nAChR signaling can activate NFkB, Akt, Src. NNN and NNK can activate several pathways as well including GATA-3, NFkB, and Stat-1. Ca^{2+} influx through nAChRs can activate protein kinase XC and the MAP kinase pathway in SCLC (Egleton et al., 2008). α7 nAChR stimulation leads to epinephrine or norepinephrine release, which increases cancer cell growth. NNK, in addition to nicotine, activates α7 nAChRs. Activation of α4β2 nAChRs stimulates GABA release, which appears to inhibit growth in some cancers (Schuller, 2009). Heteromeric nAChRs such as α4β2 are activated by NNN. Other subtypes are involved as well, but the exact subtypes of heteromeric receptors expressed in most cancers have not been characterized.

While nicotine clearly stimulates the growth of both lung cancer and pancreatic cancer, an additional effect of nicotine and nAChR signaling is that it reduces the effectiveness of cancer chemotherapy (An et al., 2012; Banerjee et al., 2013; Maneckjee & Minna, 1994). One mechanism common to its actions in several tumor types is the blockade of the apoptosis promoting action of many cancer chemotherapeutic drugs (Dasgupta, Kinkade, et al., 2006; Warren & Singh, 2013; Xu et al., 2007; Zeidler et al., 2007). Nicotine inhibits the therapeutic effects of gemcitabine on Panc-1 and BXPC-3 cells (Banerjee et al., 2013). Nicotine also inhibits gemcitabine-induced apoptosis in NSCLC (Dasgupta et al., 2006). Nicotine antagonizes cisplatin-mediated apoptosis in lung cancer cells by increasing the stability of Bcl-2 and thus reduces chemosensitivity (Nishioka et al., 2014). Nicotine also inhibited cisplatin-induced apoptosis in the SCLC cell line NCI-H446 (Zeng et al., 2012) as well as in NSCLC cells (Dasgupta et al., 2006). Nicotine reduces etoposide-induced apoptosis in A549 lung cancer cells (Zhang et al., 2009). NNK has also been shown to reduce apoptosis produced by cisplatin (Warren & Singh, 2013). Some studies indicate that α7 nAChRs may mediate anti-apoptotic effects while others show that heteromeric receptors of unknown composition do this (Egleton et al., 2008). Akt activation has been implicated in anti-apoptotic effects of nicotine in other cell lines (Dasgupta et al., 2006). Induction of X-linked inhibitor of apoptosis protein (XIAP) proteins and survivin may be mediated by α3β2 containing nAChRs in SCLC cell lines and mediate the anti-apoptotic effects of nicotine in these cells (Dasgupta et al., 2006). Nicotine has been shown to produce proliferation in normal and some cancer cells, and some nAChR antagonists reduce this effect. The specific subtypes and pathways through which this proliferation is mediated in pancreatic cancer cells are not known. Nicotine increases the level of phospho-p44/42 ERK, MEKK-1, and phosphorylated p90RSK in human mesothelioma cells (Trombino et al., 2004) leading to increased proliferation. Nicotine induces the binding of Raf-1 to Rb and phosphorylation of Rb in lung cancer cells, which was mediated by α7 nAChRs (Dasgupta, Rastogi, et al., 2006). Nicotine also induced dissociation of Rb from E2F1 in lung cancer cells (Dasgupta, Rastogi, et al., 2006).

Many patients undergoing cancer chemotherapy smoke (Burke et al., 2009; Cox et al., 2002). Even in the absence of smoking, endogenous ACh is present in a number of tissues such as the lung and pancreas that may also produce some level of chemotherapy resistance. Since nicotine and ACh are present during cancer chemotherapy, their signaling through nAChRs present on cancer cells contributes to chemoresistance.

5.5.4 Possible treatment by modulation of cholinergic signaling and nAChRs

Evidence is accumulating that targeting nAChRs can reduce tumor growth (Schuller, 2009; Spindel, 2009). Targeting specific nAChR subtypes on lung and pancreatic cancer cells is an effective route to block chemotherapy resistance mediated by nAChR signaling. Drugs targeting nAChRs such as varenicline and mecamylamine are currently being used for smoking cessation therapy. Drugs targeted to nAChR subtypes present in lung and pancreatic cancer cells may be safely used in humans at least as an adjunct to improve current drug therapies.

There are several studies that show that antagonism of nAChRs on cancer cells limits growth and proliferation. Paleari et al. (2009) used α-CbT to target α7 nAChRs and blocked growth of NSCLC cells in vivo and in vitro. Shih et al. (2010) showed that downregulation of α9-containing nAChRs reduced proliferation in nicotine-treated breast cancer cells. Antagonists of α7 nAChRs will block proliferative effects of nicotine as has been shown for some pancreatic cancer lines and other cancers (Schuller, 2009). It has been shown that nicotine promotes proliferation of lung and some pancreatic cancer cell lines and that mecamylamine (broad-spectrum nAChR antagonist) can block this effect. Mecamylamine was shown to mitigate doxorubicin-resistance induced by cigarette smoke condensate, using a head and neck cancer cell line UMSCC10B (An et al., 2012). α3β4α5 nAChRs were targeted wth α-conotoxin AuIB, which was shown to decrease SCLC cell viability (Improgo et al., 2013).

In summary, nAChRs' expression and cholinergic signaling are present in a number of cancers including lung and pancreatic. The development of subtype-specific compounds that may reduce tumor growth stimulated by normal cholinergic signaling or present or past nicotine exposure would potentially benefit treatment of these difficult to treat cancers.

References

Adams, C. E., Yonchek, J. C., Schulz, K. M., Graw, S. L., Stitzel, J., Teschke, P. U., & Stevens, K. E. (2012). Reduced Chrna7 expression in mice is associated with decreases in hippocampal markers of inhibitory function: Implications for neuropsychiatric diseases. *Neuroscience*, *207*, 274–282. https://doi.org/10.1016/j.neuroscience.2012.01.033.

Al-Wadei, M. H., Al-Wadei, H. A. N., & Schuller, H. M. (2012). Pancreatic cancer cells and normal pancreatic duct epithelial cells express an autocrine catecholamine loop that is activated by nicotinic acetylcholine receptors $\alpha 3$, $\alpha 5$, and $\alpha 7$. *Molecular Cancer Research*, *10*(2), 239–249. https://doi.org/10.1158/1541-7786.MCR-11-0332.

Al-Wadei, H. A. N., Plummer, H. K., & Schuller, H. M. (2009). Nicotine stimulates pancreatic cancer xenografts by systemic increase in stress neurotransmitters and suppression of the inhibitory neurotransmitter γ-aminobutyric acid. *Carcinogenesis*, *30*(3), 506–511. https://doi.org/10.1093/carcin/bgp010.

Alzheimer's Disease Facts and Figures. (2020). *Alzheimer's Dementia*, *16*(3), 1–91.

An, Y., Kiang, A., Lopez, J. P., Kuo, S. Z., Yu, M. A., Abhold, E. L., Chen, J. S., Wang-Rodriguez, J., Ongkeko, W. M., & Navarro, A. (2012). Cigarette smoke promotes drug resistance and expansion of cancer stem cell-like side population. *PLoS ONE*, *7*(11), e47919. https://doi.org/10.1371/journal.pone.0047919.

Arora, K., Alfulaij, N., Higa, J. K., Panee, J., & Nichols, R. A. (2013). Impact of sustained exposure to β-amyloid on calcium homeostasis and neuronal integrity in model nerve cell system expressing $\alpha 4 \beta 2$ nicotinic acetylcholine receptors. *Journal of Biological Chemistry*, *288*(16), 11175–11190. https://doi.org/10.1074/jbc.M113.453746.

Ballinger, E. C., Ananth, M., Talmage, D. A., & Role, L. W. (2016). Basal forebrain cholinergic circuits and signaling in cognition and cognitive decline. *Neuron*, *91*(6), 1199–1218. https://doi.org/10.1016/j.neuron.2016.09.006.

Banerjee, J., Al-Wadei, H. A. N., & Schuller, H. M. (2013). Chronic nicotine inhibits the therapeutic effects of gemcitabine on pancreatic cancer in vitro and in mouse xenografts. *European Journal of Cancer*, *49*(5), 1152–1158. https://doi.org/10.1016/j.ejca.2012.10.015.

Bejerot, S., & Nylander, L. (2003). Low prevalence of smoking in patients with autism spectrum disorders. *Psychiatry Research*, *119*(1–2), 177–182. https://doi.org/10.1016/S0165-1781(03)00123-9.

Bencherif, M., Stachowiak, M. K., Kucinski, A. J., & Lippiello, P. M. (2012). Alpha7 nicotinic cholinergic neuromodulation may reconcile multiple neurotransmitter hypotheses of schizophrenia. *Medical Hypotheses*, *78*(5), 594–600. https://doi.org/10.1016/j.mehy.2012.01.035.

Bordia, T., Campos, C., Huang, L., & Quik, M. (2008). Continuous and intermittent nicotine treatment reduces L-3,4-dihydroxyphenylalanine (L-DOPA)-induced dyskinesias in a rat model of Parkinson's disease. *Journal of Pharmacology and Experimental Therapeutics*, *327*(1), 239–247. https://doi.org/10.1124/jpet.108.140897.

Bordia, T., McGregor, M., Papke, R. L., Decker, M. W., Michael McIntosh, J., & Quik, M. (2015). The $\alpha 7$ nicotinic receptor agonist ABT-107 protects against nigrostriatal damage in rats with unilateral 6-hydroxydopamine lesions. *Experimental Neurology*, *263*, 277–284. https://doi.org/10.1016/j.expneurol.2014.09.015.

Bordia, T., Perez, X. A., Heiss, J. E., Zhang, D., & Quik, M. (2016). Optogenetic activation of striatal cholinergic interneurons regulates L-dopa-induced dyskinesias. *Neurobiology of Disease*, *91*, 47–58. https://doi.org/10.1016/j.nbd.2016.02.019.

Bose, C., Zhang, H., Udupa, K. B., & Chowdhury, P. (2005). Activation of p-ERK1/2 by nicotine in pancreatic tumor cell line AR42J: Effects on proliferation and secretion. *American Journal of Physiology - Gastrointestinal and Liver Physiology*, *289*(5), G926–G934. https://doi.org/10.1152/ajpgi.00138.2005.

Boyd, R. T. (2012). Neuronal nicotinic acetylcholine receptors in pancreatic cancer. Program no. 329. In *Vol. 16. Neuroscience meeting planner. New Orleans*. LA: Society for Neuroscience. Online.

Breese, C. R., Lee, M. J., Adams, C. E., Sullivan, B., Logel, J., Gillen, K. M., Marks, M. J., Collins, A. C., & Leonard, S. (2000). Abnormal regulation of high affinity nicotinic receptors in subjects with schizophrenia. *Neuropsychopharmacology*, *23*(4), 351–364. https://doi.org/10.1016/S0893-133X(00)00121-4.

Brown, K. C., Lau, J. K., Dom, A. M., Witte, T. R., Luo, H., Crabtree, C. M., Shah, Y. H., Shiflett, B. S., Marcelo, A. J., Proper, N. A., Hardman, W. E., Egleton, R. D., Chen, Y. C., Mangiarua, E. I., & Dasgupta, P. (2012). MG624, an α7-nAChR antagonist, inhibits angiogenesis via the Egr-1/FGF2 pathway. *Angiogenesis*, *15*(1), 99–114. https://doi.org/10.1007/s10456-011-9246-9.

Brown, K. C., Perry, H. E., Lau, J. K., Jones, D. V., Pulliam, J. F., Thornhill, B. A., Crabtree, C. M., Luo, H., Chen, Y. C., & Dasgupta, P. (2013). Nicotine induces the up-regulation of the α7-nicotinic receptor (α7-nAChR) in human squamous cell lung cancer cells via the Sp1/GATA protein pathway. *Journal of Biological Chemistry*, *288*(46), 33049–33059. https://doi.org/10.1074/jbc.M113.501601.

Burke, L., Miller, L. A., Saad, A., & Abraham, J. (2009). Smoking behaviors among cancer survivors: An observational clinical study. *Journal of Oncology Practice*, *5*(1), 6–9. https://doi.org/10.1200/JOP.0912001.

Chowdhury, P., MacLeod, S., Udupa, K. B., & Rayford, P. L. (2002). Pathophysiological effects of nicotine on the pancreas: An update. *Experimental Biology and Medicine*, *227*(7), 445–454. https://doi.org/10.1177/153537020222700708.

Clemens, P., Baron, J. A., Coffey, D., & Reeves, A. (1995). The short-term effect of nicotine chewing gum in patients with Parkinson's disease. *Psychopharmacology*, *117*(2), 253–256. https://doi.org/10.1007/BF02245195.

Copeland, R. L., Leggett, Y. A., Kanaan, Y. M., Taylor, R. E., & Tizabi, Y. (2005). Neuroprotective effects of nicotine against salsolinol-induced cytotoxicity: Implications for Parkinson's disease. *Neurotoxicity Research*, *8*(3–4), 289–293. https://doi.org/10.1007/BF03033982.

Cox, L. S., Patten, C. A., Ebbert, J. O., Drews, A. A., Croghan, G. A., Clark, M. M., Wolter, T. D., Decker, P. A., & Hurt, R. D. (2002). Tobacco use outcomes among patients with lung cancer treated for nicotine dependence. *Journal of Clinical Oncology*, *20*(16), 3461–3469. https://doi.org/10.1200/JCO.2002.10.085.

Dani, J. A., & Bertrand, D. (2007). Nicotinic acetylcholine receptors and nicotinic cholinergic mechanisms of the central nervous system. *Annual Review of Pharmacology and Toxicology*, *47*, 699–729. https://doi.org/10.1146/annurev.pharmtox.47.120505.105214.

Dasgupta, P., Kinkade, R., Joshi, B., DeCook, C., Haura, E., & Chellappan, S. (2006). Nicotine inhibits apoptosis induced by chemotherapeutic drugs by up-regulating XIAP and survivin. *Proceedings of the National Academy of Sciences of the United States of America*, *103*(16), 6332–6337. https://doi.org/10.1073/pnas.0509313103.

Dasgupta, P., Rastogi, S., Pillai, S., Ordonez-Ercan, D., Morris, M., Haura, E., & Chellappan, S. (2006). Nicotine induces cell proliferation by β-arrestin-mediated activation of Src and Rb-Raf-1 pathways. *Journal of Clinical Investigation*, *116*(8), 2208–2217. https://doi.org/10.1172/JCI28164.

De Luca, V., Voineskos, S., Wong, G., & Kennedy, J. L. (2006). Genetic interaction between α4 and β2 subunits of high affinity nicotinic receptor: Analysis in schizophrenia. *Experimental Brain Research*, *174*(2), 292–296. https://doi.org/10.1007/s00221-006-0458-y.

De Rubeis, S., & Buxbaum, J. D. (2015). Genetics and genomics of autism spectrum disorder: Embracing complexity. *Human Molecular Genetics*, *24*(1), R24–R31. https://doi.org/10.1093/hmg/ddv273.

Deutsch, S. I., & Burket, J. A. (2020). An evolving therapeutic rationale for targeting the α7 nicotinic acetylcholine receptor in autism spectrum disorder. In *Vol. 45. Current topics in behavioral neurosciences* (pp. 167–208). Springer Science and Business Media Deutschland GmbH. https://doi.org/10.1007/7854_2020_136.

Deutsch, S. I., Urbano, M. R., Neumann, S. A., Burket, J. A., & Katz, E. (2010). Cholinergic abnormalities in autism: Is there a rationale for selective nicotinic agonist interventions? *Clinical Neuropharmacology*, *33*(3), 114–120. https://doi.org/10.1097/WNF.0b013e3181d6f7ad.

Dineley, K. T., Bell, K. A., Bui, D., & Sweatt, J. D. (2002). β-Amyloid peptide activates α7 nicotinic acetylcholine receptors expressed in Xenopus oocytes. *Journal of Biological Chemistry*, *277*(28), 25056–25061. https://doi.org/10.1074/jbc.M200066200.

Dineley, K. T., Pandya, A. A., & Yakel, J. L. (2015). Nicotinic ACh receptors as therapeutic targets in CNS disorders. *Trends in Pharmacological Sciences*, *36*(2), 96–108. https://doi.org/10.1016/j.tips.2014.12.002.

Dineley, KT, Westerman, M, Bui, D, Bell, K, Ashe, KH, & Sweatt, JD. (2001). B-amyloid activated the mitogen-activated protein kinase cascade via hippocampal alpha7 nicotinic acetylcholine receptors: in vitro and in vivo mechanisms related to Alzheimer's disease. *Journal of Neuroscience*, *21*(12), 4125–4133.

Dougherty, J. J., Wu, J., & Nichols, R. A. (2003). B-amyloid regulation of presynaptic nicotinic receptors in rat hippocampus and neocortex. *Journal of Neuroscience*, *23*(17), 6740–6747.

Duell, E. J. (2012). Epidemiology and potential mechanisms of tobacco smoking and heavy alcohol consumption in pancreatic cancer. *Molecular Carcinogenesis*, *51*(1), 40–52. https://doi.org/10.1002/mc.20786.

Dziewczapolski, G., Glogowski, C. M., Masliah, E., & Heinemann, S. F. (2009). Deletion of the α7 nicotinic acetylcholine receptor gene improves cognitive deficits and synaptic pathology in a mouse model of Alzheimer's disease. *Journal of Neuroscience*, *29*(27), 8805–8815. https://doi.org/10.1523/JNEUROSCI.6159-08.2009.

Ebert, D. H., & Greenberg, M. E. (2013). Activity-dependent neuronal signalling and autism spectrum disorder. *Nature*, *493*(7432), 327–337. https://doi.org/10.1038/nature11860.

Egleton, R. D., Brown, K. C., & Dasgupta, P. (2008). Nicotinic acetylcholine receptors in cancer: Multiple roles in proliferation and inhibition of apoptosis. *Trends in Pharmacological Sciences*, *29*(3), 151–158. https://doi.org/10.1016/j.tips.2007.12.006.

Featherstone, R. E., & Siegel, S. J. (2015). The role of nicotine in schizophrenia. In *Vol. 124. International review of neurobiology* (pp. 23–78). Academic Press Inc. https://doi.org/10.1016/bs.irn.2015.07.002.

Fung, Y. K., Fiske, L. A., & Lau, Y. S. (1991). Chronic administration of nicotine fails to alter the MPTP-induced neurotoxicity in mice. *General Pharmacology*, *22*(4), 669–672. https://doi.org/10.1016/0306-3623(91)90075-H.

Gao, ZG, Cui, WY, Zhang, HT, & Jiu, CG. (1998). Effects of nicotine on 1-methyl-4-phenyl-1,2,5,6-tetrahydropyridine-induceddepression of striatal dopamine content and spontaneous locomotor activity in C57 black mice. *Pharmacological Research*, *38*(2), 101–106.

Gaudet, M. M., Gapstur, S. M., Sun, J., Diver, W. R., Hannan, L. M., & Thun, M. J. (2013). Active smoking and breast cancer risk: Original cohort data and meta-analysis. *JNCI: Journal of the National Cancer Institute*, *105*(8), 515–525. https://doi.org/10.1093/jnci/djt023.

Gaugler, T., Klei, L., Sanders, S. J., Bodea, C. A., Goldberg, A. P., Lee, A. B., … Buxbaum, J. D. (2014). Most genetic risk for autism resides with common variation. *Nature Genetics*, *46*(8), 881–885. https://doi.org/10.1038/ng.3039.

George, A. A., Vieira, J. M., Xavier-Jackson, C., Gee, M. T., Cirrito, J. R., Bimonte-Nelson, H. A., Picciotto, M. R., Lukas, R. J., & Whiteaker, P. (2021). Implications of oligomeric amyloid-beta (oAβ42) signaling through α7β2-nicotinic acetylcholine receptors (nAChRs) on basal forebrain cholinergic neuronal intrinsic excitability and cognitive decline. *Journal of Neuroscience*, *41*(3), 555–575. https://doi.org/10.1523/JNEUROSCI.0876-20.2020.

Getachew, B., Csoka, A. B., Aschner, M., & Tizabi, Y. (2019). Nicotine protects against manganese and iron-induced toxicity in SH-SY5Y cells: Implication for Parkinson's disease. *Neurochemistry International*, *124*, 19–24. https://doi.org/10.1016/j.neuint.2018.12.003.

Ghaleiha, A., Ghyasvand, M., Mohammadi, M. R., Farokhnia, M., Yadegari, N., Tabrizi, M., Hajiaghaee, R., Yekehtaz, H., & Akhondzadeh, S. (2014). Galantamine efficacy and tolerability as an augmentative therapy in autistic children: A randomized, double-blind, placebo-controlled trial. *Journal of Psychopharmacology*, *28*(7), 677–685. https://doi.org/10.1177/0269881113508830.

Gibbons, A., & Dean, B. (2016). The cholinergic system: An emerging drug target for schizophrenia. *Current Pharmaceutical Design*, *22*(14), 2124–2133. https://doi.org/10.2174/1381612822666160127114010.

Grassi, F., Palma, E., Tonini, R., Amici, M., Ballivet, M., & Eusebi, F. (2003). Amyloid β1-42 peptide alters the gating of human and mouse α-bungarotoxin-sensitive nicotinic receptors. *Journal of Physiology*, *547*(1), 147–157. https://doi.org/10.1113/jphysiol.2002.035436.

Hecht, S. S. (2012). Lung carcinogenesis by tobacco smoke. *International Journal of Cancer*, *131* (12), 2724–2732. https://doi.org/10.1002/ijc.27816.

Hernandez, C. M., Kayed, R., Zheng, H., Sweatt, J. D., & Dineley, K. T. (2010). Loss of α7 nicotinic receptors enhances β-amyloid oligomer accumulation, exacerbating early-stage cognitive decline and septohippocampal pathology in a mouse model of Alzheimer's disease. *Journal of Neuroscience*, *30*(7), 2442–2453. https://doi.org/10.1523/JNEUROSCI.5038-09.2010.

Hidalgo, M. (2010). Pancreatic cancer. *New England Journal of Medicine*, *362*(17), 1605–1617. https://doi.org/10.1056/nejmra0901557.

Hoskin, J. L., Al-Hasan, Y., & Sabbagh, M. N. (2019). Nicotinic acetylcholine receptor agonists for the treatment of Alzheimer's dementia: An update. *Nicotine and Tobacco Research*, *21*(3), 370–376. https://doi.org/10.1093/ntr/nty116.

Huang, L. Z., Parameswaran, N., Bordia, T., Michael McIntosh, J., & Quik, M. (2009). Nicotine is neuroprotective when administered before but not after nigrostriatal damage in rats and monkeys. *Journal of Neurochemistry*, *109*(3), 826–837. https://doi.org/10.1111/j.1471-4159.2009.06011.x.

Hurst, R., Rollema, H., & Bertrand, D. (2013). Nicotinic acetylcholine receptors: From basic science to therapeutics. *Pharmacology and Therapeutics*, *137*(1), 22–54. https://doi.org/10.1016/j.pharmthera.2012.08.012.

Improgo, MR, Schlichting, NA, Cortes, RY, Zhao-Shea, R, Tapper, AR, & Gardner, PD. (2010). ASCL1 regulates the expression of the CHRNA5/A3/B4 lung cancer susceptibility locus. *Moleular Cancer Research 8*, *(2):*, 194–203.

Improgo, M. R., Soll, L. G., Tapper, A. R., & Gardner, P. D. (2013). Nicotinic acetylcholine receptors mediate lung cancer growth. *Frontiers in Physiology*, *4*. https://doi.org/10.3389/fphys.2013.00251.

Jansen, W. J., Ossenkoppele, R., Knol, D. L., Tijms, B. M., Scheltens, P., Verhey, F. R. J., Visser, P. J., Aalten, P., Aarsland, D., Alcolea, D., Alexander, M., Almdahl, I. S., Arnold, S. E., Baldeiras, I., Barthel, H., van Berckel, B. N. M., Bibeau, K., Blennow, K., Brooks, D. J., … Zetterberg, H. (2015). Prevalence of cerebral amyloid pathology in persons without dementia. *JAMA*, *313*(19), 1924. https://doi.org/10.1001/jama.2015.4668.

Jeyarasasingam, G., Tompkins, L., & Quik, M. (2002). Stimulation of non-α7 nicotinic receptors partially protects dopaminergic neurons from 1-methyl-4-phenylpyridinium-induced toxicity in culture. *Neuroscience, 109*(2), 275–285. https://doi.org/10.1016/S0306-4522(01)00488-2.

John, D., & Berg, D. K. (2015). Long-lasting changes in neural networks to compensate for altered nicotinic input. *Biochemical Pharmacology, 97*(4), 418–424. https://doi.org/10.1016/j.bcp.2015.07.020.

Jung, Y., Lee, A. M., McKee, S. A., & Picciotto, M. R. (2017). Maternal smoking and autism spectrum disorder: Meta-analysis with population smoking metrics as moderators. *Scientific Reports, 7*(1). https://doi.org/10.1038/s41598-017-04413-1.

Jurado-Coronel, J. C., Ávila-Rodriguez, M., Capani, F., Gonzalez, J., Morán, V. E., & Barreto, G. E. (2016). Targeting the nicotinic acetylcholine receptors (nAChRs) in astrocytes as a potential therapeutic target in Parkinson's disease. *Current Pharmaceutical Design, 22*(10), 1305–1311. https://doi.org/10.2174/13816128221016030411213.

Kelton, M. C., Kahn, H. J., Conrath, C. L., & Newhouse, P. A. (2000). The effects of nicotine on Parkinson's disease. *Brain and Cognition, 43*(1–3), 274–282.

Kolodecik, T., Shugrue, C., Ashat, M., & Thrower, E. C. (2014). Risk factors for pancreatic cancer: Underlying mechanisms and potential targets. *Frontiers in Physiology, 4*. https://doi.org/10.3389/fphys.2013.00415.

Lam, D. C. L., Girard, L., Ramirez, R., Chau, W. S., Suen, W. S., Sheridan, S., Tin, V. P. C., Chung, L. P., Wong, M. P., Shay, J. W., Gazdar, A. F., Lam, W. K., & Minna, J. D. (2007). Expression of nicotinic acetylcholine receptor subunit genes in non-small-cell lung cancer reveals differences between smokers and nonsmokers. *Cancer Research, 67* (10), 4638–4647. https://doi.org/10.1158/0008-5472.CAN-06-4628.

Lamb, P. W., Melton, M. A., & Yakel, J. L. (2005). Inhibition of neuronal nicotinic acetylcholine receptor channels expressed in Xenopus oocytes by β-amyloid 1-42 peptide. *Journal of Molecular Neuroscience, 27*(1), 13–22. https://doi.org/10.1385/JMN:27:01:13.

Lee, M., Martin-Ruiz, C., Graham, A., Court, J., Jaros, E., Perry, R., Iversen, P., Bauman, M., & Perry, E. (2002). Nicotinic receptor abnormalities in the cerebellar cortex in autism. *Brain, 125*(7), 1483–1495. https://doi.org/10.1093/brain/awf160.

Leonard, S., Gault, J., Hopkins, J., Logel, J., Vianzon, R., Short, M., Drebing, C., Berger, R., Venn, D., Sirota, P., Zerbe, G., Olincy, A., Ross, R. G., Adler, L. E., & Freedman, R. (2002). Association of promoter variants in the α7 nicotinic acetylcholine receptor subunit gene with an inhibitory deficit found in schizophrenia. *Archives of General Psychiatry, 59*(12), 1085–1096. https://doi.org/10.1001/archpsyc.59.12.1085.

Leonard, S., Mexal, S., & Freedman, R. (2007). Genetics of smoking and schizophrenia. *Journal of Dual Diagnosis, 3*(3–4), 43–59. https://doi.org/10.1300/J374v03n03_05.

Leonardi-Bee, J., Ellison, T., & Bath-Hextall, F. (2012). Smoking and the risk of non-melanoma skin cancer: Systematic review and meta-analysis. *Archives of Dermatology, 148*(8), 939–946. https://doi.org/10.1001/archdermatol.2012.1374.

Lewis, A. S., van Schalkwyk, G. I., Lopez, M. O., Volkmar, F. R., Picciotto, M. R., & Sukhodolsky, D. G. (2018). An exploratory trial of transdermal nicotine for aggression and irritability in adults with autism spectrum disorder. *Journal of Autism and Developmental Disorders, 48*(8), 2748–2757. https://doi.org/10.1007/s10803-018-3536-7.

Li, S., & Selkoe, D. J. (2020). A mechanistic hypothesis for the impairment of synaptic plasticity by soluble Aβ oligomers from Alzheimer's brain. *Journal of Neurochemistry, 154*(6), 583–597. https://doi.org/10.1111/jnc.15007.

Lieberman, A., Lockhart, T. E., Olson, M. C., Smith Hussain, V. A., Frames, C. W., Sadreddin, A., McCauley, M., & Ludington, E. (2019). Nicotine bitartrate reduces falls and freezing of gait in Parkinson disease: A reanalysis. *Frontiers in Neurology, 10*(May). https://doi.org/10.3389/fneur.2019.00424.

Lilja, A. M., Porras, O., Storelli, E., Nordberg, A., & Marutle, A. (2011). Functional inter-actions of fibrillar and oligomeric amyloid-β with alpha7 nicotinic receptors in alzheimer's disease. *Journal of Alzheimer's Disease*, *23*(2), 335–347. https://doi.org/10.3233/JAD-2010-101242.

Lippiello, P. M. (2006). Nicotinic cholinergic antagonists: A novel approach for the treat-ment of autism. *Medical Hypotheses*, *66*(5), 985–990. https://doi.org/10.1016/j.mehy.2005.11.015.

Liu, Q., Huang, Y., Shen, J., Steffenson, S., & Wu, J. (2012). Functional a7b2 nicotinic ace-tylcholine receptors expressed in hippocampal interneurons exhibit high sensitivity to pathological level of amyloid b peptides. *BMC Neuroscience*, *13*.

Liu, Q., Huang, Y., Xue, F., Simard, A., DeChon, J., Li, G., Zhang, J., Lucero, L., Wang, M., Sierks, M., Hu, G., Chang, Y., Lukas, R. J., & Wu, J. (2009). A novel nicotinic acetylcholine receptor subtype in basal forebrain cholinergic neurons with high sensitiv-ity to amyloid peptides. *Journal of Neuroscience*, *29*(4), 918–929. https://doi.org/10.1523/jneurosci.3952-08.2009.

Liu, Q. S., Kawai, H., & Berg, D. K. (2001). β-Amyloid peptide blocks the response of α7-containing nicotinic receptors on hippocampal neurons. *Proceedings of the National Acad-emy of Sciences of the United States of America*, *98*(8), 4734–4739. https://doi.org/10.1073/pnas.081553598.

Lombardo, S., & Maskos, U. (2015). Role of the nicotinic acetylcholine receptor in Alzheimer's disease pathology and treatment. *Neuropharmacology*, *96*, 255–262. https://doi.org/10.1016/j.neuropharm.2014.11.018.

Ma, X., Jia, Y., Zu, S., Li, R., Jia, Y., Zhao, Y., Xiao, D., Dang, N., & Wang, Y. (2014). Alpha5 nicotinic acetylcholine receptor mediates nicotine-induced HIF-1α and VEGF expression in non-small cell lung cancer. *Toxicology and Applied Pharmacology*, *278*(2), 172–179. https://doi.org/10.1016/j.taap.2014.04.023.

Maneckjee, R., & Minna, J. D. (1994). Opioids induce while nicotine suppresses apoptosis in human lung cancer cells. *Cell Growth and Differentiation*, *5*(10), 1033–1040.

Martin-Ruiz, C. M., Lee, M., Perry, R. H., Baumann, M., Court, J. A., & Perry, E. K. (2004). Molecular analysis of nicotinic receptor expression in autism. *Molecular Brain Research*, *123*(1–2), 81–90. https://doi.org/10.1016/j.molbrainres.2004.01.003.

Maus, A. D. J., Pereira, E. F. R., Karachunski, P. I., Horton, R. M., Navaneetham, D., Macklin, K., Cortes, W. S., Albuquerque, E. X., & Conti-Fine, B. M. (1998). Human and rodent bronchial epithelial cells express functional nicotinic acetylcholine receptors. *Molecular Pharmacology*, *54*(5), 779–788. https://doi.org/10.1124/mol.54.5.779.

Meshul, CK, Kamel, D, Moore, C, Kay, TS, & Krentz, L. (2002). Nicotine alters striatal glutamate function and decreases apomorphine-induced contralateral rotations in 6-OHD-lesioned rats. *Experimental Neurology*, *175*, 257–274.

Mexal, S., Frank, M., Berger, R., Adams, C. E., Ross, R. G., Freedman, R., & Leonard, S. (2005). Differential modulation of gene expression in the NMDA postsynaptic density of schizophrenic and control smokers. *Molecular Brain Research*, *139*(2), 317–332. https://doi.org/10.1016/j.molbrainres.2005.06.006.

Momi, N., Ponnusamy, M. P., Kaur, S., Rachagani, S., Kunigal, S. S., Chellappan, S., … Batra, S. K. (2012). Nicotine/cigarette smoke promotes metastasis of pancreatic cancer through a7 nAChR-mediated MUC4 upregulation. *Oncogene*, 1–12.

Murray, T. A., Bertrand, D., Papke, R. L., George, A. A., Pantoja, R., Srinivasan, R., Liu, Q., Wu, J., Whiteaker, P., Lester, H. A., & Lukas, R. J. (2012). α7β2 nicotinic acetylcholine receptors assemble, function and are activated primarily via their α7-α7 interfaces. *Molecular Pharmacology*, *81*(2), 175–188. https://doi.org/10.1124/mol.111.074088.

Nagele, R. G., D'Andrea, M. R., Anderson, W. J., & Wang, H. Y. (2002). Intracellular accu-mulation of β-amyloid(1–42) in neurons is facilitated by the α7 nicotinic acetylcholine receptor in Alzheimer's disease. *Neuroscience*, *110*, 199–211.

Nicolson, R., Craven-Thuss, B., & Smith, J. (2006). A prospective, open-label trial of galantamine in autistic disorder. *Journal of Child and Adolescent Psychopharmacology, 16* (5), 621–629. https://doi.org/10.1089/cap.2006.16.621.

Nishioka, T., Luo, L. Y., Shen, L., He, H., Mariyannis, A., Dai, W., & Chen, C. (2014). Nicotine increases the resistance of lung cancer cells to cisplatin through enhancing Bcl-2 stability. *British Journal of Cancer, 110*(7), 1785–1792. https://doi.org/10.1038/bjc.2014.78.

Olfson, E., Saccone, N. L., Johnson, E. O., Chen, L. S., Culverhouse, R., Doheny, K., Foltz, S. M., Fox, L., Gogarten, S. M., Hartz, S., Hetrick, K., Laurie, C. C., Marosy, B., Amin, N., Arnett, D., Barr, R. G., Bartz, T. M., Bertelsen, S., Borecki, I. B., ... Bierut, L. J. (2016). Rare, low frequency and common coding variants in CHRNA5 and their contribution to nicotine dependence in European and African Americans. *Molecular Psychiatry, 21*(5), 601–607. https://doi.org/10.1038/mp.2015.105.

Olincy, A., Blakeley-Smith, A., Johnson, L., Kem, W. R., & Freedman, R. (2016). Brief report: Initial trial of Alpha7-nicotinic receptor stimulation in two adult patients with autism spectrum disorder. *Journal of Autism and Developmental Disorders, 46*(12), 3812–3817. https://doi.org/10.1007/s10803-016-2890-6.

Olivera, G, Grilli, M, Chen, J, Preda, S, Mura, E, Govoni, S, & Marchi, M. (2014). Effects of soluble β-amyloid on the release of neurotransmitters from rat brain synaptosomes. *Frontiers in Aging Neuroscience, 6,* 1–9.

Osborne, J. K., Guerra, M. L., Gonzales, J. X., McMillan, E. A., Minna, J. D., & Cobb, M. H. (2014). NeuroD1 mediates nicotine-induced migration and invasion via regulation of the nicotinic acetylcholine receptor subunits in a subset of neural and neuroendocrine carcinomas. *Molecular Biology of the Cell, 25*(11), 1782–1792. https://doi.org/10.1091/mbc.E13-06-0316.

Paleari, L., Negri, E., Catassi, A., Cilli, M., Servent, D., D'Angelillo, R., Cesario, A., Russo, P., & Fini, M. (2009). Inhibition of nonneuronal α7-nicotinic receptor for lung cancer treatment. *American Journal of Respiratory and Critical Care Medicine, 179*(12), 1141–1150. https://doi.org/10.1164/rccm.200806-908OC.

Pandya, A., & Yakel, J. L. (2011). Allosteric modulator desformylflustrabromine relieves the inhibition of α2β2 and α4β2 nicotinic acetylcholine receptors by β-amyloid 1-42 peptide. *Journal of Molecular Neuroscience, 45*(1), 42–47. https://doi.org/10.1007/s12031-011-9509-3.

Parain, K., Marchand, V., Dumery, B., & Hirsch, E. (2001). Nicotine, but not cotinine, partially protects dopaminergic neurons against MPTP-induced degeneration in mice. *Brain Research, 890*(2), 347–350. https://doi.org/10.1016/S0006-8993(00)03198-X.

Perry, E. K., Lee, M. L. W., Martin-Ruiz, C. M., Court, J. A., Volsen, S. G., Merrit, J., Folly, E., Iversen, P. E., Bauman, M. L., Perry, R. H., & Wenk, G. L. (2001). Cholinergic activity in autism: Abnormalities in the cerebral cortex and basal forebrain. *American Journal of Psychiatry, 158*(7), 1058–1066. https://doi.org/10.1176/appi.ajp.158.7.1058.

Pettit, D., Shao, Z., & Yakel, J. (2001). B-amyloid 1-42 peptide directly modulates nicotinic receptors in the rat hippocampal slice. *The Journal of Neuroscience, 21.*

Preskorn, S. H., Gawryl, M., Dgetluck, N., Palfreyman, M., Bauer, L. O., & Hilt, D. C. (2014). Normalizing effects of EVP-6124, an alpha-7 nicotinic partial agonist, on event-related potentials and cognition: A proof of concept, randomized trial in patients with schizophrenia. *Journal of Psychiatric Practice, 20*(1), 12–24. https://doi.org/10.1097/01.pra.0000442935.15833.c5.

Pym, L., Kemp, M., Raymond-Delpech, V., Buckingham, S., Boyd, C. A. R., & Sattelle, D. (2005). Subtype-specific actions of β-amyloid peptides on recombinant human neuronal nicotinic acetylcholine receptors (α7, α4β2, α3β4) expressed in *Xenopus laevis* oocytes. *British Journal of Pharmacology, 146*(7), 964–971. https://doi.org/10.1038/sj.bjp.0706403.

Quik, M, Bordia, T, Zhang, D, & Perez, XA. (2015). Nicotine and nicotinic receptor drugs: potential for Parkinson's disease and drug-induced movement disorders. *International Review of Neurobiology, 124,* 247–271.

Quik, M., Boyd, J. T., Bordia, T., & Perez, X. (2019). Potential therapeutic application for nicotinic receptor drugs in movement disorders. *Nicotine and Tobacco Research, 21*(3), 357–369. https://doi.org/10.1093/ntr/nty063.

Quik, M., Cox, H., Parameswaran, N., O'Leary, K., Langston, J. W., & Di Monte, D. (2007). Nicotine reduces levodopa-induced dyskinesias in lesioned monkeys. *Annals of Neurology, 62*(6), 588–596. https://doi.org/10.1002/ana.21203.

Quik, M., O'Neill, M., & Perez, X. A. (2007). Nicotine neuroprotection against nigrostriatal damage: Importance of the animal model. *Trends in Pharmacological Sciences, 28*(5), 229–235. https://doi.org/10.1016/j.tips.2007.03.001.

Quik, M., Perez, X. A., & Bordia, T. (2012). Nicotine as a potential neuroprotective agent for Parkinson's disease. *Movement Disorders, 27*(8), 947–957. https://doi.org/10.1002/mds.25028.

Quik, M., & Wonnacott, S. (2011). $\alpha6\beta2^*$ and $\alpha4\beta2^*$ nicotinic acetylcholine receptors as drug targets for Parkinson's disease. *Pharmacological Reviews, 63*(4), 938–966. https://doi.org/10.1124/pr.110.003269.

Quik, M., Zhang, D., McGregor, M., & Bordia, T. (2015). Alpha7 nicotinic receptors as therapeutic targets for Parkinson's disease. *Biochemical Pharmacology, 97*(4), 399–407. https://doi.org/10.1016/j.bcp.2015.06.014.

Radek, R. J., Miner, H. M., Bratcher, N. A., Decker, M. W., Gopalakrishnan, M., & Bitner, R. S. (2006). $\alpha4\beta2$ nicotinic receptor stimulation contributes to the effects of nicotine in the DBA/2 mouse model of sensory gating. *Psychopharmacology, 187*(1), 47–55. https://doi.org/10.1007/s00213-006-0394-3.

Ray, M. A., Graham, A. J., Lee, M., Perry, R. H., Court, J. A., & Perry, E. K. (2005). Neuronal nicotinic acetylcholine receptor subunits in autism: An immunohistochemical investigation in the thalamus. *Neurobiology of Disease, 19*(3), 366–377. https://doi.org/10.1016/j.nbd.2005.01.017.

Rodriguez-Diaz, R., Dando, R., Jacques-Silva, M. C., Fachado, A., Molina, J., Abdulreda, M. H., Ricordi, C., Roper, S. D., Berggren, P. O., & Caicedo, A. (2011). Alpha cells secrete acetylcholine as a non-neuronal paracrine signal priming beta cell function in humans. *Nature Medicine, 17*(7), 888–892. https://doi.org/10.1038/nm.2371.

Ryan, R. E., Ross, S. A., Drago, J., & Loiacono, R. E. (2001). Dose-related neuroprotective effects of chronic nicotine in 6-hydroxydopamine treated rats, and loss of neuroprotection in $\alpha4$ nicotinic receptor subunit knockout mice. *British Journal of Pharmacology, 132*(8), 1650–1656. https://doi.org/10.1038/sj.bjp.0703989.

Schaaf, C. P. (2014). Nicotinic acetylcholine receptors in human genetic disease. *Genetics in Medicine, 16*(9), 649–656. https://doi.org/10.1038/gim.2014.9.

Schaal, C., Padmanabhan, J., & Chellappan, S. (2015). The role of nAChR and calcium signaling in pancreatic cancer initiation and progression. *Cancers, 7*(3), 1447–1471. https://doi.org/10.3390/cancers7030845.

Schneider, J. S., Pope-Coleman, A., Van Velson, M., Menzaghi, F., & Lloyd, G. K. (1998). Effects of SIB-1508Y, a novel neuronal nicotinic acetylcholine receptor agonist, on motor behavior in parkinsonian monkeys. *Movement Disorders, 13*(4), 637–642. https://doi.org/10.1002/mds.870130405.

Schuller, H. M. (2009). Is cancer triggered by altered signalling of nicotinic acetylcholine receptors? *Nature Reviews Cancer, 9*(3), 195–205. https://doi.org/10.1038/nrc2590.

Schuller, H. M., & Al-Wadei, H. A. (2010). Neurotransmitter receptors as central regulators of pancreatic cancer. *Future Oncology (London, England), 6*(2), 221–228. https://doi.org/10.2217/fon.09.171.

Sekhon, H. S., Jia, Y., Raab, R., Kuryatov, A., Pankow, J. F., Whitsett, J. A., Lindstrom, J., & Spindel, E. R. (1999). Prenatal nicotine increases pulmonary α7 nicotinic receptor expression and alters fetal lung development in monkeys. *Journal of Clinical Investigation*, *103*(5), 637–647. https://doi.org/10.1172/jci5232.

Shih, Y. L., Liu, H. C., Chen, C. S., Hsu, C. H., Pan, M. H., Chang, H. W., Chang, C. H., Chen, F. C., Ho, C. T., Yang, Y. I. Y., & Ho, Y. S. (2010). Combination treatment with luteolin and quercetin enhances antiproliferative effects in nicotine-treated MDA-MB-231 cells by down-regulating nicotinic acetylcholine receptors. *Journal of Agricultural and Food Chemistry*, *58*(1), 235–241. https://doi.org/10.1021/jf9031684.

Shishido, E., Aleksic, B., & Ozaki, N. (2014). Copy-number variation in the pathogenesis of autism spectrum disorder. *Psychiatry and Clinical Neurosciences*, *68*(2), 85–95. https://doi.org/10.1111/pcn.12128.

Shoulson, I. (2006). Randomized placebo-controlled study of the nicotinic agonist SIB-1508Y in Parkinson's disease. *Neurology*, *66*(3), 408–410.

Song, P., Sekhon, H. S., Fu, X. W., Maier, M., Jia, Y., Duan, J., Proskosil, B. J., Gravett, C., Lindstrom, J., Mark, G. P., Saha, S., & Spindel, E. R. (2008). Activated cholinergic signaling provides a target in squamous cell lung carcinoma. *Cancer Research*, *68*(12), 4693–4700. https://doi.org/10.1158/0008-5472.can-08-0183.

Song, P., Sekhon, H. S., Jia, Y., Keller, J. A., Blusztajn, J. K., Mark, G. P., & Spindel, E. R. (2003). Acetylcholine is synthesized by and acts as an autocrine growth factor for small cell lung carcinoma. *Cancer Research*, *63*(1), 214–221.

Spindel, E. R. (2009). Is nicotine the estrogen of lung cancer? *American Journal of Respiratory and Critical Care Medicine*, *179*(12), 1081–1082. https://doi.org/10.1164/rccm.200901-0013ED.

Stefansson, H., Rujescu, D., Cichon, S., Pietiläinen, O. P. H., Ingason, A., Steinberg, S., Fossdal, R., Sigurdsson, E., Sigmundsson, T., Buizer-Voskamp, J. E., Hansen, T., Jakobsen, K. D., Muglia, P., Francks, C., Matthews, P. M., Gylfason, A., Halldorsson, B. V., Gudbjartsson, D., Thorgeirsson, T. E., … Myin-Germeys, I. (2008). Large recurrent microdeletions associated with schizophrenia. *Nature*, *455*(7210), 232–236. https://doi.org/10.1038/nature07229.

Stevens, K. E., Freedman, R., Collins, A. C., Hall, M., Leonard, S., Marks, M. J., & Rose, G. M. (1996). Genetic correlation of inhibitory gating of hippocampal auditory evoked response and α-bungarotoxin-binding nicotinic cholinergic receptors in inbred mouse strains. *Neuropsychopharmacology*, *15*(2), 152–162. https://doi.org/10.1016/0893-133X(95)00178-G.

Stuckenholz, V., Bacher, M., Balzer-Geldsetzer, M., Alvarez-Fischer, D., Oertel, W. H., Dodel, R. C., & Noelker, C. (2013). The α7 nAChR agonist PNU-282987 reduces inflammation and MPTP-induced nigral dopaminergic cell loss in mice. *Journal of Parkinson's Disease*, *3*(2), 161–172. https://doi.org/10.3233/JPD-120157.

Sullivan, J., Blair, L., Alnajar, A., Aziz, T., Chipitsynas, G., Gong, Q., Yeo, C. J., & Arafat, H. A. (2011). Expression and regulation of nicotine receptor and osteopontin isoforms in human pancreatic ductal adenocarcinoma. *Histology and Histopathology*, *26*(7), 893–904. http://www.hh.um.es/pdf/Vol_26/26_7/Sullivan-26-893-904-2011.pdf.

Talantova, M., et al. (2013). Aβ induces astrocytic glutamate release, extrasynaptic NMDA receptor activation, and synaptic loss. *Proceedings of the National Academy of Sciences of the United States of America*, *110*, E2518–E2527.

Tarroni, P., Rubboli, F., Chini, B., Zwart, R., Oortgiesen, M., Sher, E., & Clementi, F. (1992). Neuronal-type nicotinic receptors in human neuroblastoma and small-cell lung carcinoma cell lines. *FEBS Letters*, *312*(1), 66–70. https://doi.org/10.1016/0014-5793(92)81411-E.

Thies, W., & Bleiler, L. (2013). 2013 Alzheimer's disease facts and figures. *Alzheimer's & Dementia*, *9*(2), 208–245. https://doi.org/10.1016/j.jalz.2013.02.003.

Tizabi, Y., & Getachew, B. (2017). Nicotinic receptor intervention in Parkinson's disease: Future directions. *Clinical Pharmacology and Translational Medicine*, *1*, 14–19.

Tozaki, H., Matsumoto, A., Kanno, T., Nagai, K., Nagata, T., Yamamoto, S., & Nishizaki, T. (2002). The inhibitory and facilitatory actions of amyloid-β peptides on nicotinic ACh receptors and AMPA receptors. *Biochemical and Biophysical Research Communications*, *294* (1), 42–45. https://doi.org/10.1016/S0006-291X(02)00429-1.

Tregellas, J. R., Tanabe, J., Rojas, D. C., Shatti, S., Olincy, A., Johnson, L., Martin, L. F., Soti, F., Kem, W. R., Leonard, S., & Freedman, R. (2011). Effects of an alpha 7-nicotinic agonist on default network activity in schizophrenia. *Biological Psychiatry*, *69* (1), 7–11. https://doi.org/10.1016/j.biopsych.2010.07.004.

Trombino, S., Cesario, A., Margaritora, S., Granone, P. L., Motta, G., Falugi, C., & Russo, P. (2004). α7-nicotinic acetylcholine receptors affect growth regulation of human mesothelioma cells: Role of mitogen-activated protein kinase pathway. *Cancer Research*, *64*(1), 135–145. https://doi.org/10.1158/0008-5472.CAN-03-1672.

Vieregge, A., Sieberer, M., Jacobs, H., Hagenah, J. M., & Vieregge, P. (2001). Transdermal nicotine in PD: A randomized, double-blind, placebo-controlled study. *Neurology*, *57*(6), 1032–1035. https://doi.org/10.1212/WNL.57.6.1032.

Villafane, G., Cesaro, P., Rialland, A., Baloul, S., Azimi, S., Bourdet, C., Le Houezec, J., Macquin-Mavier, I., & Maison, P. (2007). Chronic high dose transdermal nicotine in Parkinson's disease: An open trial. *European Journal of Neurology*, *14*(12), 1313–1316. https://doi.org/10.1111/j.1468-1331.2007.01949.x.

Wallace, T. L., & Bertrand, D. (2015). Neuronal α7 nicotinic receptors as a target for the treatment of schizophrenia. In *Vol. 124. International review of neurobiology* (pp. 79–111). Academic Press Inc. https://doi.org/10.1016/bs.irn.2015.08.003.

Wang, L., Almeida, L. E. F., Nettleton, M., Khaibullina, A., Albani, S., Kamimura, S., Nouraie, M., & Quezado, Z. M. N. (2016). Altered nocifensive behavior in animal models of autism spectrum disorder: The role of the nicotinic cholinergic system. *Neuropharmacology*, *111*, 323–334. https://doi.org/10.1016/j.neuropharm.2016.09.013.

Wang, L., Almeida, L. E. F., Spornick, N. A., Kenyon, N., Kamimura, S., Khaibullina, A., Nouraie, M., Quezado, Z. M. N. (2015). Modulation of social deficits and repetitive behaviors in a mouse model of autism: the role of the cholinergic system. *Psychopharmacology*, *232*, 4303–4316.

Wang, H. Y., Lee, D. H. S., D'Andrea, M. R., Peterson, P. A., Shank, R. P., & Reitz, A. B. (2000). β-Amyloid 1-42 binds to α7 nicotinic acetylcholine receptor with high affinity. Implications for Alzheimer's disease pathology. *Journal of Biological Chemistry*, *275*(8), 5626–5632. https://doi.org/10.1074/jbc.275.8.5626.

Wang, H. Y., Lee, D. H. S., Davis, C. B., & Shank, R. P. (2000). Amyloid peptide Aβ1–42 binds selectively and with picomolar affinity to α7 nicotinic acetylcholine receptors. *Journal of Neurochemistry*, *75*(3), 1155–1161. https://doi.org/10.1046/j.1471-4159.2000.0751155.x.

Wang, Y., Pereira, E. F. R., Maus, A. D. J., Ostlie, N. S., Navaneetham, D., Lei, S., Albuquerque, E. X., & Conti-Fine, B. M. (2001). Human bronchial epithelial and endothelial cells express α7 nicotinic acetylcholine receptors. *Molecular Pharmacology*, *60*(6), 1201–1209. https://doi.org/10.1124/mol.60.6.1201.

Warren, G., & Singh, A. (2013). Nicotine and lung cancer. *Journal of Carcinogenesis*, *12*. https://doi.org/10.4103/1477-3163.106680.

Wessler, I., & Kirkpatrick, C. J. (2008). Acetylcholine beyond neurons: The non-neuronal cholinergic system in humans. *British Journal of Pharmacology*, *154*(8), 1558–1571. https://doi.org/10.1038/bjp.2008.185.

Wessler, I., Kirkpatrick, C. J., & Racké, K. (1998). Non-neuronal acetylcholine, a locally acting molecule, widely distributed in biological systems: Expression and function in humans. *Pharmacology and Therapeutics*, *77*(1), 59–79. https://doi.org/10.1016/S0163-7258(97)00085-5.

WHO. n.d. http://www.who.int/mediacentre/factsheets/fs297/en/.

Won, L., Ding, Y., Singh, P., & Kang, U. J. (2014). Striatal cholinergic cell ablation attenuates L-DOPA induced dyskinesia in parkinsonian mice. *Journal of Neuroscience*, *34*(8), 3090–3094. https://doi.org/10.1523/JNEUROSCI.2888-13.2014.

Wu, J, Kuo, Y-P, George, AA, Xu, L, Hu, J, & Lukas, RJ. (2004). beta-Amyloid directly inhibits human alpha4beta2-nicotinic acetylcholine receptors heterologously expressed in human SH-EP1 cells. *Journal of Biological Chemistry*, *279*(36), 37842–37851.

Xu, J., Huang, H., Pan, C., Zhang, B., Liu, X., & Zhang, L. (2007). Nicotine inhibits apoptosis induced by cisplatin in human oral cancer cells. *International Journal of Oral and Maxillofacial Surgery*, *36*(8), 739–744. https://doi.org/10.1016/j.ijom.2007.05.016.

Yoshimura, R. F., Tran, M. B., Hogenkamp, D. J., Ayala, N. L., Johnstone, T., Dunnigan, A. J., Gee, T. K., & Gee, K. W. (2017). Allosteric modulation of nicotinic and GABAA receptor subtypes differentially modify autism-like behaviors in the BTBR mouse model. *Neuropharmacology*, *126*, 38–47. https://doi.org/10.1016/j.neuropharm.2017.08.029.

Zeidler, R., Albermann, K., & Lang, S. (2007). Nicotine and apoptosis. *Apoptosis*, *12*(11), 1927–1943. https://doi.org/10.1007/s10495-007-0102-8.

Zeng, F, Li, YC, Zhang, YK, Wang, YK, Zhou, SQ, Ma, LN, … Liu, XG. (2012). Nicotine inhibits cisplatin-induced apoptosis in NCI H446 cells. *Medical Oncology*, *29*(1), 364–373.

Zhang, D., Bordia, T., Mcgregor, M., Mcintosh, J. M., Decker, M. W., & Quik, M. (2014). ABT-089 and ABT-894 reduce levodopa-induced dyskinesias in a monkey model of Parkinson's disease. *Movement Disorders*, *29*(4), 508–517. https://doi.org/10.1002/mds.25817.

Zhang, J., Kamdar, O., Le, W., Rosen, G. D., & Upadhyay, D. (2009). Nicotine induces resistance to chemotherapy by modulating mitochondrial signaling in lung cancer. *American Journal of Respiratory Cell and Molecular Biology*, *40*(2), 135–146. https://doi.org/10.1165/rcmb.2007-0277OC.

Zhang, D., McGregor, M., Decker, M. W., & Quik, M. (2014). The α7 nicotinic receptor agonist ABT-107 decreases L-dopa-induced dyskinesias in Parkinsonian monkeys. *Journal of Pharmacology and Experimental Therapeutics*, *351*(1), 25–32. https://doi.org/10.1124/jpet.114.216283.

Zhang, S., Togo, S., Minakata, K., Gu, T., Ohashi, R., Tajima, K., Murakami, A., Iwakami, S., Zhang, J., Xie, C., & Takahashi, K. (2010). Distinct roles of cholinergic receptors in small cell lung cancer cells. *Anticancer Research*, *30*(1), 97–106.

Zhou, F. M., Liang, Y., & Dani, J. A. (2001). Endogenous nicotinic cholinergic activity regulates dopamine release in the striatum. *Nature Neuroscience*, *4*(12), 1224–1229. https://doi.org/10.1038/nn769.

CHAPTER SIX

Role of neuronal nAChRs in normal development and plasticity

6.1 Acetylcholine and receptors are present and function before and early in nervous system development

Acetylcholine is expressed in a number of species even before the appearance of the nervous system. Developing organisms use morphogens of various kinds to signal cells to differentiate along specific developmental pathways. ACh is not the only neurotransmitter to act as a chemical signal during development (Lauder & Schambra, 1999). ACh acts through mAChRs as well as nAChRs, but we will focus on the actions through nAChRs. nAChRs are detected in mammalian brains earlier than mAChRs (Lauder & Schambra, 1999). In addition to ACh, other parts of the cholinergic system are present early in development such as choline acetyltransferase and acetylcholinesterase (Layer, 1983). Evidence suggests that ACh plays a key role in regulation of morphogenic cell movements, differentiation, cell proliferation (mainly through mAChRs), and growth in diverse species such as humans, rat, mouse, worms, chick, and fruit fly (Lauder, 1993; Lauder & Schambra, 1999).

ACh and nAChRs are important in reproduction in some organisms (Wessler & Kirkpatrick, 2017). ACh and α7, α3, α5, α9, and β4 nAChR subunit proteins are present in human sperm (Kumar & Meizel, 2005; Wessler & Kirkpatrick, 2017), and there is evidence that nAChRs may be involved in sperm motility. Receptors containing the α7 subunit are involved in the acrosome reaction as well, since α–Bgt can block this reaction (Wessler & Kirkpatrick, 2017). nAChRs, ChAT, and AChE are also present in the human placenta. The functional role of this cholinergic system in the placenta is not clear, but may have a role in embryogenesis (Wessler & Kirkpatrick, 2017).

Nicotinic Acetylcholine Receptors in Health and Disease
https://doi.org/10.1016/B978-0-12-819958-9.00009-8

Cell movement during gastrulation and postgastrulation is regulated, at least in part, by ACh in sea urchin embryos (Gustafson & Toneby, 1970). Neuronal nAChRs are expressed on the cell membrane of sea urchin eggs (Ivonnet & Chambers, 1997). Disruption of cholinergic signaling through nAChRs during the midblastula stage of sea urchin development produced prominent changes in cell phenotype and overall larvae structure (Buznikov et al., 2001). A role for ACh in gastrulation of chick embryos has also been suggested (Lauder & Schambra, 1999). ACh may also play a role in palate morphogenesis acting through nAChRs but not mAChRs (Lauder, 1993). Thus, nAChRs and ACh are important for embryonic development of many tissues, not just the nervous system.

Cholinergic activation is important for development of proper brain structure. Interference with cholinergic systems during development produced structural damage with effects on behavior (Bachman et al., 1994; Hohmann et al., 1988, 1991; Slotkin, 1999). Overstimulation of cholinergic pathways at the incorrect time in development also caused developmental abnormalities (Slotkin, 2004). Studies using perfusion of nicotine into cultured embryos have demonstrated that this inappropriate cholinergic signaling can produce profound effects on neural development in rats (Roy et al., 1998). The targeting of these effects follows the concentration of nAChRs, suggesting direct action on cholinergic cell development (Slotkin, 1992). The existence of cholinergic circuits in embryonic nervous tissue suggests that ACh and nAChRs may be involved in development of the rat PNS and CNS (Lauder, 1993.

During rat embryonic development, cholinergic neurons are among the first to differentiate and thought to play a role in differentiation of other neurons. Acetylcholine acting presynaptically mediates the release of several neurotransmitters (Role & Berg, 1996) including glutamate. Thus, disruption of cholinergic signaling during development may also affect development of brain regions dependent on other neurotransmitter systems. Nicotinic cholinergic agents also have been shown to exert effects on cell growth, neurite outgrowth, and synaptogenesis (Chan & Quik, 1993; Hory-Lee & Frank, 1995; Pugh & Berg, 1994; Zheng et al., 1994). In cultured rat retinal ganglion cells, ACh inhibited neurite outgrowth (Lipton et al., 1988), while ACh acted as a chemoattractant for Xenopus spinal neurons (Zheng et al., 1994). ACh inhibits neurite outgrowth in rat hippocampus (Lauder & Schambra, 1999) and can regulate neurite outgrowth also in the snail. Activation of nAChRs in primary and immortalized hippocampal progenitor cells caused apoptosis (Berger et al., 1998), suggesting that cholinergic signaling at the wrong time in a developing animal may alter normal neural development by causing inappropriate cell death. nAChRs also regulate the survival of newborn neurons in the adult olfactory bulb (Mechawar et al., 2004). Mutation of an AChR subunit in *Caenorhabditis elegans* led to neuronal degeneration (Treinin & Chalfie, 1995). Activation of nAChRs may also prevent apoptosis by downregulating molecules involved in cell death or upregulating molecules involved in cell survival (Pugh & Margiotta, 2000). Signaling through nAChRs is present in multiple animals and cells during development.

How could activation or blockade of nAChRs affect development? Signaling through nAChRs located on neurons can alter excitability of cells and control release of various neurotransmitters in addition to ACh such as glutamate or GABA. In addition, some neuronal nAChR subtypes have a high permeability to Ca^{2+} in addition to Na^+ and K^+. Entry of Ca^{2+} through nAChRs can alter various second messenger pathways directly or through activation of voltage-dependent Ca^{2+} channels or release of Ca^{2+} from intracellular stores (Dajas-Bailador & Wonnacott, 2004). Nicotine acing through nAChRs leads to the expression of several genes including immediate early genes such as c-fos (Dunckley & Lukas, 2003; Greenberg et al., 1986), thereby potentially altering expression of downstream genes involved in differentiation, cell motility, growth, or apoptosis. nAChR activation in embryonic hippocampal cells produces cell death and activates p53 and p21 (Berger et al., 1998), which could feed into apoptotic pathways to increase caspase activity. nAChR signaling can alter developmental programs in multiple ways.

6.2 Timing and location of nAChR subtypes during development

Some of the roles described above do not require an intact nervous system or even neurons. Since ACh is acting as a signal during development, it must be acting through specific receptors. When are nAChRs first expressed and what subtypes? Zebrafish nAChR α2a, β3a, and β5b mRNAs have been detected in unfertilized embryos as part of the population of maternal mRNAs (Ackerman & Boyd, 2016). Maternal RNAs are known to have a role in early events in embryogenesis (Marlow, 2010), consistent with a role for nAChRs in early developmental events. α2a and β3a nAChR subunit RNAs are also expressed at 2h postfertilization, α4 RNA is detected at 4h postfertilization (hpf) with α6 being detected at

10 hpf, thus very early in development (Ackerman et al., 2009; Ackerman & Boyd, 2016). nAChR α4 and α6 subunit RNAs are detected in specific neural cell types (i.e., spinal neurons, neural crest, forebrain, trigeminal ganglion), as early as 24 hpf. However, it has not been shown yet that these are part of functional nAChRs at this stage. Fully assembled nAChRs have been detected as early as 2 and 5 days postfertilization (dpf) in zebrafish (Zirger et al., 2003). Zebrafish α2 subunit is expressed in a time course with a large amount of expression in early stages, but much less later, consistent with a transitory role in development (Ackerman & Boyd, 2016; Zirger et al., 2003).

Zebrafish are not the only vertebrate to express cholinergic markers early in development. At the beginning of migration, quail neural crest cells synthesize ACh and acetylcholinesterase, early expression of which may play a role in migration (Smith et al., 1979). α3, α5, β2, β4, and α7 nAChRs RNAs were detected in quail neural crest cells after 2 days in culture (Howard et al., 1995). α3 nAChR protein was detected at 7 days in culture. While nAChR RNA expression has been shown to be affected by innervation, some neural crest cells expressing nAChR RNAs and proteins may do so autonomously (Howard et al., 1995) and on cells that express a neuronal marker. Some neural crest derived cells after 7–8 days in culture expressed functional responses to applied ACh, indicating the presence of assembled and functional receptors. No synapses could be detected at up to 8 days in culture using SV2 antibodies to detect synaptic vesicles (Howard et al., 1995), indicating that innervation was not required for expression. The timing and location of the appearance of functional nAChRs before innervation may support a role in development other than mediating synaptic transmission.

nAChRs are present in early embryonic day (ED) 5 chick ciliary ganglion neurons (parasympathetic); however, expression was significantly reduced in the absence of innervation (Arenella et al., 1993). Multiple nAChR RNAs (α7, α3, α5, β2, and β4) are also detected as early as ED8 in chick ciliary ganglion (parasympathetic neurons) and continue to be expressed at ED18 (Corriveau & Berg, 1993). nAChRs also increase from ED8 to ED18 (Corriveau & Berg, 1993). Chick sympathetic ganglion neurons express α3, α4, α5, and β4 RNAs as early as ED 8 (before innervation) and are also present at ED17 (Devay et al., 1994). nAChR expression is regulated by both presynaptic and postsynaptic inputs (Boyd et al., 1988; Devay et al., 1994). α3- and α4-containing nAChRs are present in ED18 embryonic chick dorsal root ganglion (DRG) and assembled, and functional nAChRs are detected at that age (Boyd et al., 1991). Electrophysiological recordings indicated that two populations of neurons with different responses to ACh were detected. ChAT was also present, but no clusters of nAChRs suggesting a diffuse mode of transmission.

Zoli et al. (1995) demonstrated the expression during development in the rat brain and PNS of multiple (α3, α4, β2, and β4) nAChR RNAs. Each subunit RNA had specific timing and localization of expression, examined by in situ hybridization. At ED11, all four subunits were detected in mesencephalon/rhombencephalon (Zoli et al., 1995), and no expression was detected before this for any subunit in any region. Please refer to Zoli et al. (1995) for detailed expression patterns.

α3 RNA levels decreased by ED15-17, but persisted in some areas such as the medial habenula. α4 expression expands from early expression in spinal cord and rhombencephalon to widespread and stable expression throughout many CNS areas (Zoli et al., 1995). β2 nAChR RNA is also widespread and stable. β4 is more transiently expressed and has a more restricted pattern later in development. Some nAChRs are expressed early and then reduced, some seem to be replaced (α3 by α4), and some are expressed at some level throughout the times examined (Zoli et al., 1995). α3, α4, β2, and β4 nAChR RNAs are expressed at ED11 in DRG in rat, consistent with the previously observed early expression in chick ED18 DRG (Boyd et al., 1991). The RNA changes could reflect the exchange or addition of subtypes in various brain regions at specific times. Multiple subtypes are present in specific brain regions in adult brain (see elsewhere), but the expression of various nAChR RNAs early in brain development is consistent with various subtypes being present in a changing pattern, which may play a role in early developmental signaling (see below). Early nAChR expression was consistent with cessation of mitosis. GABA and glutamate receptor RNA expression appears later in development than nAChRs. This earlier nAChR expression could be independent of innervation and may be regulated by differentiation signals (Zoli et al., 1995) and may in turn regulate development.

ChAT and AChE are detected early in prenatal development. Expression of RNAs doesn't mean in every case that an assembled and functional nAChR of a specific subtype is present. However, high-affinity (10 nM) nicotine binding was detected in rat in early development (Naeff, Schlumpf, & Lichtensteiger, 1992). Not just RNA, but at least some assembled receptors are expressed as detected by nicotine binding (Naeff et al., 1992). High-affinity nAChRs include α4 and β2 subunits and at ED12, 10 nM nicotine binding was detected in rat spinal cord and brainstem, a site of early synapse

formation (Naeff et al., 1992). The pattern changed over time from early prenatal to postnatal development (Naeff et al., 1992). At ED13 expression was
detected in midbrain and at ED 14 in diencephalon and in additional, more
caudal, structures. By ED 15, nicotine binding was found in the retina and
ventral telencephalon (Naeff et al., 1992). Expression continued in the midbrain and hindbrain, but more binding was also detected each day in more
rostral regions. High-affinity nicotine binding followed a caudal to rostral progression during development (Naeff et al., 1992). Beginning of expression in a
specific region coincided with differentiation (Naeff, Schlumpf and
Lichtensteiger, 1992), similar to what was seen with RNA expression (Zoli
et al., 1995). nAChR nicotine binding was high in the subplate, suggesting
a role of this early expression in development (Naeff et al., 1992).

In summary, nAChR RNA and receptors and elements of the cholinergic system are expressed early in development, in some cases before synaptic
connections are made. There is region-specific and timing-specific expression of nAChR subtypes. This would lead to the prediction that specific
subtypes may have roles in various aspects of neural development.

6.3 The role of nAChRs in specific developmental events

Multiple normal developmental events are controlled or influenced by
nAChRs and acetylcholine. Some of these occur during early development,
and some later in adolescence or beyond. Many of these are involved in nervous system development but as seen above, other systems are affected as
well. AChRs are expressed presynaptically as well as postsynaptically and
thus can control release of numerous other neurotransmitters. Thus, acetylcholine signaling can influence development in ways other than by direct
cholinergic transmission. By influencing expression of other neurotransmitters, acetylcholine can alter circuits and development. Specific neuronal
nAChR genes/subunits have been identified as having roles in various
developmental events. Some examples of this are covered below, although
these examples are by no means exhaustive.

6.3.1 Development of retinal waves

nAChRs and ACh are present in the retina and at early stages of development. $\alpha 3$ and $\beta 4$ RNAs are detected in ED13 retina and from ED15 to postnatal day (PD) 4 $\alpha 3$, $\alpha 4$, $\beta 2$, and $\beta 4$ nAChR RNAs are all present (Zoli et al.,
1995). Epibatidine binding nAChRs (assembled heteromeric receptors) and

α–Bgt binding nAChRs are present in the rat retina from PD1 (Moretti et al., 2004). Multiple subunits and subtypes are expressed in rat, mice, and chick retina in adults as well (Feller, 2002; Moretti et al., 2004). Waves of spontaneous electrical activity in retinal ganglion cells (RGC) are driven by ACh during the first days of postnatal development, even before visual experience. The expression of nAChRs is critical for proper visual system development.

The role of β2–containing nAChRs in retinal waves and visual circuit refinement was investigated using β2 knockout (KO) mice (Burbridge et al., 2014). The mice were PD 4–7, and experiments were done in vivo in un-anesthetized mice. Mice with complete KO of β2 in all cells had reduced spontaneous RGC activity and different spatial and temporal expression, but after PD 7, the activity was closer to normal. In the KO mice, spontaneous activity in each retina was more coordinated than in WT mice (Burbridge et al., 2014). It is thought that un–coordinated activity is important for eye specific axon segregation and that this coordination in KO mice may produce the altered eye specific segregation seen in β2 KO mice (Burbridge et al., 2014).

The β2 KO also affected the spontaneous activity of one target of the RGCs, the superior colliculus (SC). Normally, the spontaneous activity of the RGCs drives activity of SC neurons. In the β2 KO, spontaneous activity was increased in SC and was expressed in a different pattern, sizes, and duration (Burbridge et al., 2014). The role of β2 nAChRs was further defined by using mice with the β2 KO confined to either the RGCs or SC. Surprising, the SC wave activity was not changed in the SC β2 KO, indicating that β2 nAChRs are not involved in SC spontaneous activity or circuit development (Burbridge et al., 2014) but must act in other cells to influence SC activity.

β2 nAChRs were knocked out in some regions of the retina (which then lacked wave activity), with β2 nAChR expression intact in the central retina (normal wave activity here). Wave activity in the SC was reduced only in areas normally innervated by areas of the retina now lacking β2 nAChRs and the RGC projections to these areas of the SC were abnormal, resulting in altered retinotopic projections. (Burbridge et al., 2014). Eye-specific segregation was also disrupted in the SC and dorsal laterial geniculate nucleus (dLGN) of retina β2 KO mice. Artificially restoring retinal waves in the mice rescued eye-specific segregation in the SC and dLGN, but not the formation of the retinotopic map (Burbridge et al., 2014). These studies clearly showed a role for nAChRs containing the β2 subunit as critical for proper early visual system development. β2 subunits are often paired with α4 subunits to form functional nAChRs. nAChR antagonists D-tubucurarine and DHβE blocked retinal waves (Bansal et al., 2000). α3 nAChR RNAs and proteins have been detected in rat and chick retina. α3-containing nAChRs are also present in PDO–PD7 mouse retina. KO mice for α3 had alterations in retinal wave activity (Bansal et al., 2000). Pharmacological blockade of α7 containing nAChRs minimally affected the retinal waves. Conotoxins affecting the functions of α3β2 and α3β4 nAChRs affected retinal wave frequency. The complete subunit combinations and subtypes are yet to be determined, but multiple nAChR subtypes in addition to those containing β2 are important for normal visual system development.

6.3.2 Development of inhibitory GABAA-mediated signaling

nAChR activation also affects development by controlling the conversion of GABA-A receptor-mediated currents from excitatory to inhibitory. During early development, the chloride gradient is controlled by the embryonic chloride transporter NKCC1 (Liu et al., 2006). GABAergic currents become inhibitory due to the timed expression of the KCC2 chloride transporter, which reverses the chloride gradient. The excitatory period is important for development, but the switch to an inhibitory role must be made at the proper time to control additional development events. The role of nAChRs in this conversion was investigated in chick ciliary ganglion (CG) neurons, which receive cholinergic innervation and have outputs to smooth and striated muscle in the eye (Liu et al., 2006). In ED9 CG neurons GABAergic depolarizing responses were detected. By ED 14, the depolarizing effect of GABA was lost, due to expression of KCC2, and now an inhibitory response was observed. However, this loss of the depolarizing effect was blocked if nAChR α7 homomeric or heteromeric nAChRs were antagonized. When nAChRs were pharmacologically blocked, more of the embryonic NKCC1 was expressed and much less of KCC2. Thus, nAChR function was required for conversion of the chloride gradient through expression of KCC2 vs. embryonic NKCC1 (Liu et al., 2007). A similar block of chloride gradient conversion occurred in dissociated embryonic chick spinal cord neurons as well (Liu et al., 2007) when treated with the nAChR antagonists MLA and DHβE.

There is also a conversion that occurs early postnatally in hippocampal neurons (Liu et al., 2006, 2007). KO mice for α7 retained the excitatory response in dissociated hippocampal neurons, also showing a role for a specific nAChR subtypes in this transition (Liu et al., 2006). This was also due to maintenance of higher levels of the embryonic NKCC1. α7 and GABA receptors are clustered in dendrites of early hippocampal interneurons (Liu et al., 2007), and both can be innervated in culture by septal neurons (Liu et al., 2007). In hippocampal slices from PD 9–14 day animals, α7 nAChR activation also inhibited GABA-mediated IPSCs in stratum radiatum neurons. (Liu et al., 2007).

The switch to an inhibitory GABAergic role is important for a number of developmental events, just as the excitatory role is important for other developmental events. Neuronal morphology and synaptic contacts were determined in CG neurons by the interplay of inhibitory GABA signaling and nAChR signaling. GABA and nAChR receptors are expressed on cell bodies of CG neurons (Liu et al., 2007), and activation of nAChRs reduced GABA-induced current. Inhibitory GABA signaling promoted formation of unipolar neurons, without GABA, they become multipolar. The morphological change produced by GABA required activity of homomeric and heteromeric nAChRs. Reduction in number of SV2 (a presynaptic marker) puncta was also dependent on inhibitory GABAergic transmission and nAChR activation (Liu et al., 2006). Studies in CG neurons showed that synergism between GABA and nAChR signaling was needed for the change in SV2 puncta (Liu et al., 2006).

The inhibitory GABAergic signaling is required for gene regulation mediated by nAChR activation of CREB in CG. nAChR activation can directly affect transcription, but only if there is no large inward calcium current mediated by VGCC present (Chang & Berg, 2001). Thus, inhibitory GABA signaling is needed to suppress VGCC activation, and the developmental switch in chloride gradient allows for the nAChRs to play a new role in gene activation during development.

6.3.3 Myelination and differentiation of oligodendrocytes

The anti-inflammatory effects of nicotine may help myelination by reducing damage that occurs during conditions manifesting an inflammatory autoimmune response such as MS. But what is the role of nAChRs in development/production of oligodendrocytes or myelination? Acetylcholine has been thought to play a role in myelination, and ChAT and AChE are detected in the white matter (Fields et al., 2017). nAChRs and mAChRs are also expressed in myelinating glia. The role of nAChRs is supported by studies showing that prenatal nicotine exposure in rats reduced myelin-associated genes in PD20–21 animals. In adolescent (PD35–36) males, nicotine treatment upregulated these genes, but in females, expression was decreased (Cao, Dwyer, et al., 2013). In adults, expression was normalized in both males and females. Oligodendrocyte numbers were also increased in the prefrontal cortex of adolescent male rats after prenatal nicotine exposure. The number of oligodendrocytes was not increased in caudate putamen. No increase was noted in either region for females (Cao, Wang, et al., 2013).

O2A/OPCs are oligodendrocyte precursor cells. A2B5+ rat O2A/OPC cells in culture or a rat primary mixed culture express α4, β2, α3, α5, β4 nAChR RNAs and a small amount of α7, mostly in dying cells (Rogers et al., 2001). The corresponding proteins were also detected by immunochemistry and at various levels. The expression was mostly confined to the OPCs, but not oligodendrocytes or astrocytes. Nicotine increased the calcium response in about 2/3 of the precursor cells, which was blocked by DHβE, an antagonist of heteromeric, (especially α4β2) nAChRs and by an L-type calcium channel blocker (Rogers et al., 2001). Mecamylamine also inhibited the response to nicotine, but MLA did not. In about a third of the cells, the calcium response oscillated, and this effect was also inhibited by DHβE and the L-type calcium channel blocker nifedipine. The oscillatory response was also not blocked by MLA. Rogers et al. (2001) point out that nicotine doesn't stimulate differentiation or proliferation of O2A cells. However, given the calcium oscillations produced, the effects of nAChR activation on O2A cells may be more subtle.

Acetylcholine is broken down by AChE. Drugs such as donepezil that inhibit AChE activity can increase cholinergic signaling. Donepezil induces differentiation of neural stem cell-derived oligodendrocyte progenitor cells (NSC-OPC) isolated from ED14.5 mouse cortex cells in culture. Donepezil didn't increase apoptosis or proliferation. This activity was blocked by mecamylamine (nAChR antagonist), but not scopolamine, an mAChR antagonist (Imamura et al., 2015). This indicates a possible role for nAChRs in OPC differentiation. However, antagonists of heteromeric nAChRs α4β2*(DHβE) and α7 (MLA) didn't block the effects of donepezil. In addition, other AChE inhibitors didn't stimulate OPC differentiation (Imamura et al., 2015). Donepezil did increase expression of myelin proteins (such as myelin basic protein) and RNAs involved in transcriptional control of myelin proteins (Imamura et al., 2015). Mecamylaine had no effect on differentiation in cells not treated with donepezil. These results indicate a role of some kind for nAChRs in differentiation of OPCs, but the mechanism is unclear. A general increase in ACh didn't seem to be required as other AChE inhibitors didn't work. The nAChR subtype is also unclear. Donepezil can interact directly with nAChRs and inhibit function. Donepezil also upregulates α7nAChRs in rat cortical neurons (Takada–Takatori et al., 2008) and may upregulate other subtypes and thus produce effects on differentiation. Perhaps nAChRs at some times in normal development are involved in maintaining OPCs in an undifferentiated state. Donepezil, by interacting in some way to disrupt this nAChR control, produced unscheduled differentiation. Support for this comes from work showing that donepezil inhibits the function of α3β2 and α7nAChRs expressed in Xenopus oocytes (Shih et al., 2020).

Cholinergic signaling in ED12.5 mouse spinal cord is important for oligodendrocyte progenitor cell (OPC) development (Osterstock et al., 2018). The OPCs don't need to express nAChRs for cholinergic signaling to have an effect in their development. During embryonic development of OPCs, functional glutamatergic and GABAergic synapses are formed on the OPCs. While application of acetylcholine didn't affect the current in OPCs, acetylcholine and nicotine did increase the number of spontaneous presynaptic GABA and glutamate-mediated currents (Osterstock et al., 2018). This effect of ACh was mediated by nAChRs on the presynaptic neurons since two nAChR antagonists, mecamylamine and d-tubocurarine, blocked the increase in frequency (Osterstock et al., 2018), and tetrodotoxin (TTX) didn't block the increase in frequency produced by ACh. The amplitude of GABAergic signaling was also reduced by nAChR antagonists, but the amplitude of glutamate-mediated signaling was not. The effect of nicotine during this time also increased embryonic proliferation of OPCs (Osterstock et al., 2018). Thus, activation of presynaptic nAChRs plays an important role in OPC development.

Glial cells positive for the marker NG2 are similar to O2A oligodendrocyte precursors and are present in hippocampal slices (Velez–Fort et al., 2009). It was shown that ACh and 1,1-dimethyl-4-phenylpiperazinium iodide (DMPP), another cholinergic agonist, induce inward currents in these NG2+ cells in hippocampal CA1 slices from PD 6–14 and mecamylamine blocked these responses. Analysis of current-voltage (IV) plots also supported that the current response seen was due to nAChRs (Velez–Fort et al., 2009). However, the expression of the nAChRs appeared transient, since they could not be detected electrophysiologically in PD46 and PD 48 preparations. Pharmacological studies showed that α7 nAChRs are present on the NG2+ cells. Choline (α7 agonist) activated an inward current that was blocked by MLA. A positive allosteric modulator of α7nAChRs (PNU-120596) increased the response to choline. DMPP activation was not blocked by DHβE, but was by MLA, supporting the identity of α7 nAChRs on the NG2+ cells (Velez-Fort et al., 2009) during a time of proliferation. α7 nAChRs are highly permeable to Ca^{2+}, and the increase in Ca^{2+} seen after α7 nAChRs activation could send an important to signal to NG2+ cells at a specific time in development given the transient expression detected the hippocampal preparations (Velez-Fort et al., 2009). Concentrations of nicotine comparable with that in the brains of smokers reduced the NG2+ cells' response to choline, indicating that smoking may inhibit signaling to NG2+ cells during a critical time (Velez-Fort et al., 2009).

6.3.4 Cortical development

α5-containing nAChRs play a role in nicotinic signaling and pyramidal neuron dendritic morphology in developing mouse medial prefrontal cortex (mPFC) (Bailey et al., 2012). Cholinergic input has been shown in many brain regions to be important for driving development, and a peak of activity normally occurs in the mPFC at 3 weeks. α5 KO mice showed reduced currents in response to ACh when tested in acute slices from 2, 3, and 4-week postnatal mice (Bailey et al., 2012). This indicates a role for α5 nAChRs in this peak. Previous studies showed that nAChR activation can change neurite structure (Lipton et al., 1988; Pugh & Berg, 1994). α5 KO mice also demonstrated changes in mPFC layer VI pyramidal neuron dendritic structure. In wild-type mice at 3 weeks, the apical dendrite of most pyramidal neurons terminates in layer I, while at PD77 slightly more than half have retracted the apical dendrite and now terminate below layer I (Bailey et al., 2012). However, in the mice lacking α5, at 3 weeks, most terminate in layer I as in wild-type mice, but in adults almost all (12/13) continue to terminate in layer 1. This supports a role for α5 nAChRs in the developmental retraction of the Layer VI pyramidal neurons apical dendrite. The number of basal dendrites was also lower in wild-type neurons than for α5 KO at both 3 weeks and in adult (Bailey et al., 2012). The α5 KO mouse didn't have changes in the structure of the mPFC or change the pattern of α4 nAChR expression (Bailey et al., 2012). α5 is expressed in combination with α3 and β4 subunits in some brain regions and with α4β2 in others including the mPFC. α4 and β2 RNA expression was not changed during development in rat (Zhang et al., 1998), but the expression of α5 RNA increases in layer VI during the first few weeks (Bailey et al., 2012). This could result in the appearance during this time of more $\alpha4\beta2\alpha5$ nAChRs. $\alpha4\beta2$ nAChRs containing $\alpha5$ are more permeable to calcium and have increased cation flow compared with $\alpha4\beta2$ nAChRs (Bailey et al., 2012). The addition of more $\alpha5$-containing nAChRs could mediate the developmental changes observed in the mPFC.

6.3.5 Glutamatergic synapse development

Elegant studies by Lozada et al. (2012) showed that $\alpha7$ nAChRs were required during development for the normal production and function of glutamatergic synapses in the cortex and hippocampus. $\alpha7$ KO mice had reduced glutamatergic synapses in CA1 pyramidal neurons examined in slices from PD12 mice. In layer 5/6 neurons of the visual cortex, a reduction was also seen (Lozada et al., 2012). $\beta2$ KO mice didn't display this change, indicating a specific role for $\alpha7$ nAChRs. In slices from PD12 and PD25 hippocampus, the number of mEPSCs was reduced in the $\alpha7$ KO mice, but the amplitude was not changed. This was also shown in wild-type mice using MLA ($\alpha7$ antagonist) to produce the same effect. Again $\beta2$ was not involved since DHβE ($\beta2^*$ nAChR antagonist) did not produce this effect, nor was it seen in $\beta2$ KO mice.

Nicotine (to mimic effect of developmental cholinergic stimulation) was used to treat cultured hippocampal neurons, exposing them to 1 µM for 1 week. This produced an increase in glutamatergic synapses. Nicotine was acting at $\alpha7$ nAChRs since $\alpha7$ antagonists MLA or α-Bgt blocked this increase, while the $\beta2$ antagonist DHβE did not (Lozada et al., 2012). This was specific to glutamatergic synapses, since the nicotine treatment didn't increase GABAergic synapses. By measuring mEPSCs, it was shown that the increased glutamatergic synapses were functional, again specific for $\alpha7$ nAChRs. This increase of glutamatergic synapses was also seen after a 7-day nicotine treatment in slices from PD4 mice (Lozada et al., 2012). Using RNAi to knock down $\alpha7$ nAChR expression in PD1-PD2 mice, it was shown that the effects of $\alpha7$ were due to expression on postsynaptic neurons, a cell autonomous effect. The RNAi knockdown in both CA1 and layer 5/6 of the visual cortex reduced the number of glutamatergic synapses. $\beta2$ nAChR-targeted RNAi did not reduce the number of glutamatergic synapses (Lozada et al., 2012). These developmental changes in glutamatergic synapse function and number mediated by $\alpha7$ nAChRs, but with no changes in GABAergic synapses, showed that timely $\alpha7$ nAChR expression can mediate a change in the excitatory/inhibitory ratio. Stimulation of postsynaptic $\alpha7$

nAChRs may mediate changes in the postsynaptic cells through gene expression or altering signal transduction pathways (Lozada et al., 2012).

Activation of nAChRs early in postnatal development can cause "silent" glutamatergic synapses to become active (Maggi et al., 2003), perhaps playing an important role in development. Nicotine used at a level transiently achieved in some smokers (1 μm) was applied to PD1–PD5 hippocampal slices and shown to produce mini AMPA-mediated EPSCs in formerly silent neurons (Maggi et al., 2003). Nonsilent cells were also more active after nicotine treatment. Nicotine worked by increasing the activity of "presynaptically silent" neurons, causing an increased probability of glutamate release. α7 nAChRs were implicated in these effects since AR-17779 (selective activator of α7) converted silent synapses to active ones and MLA and α-Bgt blocked this effect. Not only applied nicotine, but endogenous ACh also converted silent synapses to functional ones (Maggi et al., 2003), and this effect was also blocked by DHβE at a level that can block multiple nAChR subtypes. Postsynaptic changes in intracellular Ca^{2+} were not required for this effect. This is consistent with a role of presynaptic α7 nAChRs that have a high Ca^{2+} permeability, in increasing glutamate release in presynaptically silent synapses. The function of nAChRs in this process may be required for normal hippocampal synaptic development.

6.4 Additional developmental effects

6.4.1 Cholinergic role in motor circuit development

Spontaneous neural activity is important for development of proper motor circuits. Cholinergic activity drives these early spontaneous rhythms (Myers et al., 2005). Mouse mutants lacking choline acetyltransferase (ChAT) have reduced spontaneous rhythms from ED12.5 to ED18.5. Blocking nAChRs with antagonists in ED12.5–14.5 wild-type mice embryos blocked spontaneous activity, but after ED 15.5 when the locomotor network is established, nAChR antagonists didn't reduce activity, showing a need for nAChR function during a specific time window. nAChRs are also crucial to proper motor neuron path finding in the zebrafish (Svoboda et al., 2002). Exposure of embryonic zebrafish to nicotine disrupts normal axon guidance.

6.4.2 Tooth development

α7 is expressed in developing mesenchyme at around 13 days of embryonic development, and expression is lost at around ED16 (Rogers & Gahring, 2012). Selective ablation of α7 nAChR expressing cells showed that these cells instructed development of dental mesenchyme and tooth patterning. At ED18, ameloblasts (contribute to enamel production) express α7 nAChRs. These had a role in tooth development in that α7 KO mice had normal tooth size, but reduced teeth enamel volume (Rogers & Gahring, 2012).

6.4.3 Adrenal gland development

α7 is detected in mouse rhombomeres 3 and 5 at ED 9 and is seen in embryonic adrenal gland in developing chromaffin cells at ED12.5 (Gahring et al., 2014). This is at a time before catecholaminergic features are detected such as tyrosine hydroxylase (TH) or dopamine beta hydroxylase (DBH) expression. As development proceeds, expression of α7 becomes more restricted. In adults, α7 is expressed in noradrenergic chromaffin cells (those expressing TH and DBH, but not phenylethanolamine N-methyltransferase (PNMT) (Gahring et al., 2014). The α7 nAChR-expressing chromaffin cells at ED 12.5 are associated with early neural processes. The early expression of α7 is consistent with a developmental role. Indeed, neonatal nicotine exposure alters expression of TH, DBH, but not PNMT, with DBH expression elevated into adulthood, consistent with a role of α7nAChRs in this effect (Gahring et al., 2014; Rosenthal & Slotkin, 1977).

6.4.4 Cochlear inner hair cell (IHC) development

α9α10 nAChRs are present at birth in rats in inner hair cells (IHC) of the cochlea (Johnson et al., 2013). ACh released from fibers of the superior olivary complex is important in shaping the development of the proper response from inner hair cells. α9 KO mice were used to show that α9 expression was required for functional maturation of inner hair cells (Johnson et al., 2013).

References

Ackerman, K. M., & Boyd, R. T. (2016). Analysis of nicotinic acetylcholine receptor (nAChR) expression in zebrafish (*Danio rerio*) by in situ hybridization and PCR. *Nicotinic Acetylcholine Receptor Technologies*, 1–31.

Ackerman, K. M., Nakkula, R., Zirger, J. M., Beattie, C. E., & Boyd, R. T. (2009). Cloning and spatiotemporal expression of zebrafish neuronal nicotinic acetylcholine receptor alpha 6 and alph4 subunit RNAs. *Developmental Dynamics*, *238*(4), 980–992.

Arenella, L. S., Oliva, J. M., & Jacob, M. H. (1993). Reduced levels of acetylcholine receptor expression in chick ciliary ganglion neurons developing in the absence of innervation. *The Journal of Neuroscience*, *13*(10), 4525–4537.

Bachman, E. S., Berger-Sweeney, J., Coyle, J. T., & Hohmann, C. F. (1994). Developmental regulation of adult cortical morphology and behavior: An animal model for mental retardation. *International Journal of Developmental Neuroscience*, *12*, 239–253.

Bailey, C. D. C., Alves, N. C., Nashmi, R., De Biasi, M., & Lambe, E. K. (2012). Nicotinic a5 subunits drive developmental changes in the activation and morphology of prefrontal cortex layer VI neurons. *Biological Psychiatry*, *71*, 120–128.

Bansal, A., Singer, J. H., Hwang, B. J., Xu, W., Beaudet, A., & Feller, M. B. (2000). Mice lacking specific nicotinic acetylcholine receptor subunits exhibit dramatically altered spontaneous activity patterns and reveal a limited role for retinal waves in forming ON and OFF circuits in the inner retina. *The Journal of Neuroscience*, *20*(20), 7672–7681. PubMed PMID: 11027228.

Berger, F., Gage, F. H., & Vijayaraghavan, S. (1998). Nicotinic receptor-induced apoptotic cell death of hippocampal progenitor cells. *The Journal of Neuroscience*, *18*, 6871–6881.

Boyd, R. T., Jacob, M. H., Couterier, S., Ballivet, M., & Berk, D. K. (1988). Expression and regulation of neuronal acetylcholine receptor mRNA in chick ciliary ganglia. *Neuron*, *6*, 495–502.

Boyd, R. T., Jacob, M. H., McEachern, A. E., Caron, S., & Berg, D. K. (1991). Nicotinic acetylcholine receptor mRNA in dorsal root ganglion neurons. *Journal of Neurobiology*, *22*(1), 1–14.

Burbridge, T. J., Xu, H.-P., Ackman, J. B., Ge, X., Zhang, Y., Ye, M.-J., Zhou, Z. J., Xu, J., Contractor, A., & Crair, M. C. (2014). Visual circuit development requires patterned activity mediated by retinal acetylcholine receptors. *Neuron*, *84*(5), 1049–1064.

Buznikov, G. A., Nikitina, L. A., Bezuglov, V. V., Lauder, J. M., Padilla, S., & Slotkin, T. A. (2001). An invertebrate model of the developmental neurotoxicity of insecticides: Effects of chlorpyrifos and dieldrin in sea urchin embryos and larvae. *Environmental Health Perspectives*, *109*, 651–661.

Cao, J., Dwyer, J. B., Gautier, N. M., Leslie, F. M., & Li, M. D. (2013). Central myelin gene expression during postnatal development in rats exposed to nicotine gestationally. *Neuroscience Letters*, *553*, 115–120.

Cao, J., Wang, J., Dwyer, J. B., Gautier, N. M., Wang, S., Leslie, F. M., & Li, M. D. (2013). Gestational nicotine exposure modifies myelin gene expression in the brains of adolescent rats with sex differences. *Translational Psychiatry*, *3*, e247.

Chan, J., & Quik, M. (1993). A role for the neuronal nicotinic α-bungarotoxin receptor in neurite outgrowth in PC12 cells. *Neuroscience*, *56*, 441–451.

Chang, K. T., & Berg, D. K. (2001). Voltage-gated channels block nicotinic regulation of CREB phosphorylation and gene expression. *Neuron*, *32*, 855–865.

Corriveau, R. A., & Berg, D. K. (1993). Coexpression of multiple acetylcholine receptor genes in neurons: Quantification of transcripts during development. *The Journal of Neuroscience*, *13*(6), 2662–2671.

Dajas-Bailador, F., & Wonnacott, S. (2004). Nicotinic acetylcholine receptors and the regulation of neuronal signaling. *Trends in Pharmacological Sciences*, *25*(6), 317–324.

Devay, P., Qu, X., & Role, L. (1994). Regulation of nAChR subunit gene expression relative to the development of pre-and postsynaptic projections of chick sympathetic neurons. *Developmental Biology*, *162*(1), 56–70.

Dunckley, T., & Lukas, R. J. (2003). Nicotinc modulates the expression of a diverse set of genes in the neuronal SH-SY5Y cell line. *The Journal of Biological Chemistry*, *278*(18), 15633–15640.

Feller, M. B. (2002). The role of nAChR-mediated spontaneous retinal activity in visual system development. *Journal of Neurobiology*, *53*, 556–567.

Fields, R. D., Dutta, D. J., Belgrad, J., & Robnett, M. (2017). Cholinergic signaling in myelination. *Glia*, *65*, 687–698.

Gahring, L. C., Myers, E., Palumbos, S., & Rogers, S. W. (2014). Nicotinic receptor alpha 7 expression during mouse adrenal gland development. *PLoS ONE*, *9*(8), e103861.

Greenberg, M. E., Ziff, E. B., & Greene, L. A. (1986). Stimulation of neuronal acetylcholine receptors induces rapid gene transcription. *Science*, *234*, 80–83.

Gustafson, T., & Toneby, M. (1970). On the role of serotonin and acetylcholine in sea urchin morphogenesis. *Experimental Cell Research*, *62*, 102–117.

Hohmann, C. F., Brooks, A. R., & Coyle, J. T. (1988). Neonatal lesions of the basal forebrain cholinergic neurons result in abnormal cortical development. *Developmental Brain Research*, *42*, 253–264.

Hohmann, C. F., Wilson, L., & Coyle, J. T. (1991). Efferent and afferent connections of mouse sensory-motor cortex following cholinergic deafferentation at birth. *Cerebral Cortex*, *1*, 1158–1172.

Hory-Lee, F., & Frank, E. (1995). The nicotinic blocking agents d-tubocurarine and α-bungarotoxin save motoneurons from naturally occurring death in the absence of neuromuscular blockade. *The Journal of Neuroscience*, *15*, 6453–6460.

Howard, M. E., Gershon, M. D., & Margiotta, J. F. (1995). Expression of nicotinic acetylcholine receptors and subunit RNA transcripts in cultures of neural crest cells. *Developmental Biology*, *170*, 479–495.

Imamura, O., Arai, M., Dateki, M., Ogata, T., Uchida, R., Tomoda, H., & Takishima, K. (2015). Nicotinic acetylcholine receptors mediate donepezil-induced oligodendrocyte differentiation. *Journal of Neurochemistry*, *135*, 1086–1098.

Ivonnet, P. I., & Chambers, E. L. (1997). Nicotinic acetylcholine receptors of the neuronal type occur in the plasma membrane of sea urchin eggs. *Zygote*, *5*, 277–287.

Johnson, S. L., Wedemeyer, C., Vetter, D. E., Adachi, R., Holley, M. C., Elgoyhen, A. B., & Marcotti, W. (2013). Cholinergic efferent synaptic transmission regulates the maturation of auditory hair cell ribbon synapses. *Open Biology*, *3*, 130163. https://doi.org/10.1098/rsob.130163.

Kumar, P., & Meizel, S. (2005). Nicotinic acetylcholine receptor subunits and associated proteins in human sperm. *The Journal of Biological Chemistry*, *280*, 25928–25935.

Lauder, J. M. (1993). Neurotransmitters as growth regulatory signals: Role of receptors and second messengers. *Trends in Neurosciences*, *16*, 233–240.

Lauder, J. M., & Schambra, U. B. (1999). Morphogenetic roles of acetylcholine. *Environmental Health Perspectives*, *107*(Suppl 1), 65–69.

Layer, P. G. (1983). Comparative localization of acetylcholinesterase and pseudocholinesterase during morphogenesis of the chicken brain. *Proceedings of the National Academy of Sciences of the United States of America*, *80*(20), 6413–6417.

Lipton, S. A., Frosch, M. P., Phillips, M. D., Tauck, D. L., & Aizenman, E. (1988). Nicotinic antagonists enhance process outgrowth by rat retinal ganglion cells in culture. *Science*, *239*, 1293–1296.

Liu, Z., Neff, R. A., & Berg, D. K. (2006). Sequential interplay of nicotinic and GABAergic signaling guides neuronal development. *Science*, *314*, 1610–1613.

Liu, Z., Zhang, J., & Berg, D. K. (2007). Role of endogenous nicotinic signaling in guiding neuronal development. *Biochemical Pharmacology*, *74*(8), 1112–1119.

Lozada, A. F., Wang, X., Gounko, N. V., Massey, K. A., Duan, J., Liu, Z., & Berg, D. K. (2012). Glutamatergic synapse formation is promoted by α7-containing nicotinic acetylcholine receptors. *The Journal of Neuroscience*, *32*(22), 7651–7661.

Maggi, L., LeMagueresse, C., Changeux, J.-P., & Cherubini, E. (2003). Nicotine activates immature "silent" connections in the developing hippocampus. *Proceedings of the National Academy of Sciences*, *100*(4), 2059–2064.

Marlow, F. L. (2010). Maternal control of development in vertebrates. *Colloquium Series on Developmental Biology*, *1*(1), 1–196. https://doi.org/10.4199/C00023ED1V01Y201012DEB005.

Mechawar, N., Saghatelyan, A., Grailhe, R., Scoriels, L., Gheusi, G., Gabellec, M.-M., Lledo, P.-M., & Changeux, J. P. (2004). Nicotinic receptors regulate the survival of newborn neurons in the adult olfactory bulb. *Proceedings of the National Academy of Sciences of the United States of America*, *101*, 9822–9826.

Moretti, M., Vailati, S., Zoli, M., Lippi, G., Riganti, L., Longhi, R., Viegi, A., Clementi, F., & Gotti, C. (2004). Nicotinic acetylcholine receptor subtypes expression during rat retina development and their regulation by visual experience. *Molecular Pharmacology*, *66*, 85–96.

Myers, C. P., Lewcock, J. W., Hanson, M. G., Gosgnach, S., Almone, J. B., Gage, F. H., Lee, K.-F., Landmesser, L. T., & Pfaff, S. L. (2005). Cholinergic input is required during embryonic development to mediate proper assembly of spinal locomotor circuits. *Neuron*, *46*, 37–49.

Naeff, B., Schlumpf, M., & Lichtensteiger, W. (1992). Pre- and postnatal development of high-affinity [3H] nicotine binding sites in rat brain regions: An autoradiographic study. *Developmental Brain Research*, *68*, 163–174.

Osterstock, G., Le Bras, B., Arulkandarajah, K. H., Le Corronc, H., Czarnecki, A., Mouffle, C., Bullier, E., Legendre, P., & Mangin, J.-M. (2018). Axoglial synapses are formed onto pioneer oligodendrocyte precursor cells at the onset of spinal cord gliogenesis. *Glia*, *66*, 1678–1694.

Pugh, P. C., & Berg, D. K. (1994). Neuronal acetylcholine receptors that bind α-bungarotoxin mediate neurite retraction in a calcium-dependent manner. *The Journal of Neuroscience*, *14*, 889–896.

Pugh, P. C., & Margiotta, J. F. (2000). Nicotinic acetylcholine receptor agonists promote survival and reduce apoptosis of chick ciliary ganglion neurons. *Molecular and Cellular Neuroscience*, *15*, 113–122.

Rogers, S. W., & Gahring, L. C. (2012). Nicotinic receptor α7 expression during tooth morphgenesis reveals functional pleiotropy. *PLoS ONE*, *7*(5), e36467.

Rogers, S. W., Gregori, N. Z., Carlson, N., Gahring, L. C., & Noble, M. (2001). Neuronal nicotinic acetylcholine receptor expression by O2A/oligodendrocyte progenitor cells. *Glia*, *33*, 306–313.

Role, L. W., & Berg, D. K. (1996). Nicotinic receptors in the development and modulation of CNS synapses. *Neuron*, *16*, 1077–1085.

Rosenthal, R. N., & Slotkin, T. A. (1977). Development of nicotinic responses in the rat adrenal medulla and long-term effects of neonatal nicotine administration. *British Journal of Pharmacology*, *60*, 59–64.

Roy, T. S., Andrews, J. E., Seidler, F. J., & Slotkin, T. A. (1998). Nicotine evokes cell death in embryonic rat brain during neurulation. *The Journal of Pharmacology and Experimental Therapeutics*, *287*, 1136–1144.

Shih, C.-C., Chen, P.-Y., Chen, M.-F., & Lee, T. J. F. (2020). Differential blockade by huperzine A and donepezil of sympathetic nicotinic acetylcholine receptor-mediated nitrergic neurogenic dilations in porcine basilar arteries. *European Journal of Pharmacology*, *868*, 172851.

Slotkin, T. A. (1992). Prenatal exposure to nicotine: What can we learn from animal models? In I. S. Zagon, & T. A. Slotkin (Eds.), *Maternal substance abuse and the developing nervous system* (pp. 97–124). San Diego, CA: Academic Press.

Slotkin, T. A. (1999). Developmental cholinotoxicants: Nicotine and chlorpyrifos. *Environmental Health Perspectives*, *107*, 71–80.

Slotkin, T. A. (2004). Cholinergic systems in brain development and disruption by neurotoxicants: Nicotine, environmental tobacco smoke, organophosphates. *Toxicology and Applied Pharmacology*, *198*, 132–151.

Smith, J., Fauquet, M., Ziller, C., & LeDouarin, N. M. (1979). Acetylcholine synthesis by mesencephalic neural crest cells in the process of migration in vivo. *Nature*, *282*, 853–855.

Svoboda, K. R., Vijayaraghavan, S., & Tanguay, R. L. (2002). Nicotinic receptors mediate changes in spinal motoneuron development and axonal pathfinding in embryonic zebrafish exposed to nicotine. *The Journal of Neuroscience*, *22*, 10731–10741.

Takada-Takatori, Y., Kume, T., Ohgi, Y., Fujii, T., Niidome, T., Sugimoto, H., & Akaike, A. (2008). Mechanisms of α7-nicotinic receptor up-regulation and sensitization to donepezil induced by chronic donepezil treatment. *European Journal of Pharmacology*, *590*, 150–156.

Treinin, M., & Chalfie, M. (1995). A mutated acetylcholine receptor subunit causes neuronal degeneration in *C. elegans*. *Neuron*, *14*, 871–877.

Velez-Fort, M., Audinat, E., & Angulo, M. C. (2009). Functional alpha 7-containing nicotinic receptors of NG2-expressing cells in the hippocampus. *Glia*, *57*, 1104–1114.

Wessler, I. K., & Kirkpatrick, C. J. (2017). Non-neuronal acetylcholine involved in reproduction in mammals and honeybees. *Journal of Neurochemistry*, *142*(suppl.2), 144–150.

Zhang, X., Liu, C., Miao, H., Gong, Z. H., & Nordberg, A. (1998). Postnatal changes of nicotinic acetylcholine receptor αα2, α3, α4, 7 and β2 subunits genes expression in rat brain. *International Journal of Developmental Neuroscience*, *16*, 507–518.

Zheng, J. Q., Felder, M., Connor, J. A., & Poo, M. M. (1994). Turning of nerve growth cones induced by neurotransmitters. *Nature*, *368*(6467), 140–144.

Zirger, J. M., Beattie, C. E., McKay, D. B., & Boyd, R. T. (2003). Cloning and expression of zebrafish neuronal nicotinic acetylcholine receptors. *Gene Expression Patterns*, *3*(6), 747–754.

Zoli, M., Le Novere, N., Hill, J. A., & Changeux, J.-P. (1995). Developmental regulation of nicotinic ACh receptor subunit mRNAs in the rat central and peripheral nervous system. *The Journal of Neuroscience*, *15*(3), 1912–1939.

CHAPTER SEVEN

Developmental, epigenetic, and possible transgenerational effects of smoking and vaping

7.1 Effects of maternal nicotine exposure on various organ systems

Although the percentage of smokers in the United States has decreased over the last few years, unfortunately millions still do. Approximately 13%–24% of woman smoke, up to 12% while pregnant (Holbrook, 2016). Many women continue or resume smoking after birth, continuing to expose the baby to second-hand smoke. Nicotine can reach levels in the fetus 15% higher than that in the mother. In humans, it appears to be more harmful to the fetus in the third trimester (Holbrook, 2016). This corresponds to the early postnatal period in animals. This chapter will describe work from both human and animal studies. Many studies in humans and animals have shown that fetal exposure to nicotine can produce long-term effects in offspring, which last into adulthood.

Smoking leads to a higher level of fetal death, small birth weight, and premature births (Holbrook, 2016). Smoking can restrict blood flow and thus oxygen to the fetus; smoke has numerous compounds in addition to nicotine. Low birth weight may be due to the effects of nicotine on placental nAChRs (Greene & Pisano, 2019). Smokers also have increased carbon monoxide levels that can affect the fetus. Fetuses are exposed to the large amount of chemicals present in smoke, in addition to nicotine. However, many of the effects may be attributable to nicotine alone. Maternal smoking during pregnancy is a major risk factor for Sudden Infant Death Syndrome (SIDS) (Bruin et al., 2010; Holbrook, 2016). Animal studies show that nicotine exposure (not smoke) produce mice that have reduced catecholamine expression that may affect the stress response and produce a reduced response to hypoxia (Cohen et al., 2005). These mice also showed unstable breathing and small birth weight. KO mice for the nAChR β2 subunit showed the same deficits, indicating that nicotine may be acting to desensitize or somehow decrease the function of β2-containing nAChRs to mediate these effects (Cohen et al., 2005). Newborns from a smoking mother exhibit increased irritability, excitability, and withdrawal during the first 5 days after birth (England et al., 2017).

Effects of maternal smoking may be long-lasting. Orofacial clefts have been correlated with human maternal smoking (Holbrook, 2016). Children of mothers who smoked exhibit increased obesity that may extend into adulthood and an increased incidence of type 2 diabetes (Holbrook, 2016). Children also have reduced lung function and are more likely to have respiratory infections after maternal smoking and asthma. Children greater than age 16 were reported to have increased wheezing (Blacquiere et al., 2009). Goblet cell hypertrophy was seen in animals exposed to prenatal smoke (Blacquiere et al., 2009). nAChRs are expressed in the lung. Nicotine (in the absence of other components of cigarette smoke) exposure in utero led to reduced lung weight and volume (Sekhon et al., 2001) in monkeys with a resulting reduced tidal expiration and increased pulmonary resistance. Smokeless tobacco (nicotine without combustion products) use in pregnant women is correlated with still births and preterm births (England et al., 2017). Cigarettes produce low birth weight, but not the use of nicotine alone indicating different effects of nicotine and other components of smoke (England et al., 2017). Human offspring are hospitalized at an increased rate for problems with respiratory infections or difficulty breathing at altitude (Holbrook, 2016) Metabolic changes are also produced in human offspring of smoking mothers, in addition to type 2 diabetes including obesity and hypertension. Nicotine produces similar outcomes in animals. There is also evidence from human studies showing that maternal smoking reduces fertility in females, while having smaller or no effects in males (Jensen et al., 2006). Cardiovascular effects such as hypertension are also present in the offspring of smoking mothers into adulthood (Holbrook, 2016 and references therein). Infants exposed to nicotine during development showed reduced cortisol response to stress that persisted for a month (Stroud et al., 2014). Seven-month-olds demonstrated a higher cortisol response after stress (Schuetze et al., 2008). Adolescents once exposed to fetal nicotine show a higher propensity of low-frequency hearing loss that is consistent with studies in rats showing disrupted synaptic development in the auditory cortex (England et al., 2017).

7.2 Nervous system effects of prenatal nicotine exposure

As described in another chapter, signaling via nAChRs is important at multiple steps of development. This may be due to activation of nAChRs at specific times in specific cells. Disruption of normal acetylcholine-mediated signaling by nicotine can activate or desensitize various nAChR subtypes. Prenatal nicotine exposure in rats reduced expression of several elements of the cholinergic system including ChAT RNA, VAChT RNA, and choline transporters RNA (Mao et al., 2008). nAChRs and ACh are present early in neural development and thus are targets for fetal nicotine exposure. A large body of work with animals documents many changes in brain structure occurring as a result of fetal nicotine exposure. Even at low levels, nicotine exposure leads to developmental changes. Some of these seem to be long-lasting even until adulthood.

Both human and animal studies have shown the long-term effects of maternal smoking on the nervous system both structurally and functionally. Nicotine can impact all stages of development with nicotine alone producing 36%–46% of the effects in male rats (England et al., 2017). Other components of tobacco smoke also have major effects (Slotkin et al., 2015). Embryonic exposure to $1\,\mu M$ nicotine in rat embryos at a time equivalent to first trimester humans didn't produce large changes or a developmental delay but still showed increased intercellular spaces and increased numbers of pyknotic or apoptotic cells (Roy, Andrews, Seidler, & Slotkin, 1998). Adult animals may also display grossly normal brain structure after fetal exposure, but show subtle changes in synaptic architecture that may underlie functional changes (England et al., 2017). This was shown when rats exposed prenatally to nicotine at a level seen in the plasma of smokers had smaller cell sizes and thinner cell layers in the hippocampal CA3 region and dentate gyrus (Roy et al., 2002). In the somatosensory cortex, the proportion of medium-size pyramidal cells was decreased, and the proportion of smaller nonpyramidal cells increased.

Apoptotic death is induced in rat primary hippocampal progenitor cells exposed to $0.5\,\mu M$ nicotine, and this effect is mediated by $\alpha 7$-nAChRs since α-Bgt treatment blocked the effect of nicotine (Berger et al., 1998). Nicotine also induced cell death in a hippocampal progenitor cell line (HC2S2) when it was undifferentiated, but not when it was differentiated (Berger et al., 1998). This indicates that exposure to nicotine during different developmental windows may produce different outcomes. Numerous other animal studies demonstrate the effects of nicotine on fetal development. These include deficits in learning and memory, cognitive development, changes in affective behavior, increased anxiety, and decreased novelty-seeking (England et al., 2017). Prenatal nicotine exposure increased or decreased multiple genes in various brain regions in rodent brain (Sherafat et al., 2021). Genes involved include those involved in the immune response, serotonin transporter, GABA-A receptors, and cholinergic signaling. Many effects seen in animals may manifest in humans as well. A study using human postmortem fetal brains exposed in utero to nicotine showed differential expression of 14 genes in the prefrontal cortex (Semick et al., 2020). These genes included a GABA-A receptor alpha 4 subunit, a neuronal cell adhesion molecule, and a Ca^{2+}-activated K^+ channel (Semick et al., 2020). There were major changes in the gene expression profile comparing the fetal exposure changes with those genes affected in adult smokers (Semick et al., 2020). This indicates that nicotine has differential effects on the developing brain versus the adult brain.

Human epidemiological studies have shown an association between maternal smoking and ADHD, learning disabilities, behavioral issues, and increased risk of nicotine addiction (Bruin et al., 2010; Nomura et al., 2010; Schneider et al., 2011). Animal studies in rats or mice are consistent with these changes showing that prenatal nicotine exposure produces cognitive impairment, changes in nicotine sensitivity, increased anxiety, hyperactivity, and increases in nicotine self-administration (Bruin et al., 2010; Levin et al., 2006; Winzer-Serhan, 2008). Several population-based studies in humans show that maternal smoking increases the risk for ADHD in offspring (Linnet et al., 2003). Epidemiological studies also support a role for prenatal maternal smoking in conduct disorders and behavioral problems in children (Winzer-Serhan, 2008). Some studies indicate an increased risk of tobacco dependence among children exposed to prenatal nicotine (Buka et al., 2003). Studies in humans indicate that adolescents whose mothers smoked are twice as likely to smoke and continue smoking (DiFranza et al., 2004). Animal studies also support the role of nicotine in these effects. (Bruin et al., 2010). Children of smoking mothers also had an increased risk of oppositional defiant disorder (ODD). Father's smoking was not associated with increased risk (Nomura et al., 2010). A recent prospective birth cohort study showed that children of mothers who smoked are more likely to be dependent on nicotine or attempting withdrawal at age 21 than children of nonsmokers (O'Callaghan et al., 2009). Prenatal smoking is associated with a number of outcomes extending into preschool and early school-age humans including inattention, hyperactivity, and impulse control (Espy et al., 2011). Studies controlling for variables such a maternal education, parenting practices, comparison of children raised by biological mothers versus adoption, and others show disruption in conduct in children aged 4–10 after fetal exposure to nicotine (Gaysina et al., 2013). ODD and conduct disorders were demonstrated in 5-year-olds in a well-controlled study (Estabrook et al., 2016) after fetal nicotine exposure via maternal smoking.

Brain structural changes also appear to be associated with prenatal nicotine exposure. Adolescents (13–19 years old) exposed to prenatal smoking demonstrated by MRI a smaller volume of the amygdala, but not the orbitofrontal cortex or nucleus accumbans (Haghighi et al., 2013). Brain morphology changes were also evident in 6–8 year-olds exposed to nicotine via maternal smoking. MRI analysis showed a decrease in total brain volume, thinning of some cortical areas, and smaller gray and white matter volumes (El Marroun et al., 2014). Twenty-seven-year-old males exposed to nicotine in utero exhibited reduced serotonin synthesis in the medial orbitofrontal cortex (Booij et al., 2012). Using fMRI, Longo et al. (2013) demonstrated greater activity in a response inhibition task in several brain areas including cerebellum, inferior frontal gyrus and thalamus in young adults. Young adults previously exposed to nicotine in utero also demonstrated greater activity in brain regions involved in processing verbal working memory (Longo et al., 2014). These two fMRI studies indicated that to produce similar performance, brain areas worked harder and could indicate that changes in neural circuitry were produced as a result of prenatal nicotine exposure. Exposure to nicotine in utero was also shown to increase white matter density in anterior cortical white matter (Jacobsen et al., 2007). Mice exposed to nicotine from conception until postnatal day 21 showed increased dendritic spine density across multiple cortical regions (Jung et al., 2016). Mice exposed only prenatally also showed increased dendritic complexity (Jung et al., 2016).

These results support the hypothesis that nicotine exposure in utero can potentially produce long-term changes in brain function and thus influence behavior in offspring. Clearly long-term changes in behavior (ADHD, anxiety, response to subsequent nicotine exposure) and metabolism, as well as the effects of nicotine on adult animals, can be the result of fetal nicotine exposure. However, the downstream steps between nAChR signaling and long-term behavioral changes are not very well understood.

7.3 Nervous system effects of adolescent nicotine exposure

Adolescent smoking has also been associated with multiple adverse effects. Brain development continues for many years after birth, and during the adolescent years, the brain may be more plastic and thus more susceptible to the effects of drugs. Most adult smokers started before age 18 (Yuan et al., 2015). Brain circuits involved in decision-making are forming. During adolescence, cortical white matter volume increases while gray matter density and volume decrease in several brain areas including the frontal cortex (Yuan et al., 2015). Functional maturation of circuits occurs. Executive function increases during this time as well with maturation of neurotransmitter systems including dopamine, glutamate, and GABA. nAChR activation is key to signaling VTA dopaminergic neurons especially through α4β2* receptors (Yuan et al., 2015). α7 nAChRs regulate glutamate release onto dopaminergic neurons. As we have seen elsewhere in this book, nAChRs act to modulate expression of a number of neurotransmitters including glutamate, dopamine, GABA, and serotonin. Chronic nicotine treatment during adolescence alters the response in adults to subsequent nicotine exposure, indicating a long-term change in the serotonergic system induced by adolescent nicotine exposure (Slotkin et al., 2014).

nAChRs are differentially expressed throughout development relative to timing and brain region. Thus, the effects of nicotine signaling through nAChRs may vary with development and brain regions. Nicotine stimulates neurons in reward areas of the brain (i.e., VTA and basolateral amygdala) to a higher degree in adolescents than in adult rats (Yuan et al., 2015). Adolescent α4β2* and α7 nAChRs are more highly expressed in some brain regions than in adults, although α6* nAChRs are present at similar levels. (Doura et al., 2008). Several nAChR RNAs expressed in VTA and SN changed during postnatal development during a time of dopaminergic circuit maturation (Azam et al., 2007). Activity-regulated cytoskeleton–associated protein (Arc) gene expression in cortical regions is increased by nicotine or an α7-nAChR agonist in adolescent rats, but to a much lesser degree in adults (Yuan et al., 2015). Arc gene expression is regulated by activity and is an important regulator of synaptic plasticity. This suggests a mechanism by which nicotinic signaling at an inappropriate time/place in the brain may influence synaptic structure.

The mechanisms by which changes in cholinergic signaling are affected by nicotine during adolescence may be different than those operating during fetal and neonatal development. Neurotoxic effects of nicotine are more severe as a result of fetal exposure versus adolescent (Jacobsen et al., 2005). Adolescents are more at risk for addiction, in part due the plasticity and vulnerability to activation of reward circuits (England et al., 2017). This is consistent with studies showing that people smoking during the teenage years rather than those who start in their 20s are more likely to become long-term smokers (England et al., 2017). Animal and human studies show that adolescents are more sensitive to the effects of nicotine than adults (England et al., 2017). A study of 17–21 years old smokers showed deficits in several cognitive tests (oral arithmetic, vocabulary, and auditory memory) compared with the performance of the same people at ages 9–12, before they started smoking (Fried et al., 2006). Most of these deficits were reversed 6 months after smoking cessation. Adolescent smokers show deficits in working memory. The earlier smoking started, the worse the deficits displayed (Jacobsen et al., 2005). Attentional deficits were also seen (England et al., 2017). Even exposure to second-hand smoke during childhood and adolescence produces cognitive deficits (England et al., 2017). Adolescent tobacco use has been associated with reduced response to non-drug rewards (England et al., 2017). Clearly changes in communication and signaling are affected by adolescent nicotine exposure in humans. Adolescent nicotine exposure may also sensitize an adult to the effects of nicotine.

Animal studies also show more direct evidence of the effects of adolescent nicotine exposure. Lost-lasting deficits in cognition and learning and deficits in fear condition have been observed in adults after adolescent exposure (England et al., 2017). Mice exposed to nicotine via pump during 12 days of adolescence and then tested 30 days after cessation showed increased depressive responses to a forced swim test and the elevated plus maze (Holliday et al., 2016). Adult mice exposed during adolescence also showed reduced length and complexity of dendrites in the hippocampal CA1 region (Holliday et al., 2016). Adolescent exposure to nicotine resulted in adults with increased impulsivity and attention deficits (England et al., 2017). In rodents, adolescents show an increased motor activity in response to nicotine while adults have decreased activity (Yuan et al., 2015). Adolescents are more sensitive to the rewarding effects while adults are more sensitive to the aversive effects of nicotine. Higher doses of nicotine were better tolerated by adolescents, making it easier to become addicted (Yuan et al., 2015).

7.4 Role of specific nAChR subtypes and possible mechanisms

Multiple studies show changes in brain structure and in behavioral and functional activity in adolescent and adults exposed to nicotine prenatally or during adolescence. It must keep in mind that the many effects on fetal development, some of which persist into adulthood, are due to activation or in some cases, desensitization of nAChRs. What subtypes are involved and how may they mediate these effects? In most cases the specific nAChR subtypes involved are not known although α4β2*, α7, and α6* nAChRs are most likely important in mediating the effects of neonatal and adolescent nicotine exposure. Since the nAChRs are ion channels, their activation influences cell excitability that may have many effects on neurons. Mechanisms may involve effects on gene expression, cell motility, cell death, synaptic connectivity, and presynaptic control of neurotransmitter release such as dopamine, glutamate, and serotonin. α7 nAChRs can also couple to G-proteins under some conditions and induce long-term changes in cellular signaling (Kabbani & Nichols, 2018). Expression of nAChRs themselves is mediated by nicotine. Rats exposed to nicotine during gestational days (GD) 3–21, GD 8–21, and GD 15–21 showed higher levels of α2, α4, α7, and β2 RNAs in the forebrain and hindbrain. α3, α5, β3, and β4 were not changed (Lv et al., 2008). Rat exposed to oral nicotine on postnatal days (PD) 1–8 (similar to third human trimester) demonstrated increased epibatidine binding (detects heteromeric nAChRs especially α4β2) in the cortex, thalamus, and hippocampus (Huang & Winzer-Serhan, 2006). Epibatidine binding was not increased in the medial habenula. α-Btx binding (detects α7 homomeric receptors) was not increased by chronic neonatal nicotine. The α4β2 antagonist DHβE also caused increased epibatidine binding. There is evidence that these nAChRs remain functional after upregulation and thus could influence developmental events during this crucial period (Huang & Winzer-Serhan, 2006). Upregulation of nicotine binding (α4β2* nAChRs) was observed in 20–30-day-old mice following neonatal exposure. This is in contrast to chronic nicotine treatment in adults that upregulate nicotine binding, but the levels return to baseline with 7 days following cessation of treatment (Van de Kamp & Collins, 1994). α4β2* nAChRs are implicated in nicotine-induced apoptosis in the embryonic zebrafish (Matta et al., 2007). β2-containing nAChRs were implicated in the effects of prenatal nicotine exposure. β2 KO mice showed similar deficits (unstable breathing, impaired arousal and catecholamine synthesis) as seen in wild-type mice exposed to nicotine from GD 14 until PD5 (Cohen et al., 2005). Chronic nicotine may act by desensitizing β2-containing nAChRs and thus essentially "knocking" them out.

7.5 Vaping: Prenatal and adolescent effects

Use of vaping products containing nicotine is on the rise. There was an increase of 900% among high school students between 2011 and 2015. Four million high school students use e-cigarettes (Cooper & Henderson, 2020). In total, 6.5%–31% of adolescents have used e-cigarettes, sometimes referred to as electronic nicotine delivery systems (ENDS), at least once (Greenhill et al., 2016). Teenagers using ENDS have a greater risk (7×) of using traditional cigarettes (Sailer et al., 2019), even if ENDS are promoted by some as a smoking cessation product. Adult e-cigarette use is increasing as well with 13% of adults using e-cigarettes in 2013 (Greenhill et al., 2016). People around ENDS users are exposed to second-hand smoke and have the nicotine metabolite cotinine in their blood at levels similar to those exposed to second-hand smoke from traditional cigarettes (Smith et al., 2015). Some product labels don't include the actual nicotine content (cancer.org), and many products have an incomplete list of ingredients. Some vaping products are counterfeits that contain large amounts of nicotine (Sailer et al., 2019). Vaping products also contain some of the same chemicals in cigarettes including formaldehyde. Vaping solvents can include toxic metals, acrolein, 1,2-propanediol (1,2PDO), vegetable glycerin, and propylene oxide, some of which produce lung irritation. Some flavoring additives also contain compounds such as menthol that alone can affect expression and trafficking of nAChRs (Henderson et al., 2016). While cigarettes may contain menthol as the only additive, thousands of flavors are available with ENDS (Cooper & Henderson, 2020). Many ENDS users now prefer flavorants (Cooper & Henderson, 2020). Some contents become harmful after they are heated in the vaping device. Unfortunately, e-cigarette use is increasing in pregnant woman, under the mistaken belief that it is a safer alternative to cigarette smoking. This may not be the case as the nicotine contents of vaping products and cigarettes are similar (Whittington et al., 2018). Newer ENDSs have been shown to deliver doses of nicotine similar to that achieved by smoking cigarettes (Sailer et al., 2019), but over a slower time course. A popular e-cigarette in use is JUUL (Greene & Pisano, 2019). These use nicotine salt with benzoic acid allowing for rapid absorption. The concentration of nicotine per JUUL pod is $1.5\times$ that in a pack of cigarettes (Cooper & Henderson, 2020) delivering a higher dose of nicotine than most ENDSs. The popularity of JUUL makes them especially dangerous. blu e-cigarettes are also popular and come with flavorants such as menthol, polar mint, and vanilla.

As of 2019, there were no studies in humans of how the use of ENDS during pregnancy may affect fetal development (Sailer et al., 2019). A major concern is how do nicotine and other components of vaping affect human brain development? Smoking traditional cigarettes clearly does. It is reasonable to think that since vaping products containing nicotine can deliver doses comparable to that achieved in cigarette smokers, that effects similar to those described above on neural development may also occur with ENDS.

Some work with vaping products has been done in animals, however. Mice were exposed to e-cigarette vapors (2.4% nicotine in polypropylene glycol) from GD 15–19 and for PD 2–16 (Smith et al., 2015). The mice had increased activity in open field tests and a zero maze when tested at 14 weeks. Animals exposed to nicotine and those that weren't performed similarly on the rotarod test. The animals had cotinine levels comparable with human offspring of smoking mothers (Smith et al., 2015). This suggests that animals exposed to nicotine via vaping exhibited locomotor hyperactivity. This study didn't distinguish prenatal versus postnatal exposure effects.

The offspring of mice exposed to e-cigarette aerosols with nicotine for 6 weeks before pregnancy, during gestation and during lactation, were tested at 12 weeks after birth using several behavioral assays (Nguyen et al., 2018). Mice exposed to nicotine during the neonatal and postnatal periods showed short-term memory deficits in a novel object recognition task. It was also shown that exposure to the aerosol (50% propylene glycol and 50% vegetable glycerin) with or without nicotine produced reduced anxiety and increased hyperactivity (Nguyen et al., 2018). In addition, DNA hypermethylation was also observed after aerosol exposure without nicotine at 1 day and 20 days after birth. The RNA levels of numerous chromatin-modifying genes were altered as well using aerosols with or without nicotine (Nguyen et al., 2018). This indicates that components of ENDS without nicotine can produce harmful effects and highlights the potential dangers of ENDS use in pregnant woman.

When mice were exposed to e-cigarettes with nicotine prenatally and through lactation, multiple changes in the transcriptome of frontal cortex were detected in both sexes (Lauterstein et al., 2016). Brains were collected at PD 25–31 and changes determined by RNA sequencing. Surprisingly, females exposed to ENDS without nicotine demonstrated the greatest number of genes with altered expression, while male exposure to nicotine has the lowest number of expression changes (Lauterstein et al., 2016). The levels of a large number of genes were changed in females also with nicotine exposure (blu, classic tobacco). Genes in pathways involved with neurological disorders or functions including seizures, memory, cognition, and hyperactivity were among those changes as well as those that affect cancer and multiple organs (Lauterstein et al., 2016). Note that ENDS with or without nicotine exposure caused changes in the transcriptome of brain neurons.

In addition, another danger of ENDS may come from the flavors that are added to nicotine. These may produce developmental effects to a fetus (as seen with rodents above), but some such as green apple (produced by farnesol and farnesene) also demonstrate rewarding properties alone (Cooper & Henderson, 2020). Flavors such as menthol may cover up the bad taste of nicotine and thus increase use. Menthol has also been shown to increase the reinforcing properties of nicotine (Cooper & Henderson, 2020). Farnesol also increases $\alpha6^*$-containing nAChR expression on VTA dopaminergic neurons in adult mice, promotes a reward behavior (conditioned place preference), and increases locomotor activity (Avelar et al., 2019). A study using adult male mice showed that they increase self-administration behavior after exposure to menthol + nicotine, green apple + nicotine and maintained this behavior (Cooper et al., 2021). Surprisingly, the mice also demonstrate this rate of self-administration when exposed to vapors containing green apple only e-liquids (Cooper et al., 2021). $\alpha4\beta2^*$ nAChRs were involved as pretreatment with DHβE (antagonist) reduced the self-administration response to nicotine, green apple, nicotine + green apple, and nicotine + menthol (Cooper et al., 2021). DHβE didn't reduce the response to menthol alone, indicating that possibly another subtype of nAChR is involved with this response such as $\alpha6^*$ nAChRs. Not much is known about the effects of ENDS in human fetal development. However, cigarettes have been shown to affect placental blood flow. E-cigarette vapor, without nicotine, was shown to reduce function of a cell line derived from chorionic villi from first trimester human placenta (Greene & Pisano, 2019). This study supports the hypothesis that e-cigarettes, even without nicotine, can affect placental function.

Much more needs to be done examining the effects of ENDS on human fetal and adolescent development, but what has been done so far in animals indicates that significant effects in humans on development and adult behaviors may be produced by ENDS with or without nicotine. The presence of flavorants and nicotine in ENDS could make them more addictive and thus promote use among pregnant woman and adolescents. ENDS can be thought of conservatively as dangerous to fetal development and may be shown to reproduce many of the detrimental effects of maternal smoking on fetal and adolescent development.

7.6 Epigenetic effects of nicotine exposure

Human and animal studies show that nicotine exposure has deleterious effects on babies, youth, and adults that can be long-lasting. How might these long-term effects occur? Clearly as described above, nicotine exposure alters signaling through nAChRs. In addition, some studies show that nicotine can have epigenetic effects that may underlie its long-term actions.

In order to examine gene expression changes during nicotine exposure, mice were exposed to nicotine pre- and postnatally until PD21 and then remained nicotine free until 3 months. qPCR was used to identify nine genes induced at PD21, and expression of five of these were still induced at 3 months (Jung et al., 2016). The gene Ash2l was highly induced by the nicotine treatment. This is interesting in that this gene is a known transcriptional regulator that acts through trimethylation of histone 3 (H3K4me3). Multiple sites in the mice were enriched in H3K4me3 methylation after both pre- and postnatal exposure (Jung et al., 2016). Many of these were associated with genes involved in synapse formation or plasticity. In total, 39 genes were affected by both prenatal and postnatal or just postnatal exposure. A major site of increased histone methylation after nicotine exposure was the Mef2c locus (Jung et al., 2016). Mef2c RNA levels were also increased after nicotine treatment during development. H3K4me3 methylation remained increased at the Mef2c locus 3 months after cessation of nicotine exposure. Nicotine treatment of cultured neural progenitors increased expression of both Ash2l and Mef2c proteins. Most importantly, this increase required nAChR activity and an $\alpha4\beta2$* antagonist reduced the effect, but $\alpha7$ nAChR antagonists didn't (Jung et al., 2016). Reduction of Ash2l and Mef2c RNA expression by shRNA reduced the increase of dendritic spines seen with nicotine exposure. Nicotine exposure for 3 postnatal weeks was shown to produce a hypersensitive avoidance response that was mediated by $\beta2$ nAChR subunit containing receptors. Knockdown of Ash2l and Mef2c in utero reduced the increased hypersensitive response produced by nicotine (Jung et al., 2016). Overall this work implicates $\alpha4\beta2$* nAChR activity due to developmental nicotine exposure producing long-term changes in the epigenome. These changes produced long-term effects on gene expression and behavior. Other studies show that epigenetic changes are produced in specific regions of the epigenome in response to fetal nicotine exposure. These changes can occur in somatic cells and germ cells. These changes influence behaviors, metabolism, and physiology through histone modifications and altered gene regulation due to changes in DNA methylation patterns or chromatin remodeling in offspring.

Levine et al. (2011) found that pretreatment of 10–12-week-old mice with nicotine increased the responses to subsequent cocaine exposure, while cocaine pretreatment did not alter responsiveness to nicotine. Nicotine led to increased FosB expression after cocaine exposure by inhibiting histone deacetylase (HDAC). Acetylation of both histone H3 and histone H4 was increased at the FosB promoter after nicotine pretreatment. The effect of nicotine could be simulated by pretreatment with a HDAC inhibitor. HDAC inhibition also decreased nicotine-conditioned place preference without affecting aversion for nicotine in 30–35-day-old rats (Pastor et al., 2011). In another study, nicotine pretreatment in mice was shown to reduce DNA methyltransferase (DNMT) 1 activity with subsequent reduction in GAD67 promoter methylation and increased GAD67 expression in cortical or hippocampal neurons (Satta et al., 2008). Varenicline, an $\alpha 4 \beta 2$ nAChR partial agonist and $\alpha 7$ nAChR full agonist, reduced methylation of GAD 67 promoters and increased expression of GAD67 mRNA. Another $\alpha 4 \beta 2$ nAChR agonist also produced this effect, but an $\alpha 7$ nAChR agonist did not (Maloku et al., 2011). Epigenetic and gene expression changes have been noted in adult rats as well with nicotine administration (Castino et al., 2018). Nicotine self-administration reduced methylation at two chromatin sites in the Cdk-5 and BDNF promoters. Inhibition of HDAC increased extinction of nicotine-seeking behavior (Castino et al., 2018).

A human study of 754 woman and 230 men exposed to nicotine in utero showed that even as adults (48 years old) changes in methylation at 15 CpG sites could be observed (Richmond et al., 2018). The DNA methylation scores predicted prenatal exposure to nicotine. A large study of newborns exposed to maternal smoking also demonstrated a large number of changes in CpG methylation patterns, many of which were correlated with corresponding changes in gene expression and in some genes important for development (Joubert et al., 2016). A large study of humans between ages 16 and 48 years previously exposed to nicotine in utero showed

CpG methylation changes in 36 genomic regions. Multiple changes were correlated with increased incidence of human disease including schizophrenia (Wiklund et al., 2019). Some adult diseases are due to fetal insults, and some of these have an epigenetic basis (Vo & Hardy, 2012). The frequency by which fetal nicotine exposure produces epigenetic changes that ultimately can lead to disease, changes in behavior or in responsiveness to subsequent nicotine exposure is not known. However, epigenetic changes seen in animals also occur in humans.

7.7 Intergenerational and transgenerational effects

Epigenetic changes seem to be transmitted to the F1 and F2 generation after male mice are exposed to nicotine for 8 weeks before mating (Goldberg et al., 2021). Transmission of changes to F1 may be considered intergenerational since the germ cells of the F1 mice were exposed to nicotine. F2 may be considered transgenerational since they were conceived from sperm never exposed to nicotine. Nicotine was removed in time so that no nicotine was left in the male mice at the time of mating. F1 mice sired by the nicotine-exposed males (FO) showed enhanced contextual fear and reduced acute nicotine enhancement of fear conditioning (Goldberg et al., 2021). F2 mice bread from mice (F1) that had not been exposed to nicotine showed also increased fear conditioning. Paternal exposure before conception produced F1 and F2 mice with stronger fear memories (Goldberg et al., 2021). Other effects of paternal nicotine exposure on F1 mice were noted such as decreased nicotine self-administration and increased binding of epibatidine to $\alpha4\beta2^*$ nAChRs in 8-week-old mice (Goldberg et al., 2021). Nicotine applied to the hippocampus of nicotine-sired mice (F1) 10–20 weeks old resulted in a reduction of cholinergic signaling determined by amperometry in the ventral and dorsal hippocampus. In addition, a large number of genes were shown to be differentially expressed in the F1 mice sired by nicotine-exposed males versus those fathered by males exposed only to saline. Some of these were noted to be involved in neurological disorders and nervous system development (Goldberg et al., 2021). Finally, epigenetic changes in methylation patterns were noted in F1 mice after paternal nicotine exposure. Clearly paternal nicotine exposure can produce changes in the genome or epigenome that can be transmitted across generations and produce changes in behavior and cholinergic signaling in mice.

Paternal exposure also reduced nicotine-induced hypothermia in male F1 mice sired by males exposed to nicotine, increased serum nicotine concentration in females after nicotine exposure, and reduced baseline levels of corticosterone in both male and female F1 mice (Zeid et al., 2021). Female F1 mice born of parents (FO) also exposed to nicotine during adolescence performed more poorly on a learning task than mice born of parents not exposed to nicotine (Renaud & Fountain, 2016). Nicotine exposure of the parents caused a cognitive change in the offspring. Paternal exposure to nicotine also produced transgenerational changes in response to subsequent nicotine exposure and stress including sex differences in the responses (Yohn et al., 2019).

The mechanisms of intergenerational or transgenerational transmission involve epigenetic changes that can include methylation, changes to chromatin, and changes in microRNAs. Work with *Caenorhabditis elegans* demonstrated that expression of specific miRNAs is altered in F1 and F2 worms after FO nicotine exposure. The genes targeted by the regulated miRNAs were shown to be involved in nicotine-mediated behaviors in *C. elegans* (Taki et al., 2014). While the work was done in the worm *C. elegans*, the worm and human genomes are about 80% homologous, and nicotine does mediate some behaviors in worms as in rodents (see Chapter 10). Nicotine exposure to FO worms was previously shown to produce transgenerational changes in behavior (Taki et al., 2013). This work indicates another possible mechanism of intergenerational and transgenerational effects of nicotine.

Is there evidence for intergenerational and transgenerational effects of nicotine exposure in humans? Some studies indicate that grandparental and parental smoking was correlated with increased incidence of autism spectrum disorders (ASD), asthma, and smoking in offspring (Baratta et al., 2021). Research has only touched the surface of understanding the epigenetic and transgenerational effects of nicotine exposure by maternal smoking or vaping. These are important topics given the long-term effects of nicotine on addiction, various behaviors, and the potential for prenatal or adolescent nicotine exposure to increase the chances of addiction to other drugs of abuse.

References

Avelar, A. J., Akers, A. T., Baumgard, Z. J., Cooper, S. Y., Casinelli, G. P., & Henderson, B. J. (2019). Why flavored vape products may be attractive: Green apple tobacco flavor elicits reward-related behavior, upregulates nAChRs on VTA dopamine neurons, and alters midbrain dopamine and GABA neuron function. *Neuropharmacology, 158*. https://doi.org/10.1016/j.neuropharm.2019.107729.

Azam, L., Chen, Y., & Leslie, F. M. (2007). Developmental regulation of nicotinic acetylcholine receptors within midbrain dopamine neurons. *Neuroscience, 144*(4), 1347–1360. https://doi.org/10.1016/j.neuroscience.2006.11.011.

Baratta, A. M., Rathod, R. S., Plasil, S. L., Seth, A., & Homanics, G. E. (2021). Exposure to drugs of abuse induce effects that persist across generations. *International Review of Neurobiology*, *156*, 217–277. https://doi.org/10.1016/bs.irn.2020.08.003.

Berger, F., Gage, F. H., & Vijayaraghavan, S. (1998). Nicotinic receptor-induced apoptotic cell death of hippocampal progenitor cells. *The Journal of Neuroscience*, *18*(17), 6871–6881. https://doi.org/10.1523/jneurosci.18-17-06871.1998.

Blacquiere, M. J., Timens, W., Melgert, B. N., Geerlings, M., Postma, D. S., & Hylkema, M. N. (2009). Maternal smoking during pregnancy induces airway remodelling in mice offspring. *European Respiratory Journal*, *33*(5), 1133–1140. https://doi.org/10.1183/09031936.00129608.

Booij, L., Benkelfat, C., Leyton, M., Vitaro, F., Gravel, P., Lévesque, M. L., Arseneault, L., Diksic, M., & Tremblay, R. E. (2012). Perinatal effects on in vivo measures of human brain serotonin synthesis in adulthood: A 27-year longitudinal study. *European Neuropsychopharmacology*, *22*(6), 419–423. https://doi.org/10.1016/j.euroneuro.2011.11.002.

Bruin, J. E., Gerstein, H. C., & Holloway, A. C. (2010). Long-term consequences of fetal and neonatal nicotine exposure: A critical review. *Toxicological Sciences*, *116*(2), 364–374. https://doi.org/10.1093/toxsci/kfq103.

Buka, S. L., Shenassa, E. D., & Niaura, R. (2003). Elevated risk of tobacco dependence among offspring of mothers who smoked during pregnancy: A 30-year prospective study. *American Journal of Psychiatry*, *160*(11), 1978–1984. https://doi.org/10.1176/appi.ajp.160.11.1978.

Castino, M. R., Baker-Andresen, D., Ratnu, V. S., Shevchenko, G., Morris, K. V., Bredy, T. W., Youngson, N. A., & Clemens, K. J. (2018). Persistent histone modifications at the BDNF and Cdk-5 promoters following extinction of nicotine-seeking in rats. *Genes, Brain and Behavior*, *17*(2), 98–106. https://doi.org/10.1111/gbb.12421.

Cohen, G., Roux, J. C., Grailhe, R., Malcolm, G., Changeux, J. P., & Lagercrantz, H. (2005). Perinatal exposure to nicotine causes deficits associated with a loss of nicotinic receptor function. *Proceedings of the National Academy of Sciences of the United States of America*, *102*(10), 3817–3821. https://doi.org/10.1073/pnas.0409782102.

Cooper, S. Y., Akers, A. T., & Henderson, B. J. (2021). Flavors enhance nicotine vapor self-administration in male mice. *Nicotine and Tobacco Research*, *23*(3), 566–572. https://doi.org/10.1093/ntr/ntaa165.

Cooper, S., & Henderson, B. J. (2020). The impact of nicotine delivery system (ENDS) flavors on nicotinic acetylcholine receptors and nicotine addiction related behaviors. *Molecules*, *25*.

DiFranza, J. R., Aligne, C. A., & Weitzman, M. (2004). Prenatal and postnatal environmental tobacco smoke exposure and children's health. *Pediatrics*, *113*(4), 1007–1015.

Doura, M. B., Gold, A. B., Keller, A. B., & Perry, D. C. (2008). Adult and periadolescent rats differ in expression of nicotinic cholinergic receptor subtypes and in the response of these subtypes to chronic nicotine exposure. *Brain Research*, *1215*, 40–52. https://doi.org/10.1016/j.brainres.2008.03.056.

El Marroun, H., Schmidt, M. N., Franken, I. H. A., Jaddoe, V. W. V., Hofman, A., Van Der Lugt, A., Verhulst, F. C., Tiemeier, H., & White, T. (2014). Prenatal tobacco exposure and brain morphology: A prospective study in young children. *Neuropsychopharmacology*, *39*(4), 792–800. https://doi.org/10.1038/npp.2013.273.

England, L. J., Aagaard, K., Bloch, M., Conway, K., Cosgrove, K., Grana, R., Gould, T. J., Hatsukami, D., Jensen, F., Kandel, D., Lanphear, B., Leslie, F., Pauly, J. R., Neiderhiser, J., Rubinstein, M., Slotkin, T. A., Spindel, E., Stroud, L., & Wakschlag, L. (2017). Developmental toxicity of nicotine: A transdisciplinary synthesis and implications for emerging tobacco products. *Neuroscience and Biobehavioral Reviews*, *72*, 176–189. https://doi.org/10.1016/j.neubiorev.2016.11.013.

Espy, K. A., Fang, H., Johnson, C., Stopp, C., Wiebe, S. A., & Respass, J. (2011). Prenatal tobacco exposure: Developmental outcomes in the neonatal period. *Developmental Psychology*, *47*(1), 153–169. https://doi.org/10.1037/a0020724.

Estabrook, R., Massey, S. H., Clark, C. A. C., Burns, J. L., Mustanski, B. S., Cook, E. H., O'Brien, T. C., Makowski, B., Espy, K. A., & Wakschlag, L. S. (2016). Separating family-level and direct exposure effects of smoking during pregnancy on offspring externalizing symptoms: Bridging the behavior genetic and behavior teratologic divide. *Behavior Genetics*, *46*(3), 389–402. https://doi.org/10.1007/s10519-015-9762-2.

Fried, P. A., Watkinson, B., & Gray, R. (2006). Neurocognitive consequences of cigarette smoking in young adults—A comparison with pre-drug performance. *Neurotoxicology and Teratology*, *28*(4), 517–525. https://doi.org/10.1016/j.ntt.2006.06.003.

Gaysina, D., Fergusson, D. M., Leve, L. D., Horwood, J., Reiss, D., Shaw, D. S., Elam, K. K., Natsuaki, M. N., Neiderhiser, J. M., & Harold, G. T. (2013). Maternal smoking during pregnancy and offspring conduct problems: Evidence from 3 independent genetically sensitive research designs. *JAMA Psychiatry*, *70*(9), 956–963. https://doi.org/10.1001/jamapsychiatry.2013.127.

Goldberg, L. R., Zeid, D., Kutlu, M. G., Cole, R. D., Lallai, V., Sebastian, A., Albert, I., Fowler, C. D., Parikh, V., & Gould, T. J. (2021). Paternal nicotine enhances fear memory, reduces nicotine administration, and alters hippocampal genetic and neural function in offspring. *Addiction Biology*, *26*(1). https://doi.org/10.1111/adb.12859.

Greene, R. M., & Pisano, M. M. (2019). Developmental toxicity of e-cigarette aerosols. *Birth Defects Research*, *111*(17), 1294–1301. https://doi.org/10.1002/bdr2.1571.

Greenhill, R., Dawkins, L., Notley, C., Finn, M. D., & Turner, J. J. D. (2016). Adolescent awareness and use of electronic cigarettes: A review of emerging trends and findings. *Journal of Adolescent Health*, *59*(6), 612–619. https://doi.org/10.1016/j.jadohealth.2016.08.005.

Haghighi, A., Schwartz, D. H., Abrahamowicz, M., Leonard, G. T., Perron, M., Richer, L., Veillette, S., Gaudet, D., Paus, T., & Pausova, Z. (2013). Prenatal exposure to maternal cigarette smoking, amygdala volume, and fat intake in adolescence. *JAMA Psychiatry*, *70*(1), 98. https://doi.org/10.1001/archgenpsychiatry.2012.1101.

Henderson, B. J., Wall, T. R., Henley, B. M., Kim, C. H., Nichols, W. A., Moaddel, R., Xiao, C., & Lester, H. A. (2016). Menthol alone upregulates midbrain nAChRs, alters nAChR subtype stoichiometry, alters dopamine neuron firing frequency, and prevents nicotine reward. *Journal of Neuroscience*, *36*(10), 2957–2974. https://doi.org/10.1523/JNEUROSCI.4194-15.2016.

Holbrook, B. D. (2016). The effects of nicotine on human fetal development. *Birth Defects Research Part C—Embryo Today: Reviews*, *108*(2), 181–192. https://doi.org/10.1002/bdrc.21128.

Holliday, E. D., Nucero, P., Kutlu, M. G., Oliver, C., Connelly, K. L., Gould, T. J., & Unterwald, E. M. (2016). Long-term effects of chronic nicotine on emotional and cognitive behaviors and hippocampus cell morphology in mice: Comparisons of adult and adolescent nicotine exposure. *European Journal of Neuroscience*, *44*(10), 2818–2828. https://doi.org/10.1111/ejn.13398.

Huang, L. Z., & Winzer-Serhan, U. H. (2006). Chronic neonatal nicotine upregulates heteromeric nicotinic acetylcholine receptor binding without change in subunit mRNA expression. *Brain Research*, *1113*(1), 94–109. https://doi.org/10.1016/j.brainres.2006.06.084.

Jacobsen, L. K., Krystal, J. H., Mencl, W. E., Westerveld, M., Frost, S. J., & Pugh, K. R. (2005). Effects of smoking and smoking abstinence on cognition in adolescent tobacco smokers. *Biological Psychiatry*, *57*(1), 56–66. https://doi.org/10.1016/j.biopsych.2004.10.022.

Jacobsen, L. K., Picciotto, M. R., Heath, C. J., Frost, S. J., Tsou, K. A., Dwan, R. A., Jackowski, M. P., Constable, R. T., & Mencl, W. E. (2007). Prenatal and adolescent exposure to tobacco smoke modulates the development of white matter microstructure. *Journal of Neuroscience*, *27*(49), 13491–13498. https://doi.org/10.1523/JNEUROSCI.2402-07.2007.

Jensen, TK, Joffe, M, Scheike, T, Skytthe, A, Gaist, D, Petersen, I, & Christensen, K. (2006). Early exposure to smoking and future fecundity among Danish twins. *International Journal of Andrology*, *29*, 603–613.

Joubert, B. R., Felix, J. F., Yousefi, P., Bakulski, K. M., Just, A. C., Breton, C., Reese, S. E., Markunas, C. A., Richmond, R. C., Xu, C. J., Küpers, L. K., Oh, S. S., Hoyo, C., Gruzieva, O., Söderhäll, C., Salas, L. A., Baïz, N., Zhang, H., Lepeule, J., ... London, S. J. (2016). DNA methylation in newborns and maternal smoking in pregnancy: Genome-wide consortium meta-analysis. *American Journal of Human Genetics*, *98*(4), 680–696. https://doi.org/10.1016/j.ajhg.2016.02.019.

Jung, Y., Hsieh, L. S., Lee, A. M., Zhou, Z., Coman, D., Heath, C. J., Hyder, F., Mineur, Y. S., Yuan, Q., Goldman, D., Bordey, A., & Picciotto, M. R. (2016). An epigenetic mechanism mediates developmental nicotine effects on neuronal structure and behavior. *Nature Neuroscience*, *19*(7), 905–914. https://doi.org/10.1038/nn.4315.

Kabbani, N., & Nichols, R. A. (2018). Beyond the channel: Metabotropic signaling by nicotinic receptors. *Trends in Pharmacological Sciences*, *39*(4), 354–366. https://doi.org/10.1016/j.tips.2018.01.002.

Lauterstein, D. E., Tijerina, P. B., Corbett, K., Oksuz, B. A., Shen, S. S., Gordon, T., Klein, C. B., & Zelikoff, J. T. (2016). Frontal cortex transcriptome analysis of mice exposed to electronic cigarettes during early life stages. *International Journal of Environmental Research and Public Health*, *13*(4). https://doi.org/10.3390/ijerph13040417.

Levin, E. D., Lawrence, S., Petro, A., Horton, K., Seidler, F. J., & Slotkin, T. A. (2006). Increased nicotine self-administration following prenatal exposure in female rats. *Pharmacology, Biochemistry, and Behavior*, *85*, 669–674.

Levine, A., Huang, Y. Y., Drisaldi, B., Griffin, E. A., Pollak, D. D., Xu, S., Yin, D., Schaffran, C., Kandel, D. B., & Kandel, E. R. (2011). Molecular mechanism for a gateway drug: Epigenetic changes initiated by nicotine prime gene expression for cocaine. *Science Translational Medicine*, *3*(107). https://doi.org/10.1126/scitranslmed.3003062.

Linnet, K. M., Dalsgaard, S., Obel, G., Wisborg, K., Henriksen, T. B., Rodriguez, A., Kotimaa, A., Moilanen, I., Thomsen, P. H., Olsen, J., & Jarvelin, M. R. (2003). Maternal lifestyle factors in pregnancy risk of attention deficit hyperactivity disorder and associated behaviors: Review of the current evidence. *American Journal of Psychiatry*, *160*(6), 1028–1040. https://doi.org/10.1176/appi.ajp.160.6.1028.

Longo, C. A., Fried, P. A., Cameron, I., & Smith, A. M. (2013). The long-term effects of prenatal nicotine exposure on response inhibition: An fMRI study of young adults. *Neurotoxicology and Teratology*, *39*, 9–18. https://doi.org/10.1016/j.ntt.2013.05.007.

Longo, C. A., Fried, P. A., Cameron, I., & Smith, A. M. (2014). The long-term effects of prenatal nicotine exposure on verbal working memory: An fMRI study of young adults. *Drug and Alcohol Dependence*, *144*, 61–69. https://doi.org/10.1016/j.drugalcdep.2014.08.006.

Lv, J., Mao, C., Zhu, L., Zhang, H., Pengpeng, H., Xu, F., Liu, Y., Zhang, L., & Xu, Z. (2008). The effect of prenatal nicotine on expression of nicotine receptor subunits in the fetal brain. *Neurotoxicology*, *29*(4), 722–726. https://doi.org/10.1016/j.neuro.2008.04.015.

Maloku, E., Kadriu, B., Zhubi, A., Dong, E., Pibiri, F., Satta, R., & Guidotti, A. (2011). Selective $\alpha 4 \beta 2$ nicotinic acetylcholine receptor agonists target epigenetic mechanisms in cortical GABAergic neurons. *Neuropsychopharmacology*, *36*(7), 1366–1374. https://doi.org/10.1038/npp.2011.21.

Mao, C., Yuan, X., Zhang, H., Lv, J., Guan, J., Miao, L., Chen, L., Zhang, Y., Zhang, L., & Xu, Z. (2008). The effect of prenatal nicotine on mRNA of central cholinergic markers and hematological parameters in rat fetuses. *International Journal of Developmental Neuroscience, 26*(5), 467–475. https://doi.org/10.1016/j.ijdevneu.2008.02.007.

Matta, S. G., Balfour, D. J., Benowitz, N. L., Boyd, R. T., Buccafusco, J. J., Caggiula, A. R., Craig, C. R., Collins, A. C., Damaj, M. I., Donny, E. C., Gardiner, P. S., Grady, S. R., Heberlein, U., Leonard, S. S., Levin, E. D., Lukas, R. J., Markou, A., Marks, M. J., McCallum, S. E., ... Zirger, J. M. (2007). Guidelines on nicotine dose selection for in vivo research. *Psychopharmacology, 190*(3), 269–319. https://doi.org/10.1007/s00213-006-0441-0.

Nguyen, T., Li, G. E., Chen, H., Cranfield, C. G., McGrath, K. C., & Gorrie, C. A. (2018). Maternal E-cigarette exposure results in cognitive and epigenetic alterations in offspring in a mouse model. *Chemical Research in Toxicology, 31*(7), 601–611. https://doi.org/10.1021/acs.chemrestox.8b00084.

Nomura, Y., Marks, D. J., & Halperin, J. M. (2010). Prenatal exposure to maternal and paternal smoking on attention deficit hyperactivity disorders symptoms and diagnosis in offspring. *Journal of Nervous and Mental Disease, 198*(9), 672–678. https://doi.org/10.1097/NMD.0b013e3181ef3489.

O'Callaghan, F. V., Al Mamun, A., O'Callaghan, M., Alati, R., Najman, J. M., Williams, G. M., & Bor, W. (2009). Maternal smoking during pregnancy predicts nicotine disorder (dependence or withdrawal) in young adults—A birth cohort study. *Australian and New Zealand Journal of Public Health, 33*(4), 371–377. https://doi.org/10.1111/j.1753-6405.2009.00410.x.

Pastor, V., Host, L., Zwiller, J., & Bernabeu, R. (2011). Histone deacetylase inhibition decreases preference without affecting aversion for nicotine. *Journal of Neurochemistry, 116*(4), 636–645. https://doi.org/10.1111/j.1471-4159.2010.07149.x.

Renaud, S. M., & Fountain, S. B. (2016). Transgenerational effects of adolescent nicotine exposure in rats: Evidence for cognitive deficits in adult female offspring. *Neurotoxicology and Teratology, 56*, 47–54. https://doi.org/10.1016/j.ntt.2016.06.002.

Richmond, R. C., Suderman, M., Langdon, R., Relton, C. L., & Smith, G. D. (2018). DNA methylation as a marker for prenatal smoke exposure in adults. *International Journal of Epidemiology, 47*(4), 1120–1130. https://doi.org/10.1093/ije/dyy091.

Roy, T. S., Andrews, J. E., Seidler, F. J., & Slotkin, T. A. (1998). Nicotine evokes cell death in embryonic rat brain during neurulation. *Journal of Pharmacology and Experimental Therapeutics, 287*, 1136–1144.

Roy, T. S., Seidler, F. J., & Slotkin, T. A. (2002). Prenatal nicotine exposure evokes alterations of cell structure in hippocampus and somatosensory cortex. *Journal of Pharmacology and Experimental Therapeutics, 300*(1), 124–133. https://doi.org/10.1124/jpet.300.1.124.

Sailer, S., Sebastiani, G., Andreu-Férnández, V., & García-Algar, O. (2019). Impact of nicotine replacement and electronic nicotine delivery systems on fetal brain development. *International Journal of Environmental Research and Public Health, 16*(24). https://doi.org/10.3390/ijerph16245113.

Satta, R., Maloku, E., Zhubi, A., Pibiri, F., Hajos, M., Costa, E., & Guidotti, A. (2008). Nicotine decreases DNA methyltransferase 1 expression and glutamic acid decarboxylase 67 promoter methylation in GABAergic interneurons. *Proceedings of the National Academy of Sciences of the United States of America, 105*(42), 16356–16361. https://doi.org/10.1073/pnas.0808699105.

Schneider, T., Ilott, N., Brolese, G., Bizarro, L., Asherson, P. J. E., & Stolerman, I. P. (2011). Prenatal exposure to nicotine impairs performance of the 5-choice serial reaction time task in adult rats. *Neuropsychopharmacology, 36*(5), 1114–1125. https://doi.org/10.1038/npp.2010.249.

Schuetze, P., Lopez, F. A., Granger, D. A., & Eiden, R. D. (2008). The association between prenatal exposure to cigarettes and cortisol reactivity and regulation in 7-month-old infants. *Developmental Psychobiology*, *50*(8), 819–834. https://doi.org/10.1002/dev.20334.

Sekhon, H. S., Keller, J. A., Benowitz, N. L., & Spindel, E. R. (2001). Prenatal nicotine exposure alters pulmonary function in newborn rhesus monkeys. *American Journal of Respiratory and Critical Care Medicine*, *164*(6), 989–994. https://doi.org/10.1164/ajrccm.164.6.2011097.

Semick, S. A., Collado-Torres, L., Markunas, C. A., Shin, J. H., Deep-Soboslay, A., Tao, R., Huestis, M. A., Bierut, L. J., Maher, B. S., Johnson, E. O., Hyde, T. M., Weinberger, D. R., Hancock, D. B., Kleinman, J. E., & Jaffe, A. E. (2020). Developmental effects of maternal smoking during pregnancy on the human frontal cortex transcriptome. *Molecular Psychiatry*, *25*(12), 3267–3277. https://doi.org/10.1038/s41380-018-0223-1.

Sherafat, Y., Bautista, M., & Fowler, C. D. (2021). Multidimensional intersection of nicotine, gene expression, and behavior. *Frontiers in Behavioral Neuroscience*, *15*. https://doi.org/10.3389/fnbeh.2021.649129.

Slotkin, T. A., Card, J., & Seidler, F. J. (2014). Nicotine administration in adolescence reprograms the subsequent response to nicotine treatment and withdrawal in adulthood: Sex-selective effects on cerebrocortical serotonergic function. *Brain Research Bulletin*, *102*, 1–8. https://doi.org/10.1016/j.brainresbull.2014.01.004.

Slotkin, T. A., Skavicus, S., Card, J., Stadler, A., Levin, E. D., & Seidler, F. J. (2015). Developmental neurotoxicity of tobacco smoke directed toward cholinergic and serotonergic systems: More than just nicotine. *Toxicological Sciences*, *147*(1), 178–189. https://doi.org/10.1093/toxsci/kfv123.

Smith, D., Aherrera, A., Lopez, A., Neptune, E., Winickoff, J. P., Klein, J. D., Chen, G., Lazarus, P., Collaco, J. M., & McGrath-Morrow, S. A. (2015). Adult behavior in male mice exposed to E-cigarette nicotine vapors during late prenatal and early postnatal life. *PLoS ONE*, *10*(9). https://doi.org/10.1371/journal.pone.0137953.

Stroud, L. R., Papandonatos, G. D., Rodriguez, D., McCallum, M., Salisbury, A. L., Phipps, M. G., Lester, B., Huestis, M. A., Niaura, R., Padbury, J. F., & Marsit, C. J. (2014). Maternal smoking during pregnancy and infant stress response: Test of a prenatal programming hypothesis. *Psychoneuroendocrinology*, *48*, 29–40. https://doi.org/10.1016/j.psyneuen.2014.05.017.

Taki, F. A., Pan, X., Lee, M. H., & Zhang, B. (2014). Nicotine exposure and transgenerational impact: A prospective study on small regulatory microRNAs. *Scientific Reports*, *4*. https://doi.org/10.1038/srep07513.

Taki, F. A., Pan, X., & Zhang, B. (2013). Nicotine exposure caused significant transgenerational heritable behavioral changes in *Caenorhabditis elegans*. *EXCLI Journal*, *12*, 793–806. http://www.excli.de/vol12/Zhang_10092013_proof.pdf.

Van de Kamp, J. L., & Collins, A. C. (1994). Prenatal nicotine alters nicotinic receptor development in the mouse brain. *Pharmacology, Biochemistry and Behavior*, *47*(4), 889–900. https://doi.org/10.1016/0091-3057(94)90293-3.

Vo, T., & Hardy, D. B. (2012). Molecular mechanisms underlying the fetal programming of adult disease. *Journal of Cell Communication and Signaling*. https://doi.org/10.1007/s12079-012-0165-3.

Whittington, J. R., Simmons, P. M., Phillips, A. M., Gammill, S. K., Cen, R., Magann, E. F., & Cardenas, V. M. (2018). The use of electronic cigarettes in pregnancy: A review of the literature. *Obstetrical and Gynecological Survey*, *73*(9), 544–549. https://doi.org/10.1097/OGX.0000000000000595.

Wiklund, P., Karhunen, V., Richmond, R. C., Parmar, P., Rodriguez, A., De Silva, M., Wielscher, M., Rezwan, F. I., Richardson, T. G., Veijola, J., Herzig, K. H., Holloway, J. W., Relton, C. L., Sebert, S., & Järvelin, M. R. (2019). DNA methylation links prenatal smoking exposure to later life health outcomes in offspring. *Clinical Epigenetics*, *11*(1). https://doi.org/10.1186/s13148-019-0683-4.

Winzer-Serhan, U. H. (2008). Long-term consequences of maternal smoking and developmental chronic nicotine exposure. *Frontiers in Bioscience, 13,* 636–649.

Yohn, N. L., Caruso, M. J., & Blendy, J. A. (2019). Effects of nicotine and stress exposure across generations in C57BL/6 mice. *Stress, 22*(1), 142–150. https://doi.org/10.1080/10253890.2018.1532991.

Yuan, M., Cross, S. J., Loughlin, S. E., & Leslie, F. M. (2015). Nicotine and the adolescent brain. *Journal of Physiology, 593*(16), 3397–3412. https://doi.org/10.1113/JP270492.

Zeid, D., Goldberg, L. R., Seemiller, L. R., Mooney-Leber, S., Smith, P. B., & Gould, T. J. (2021). Multigenerational nicotine exposure affects offspring nicotine metabolism, nicotine-induced hypothermia, and basal corticosterone in a sex-dependent manner. *Neurotoxicology and Teratology, 85.* https://doi.org/10.1016/j.ntt.2021.106972.

CHAPTER EIGHT

Expression and function of nonneuronal nAChRs

8.1 Expression of nAChRs and a cholinergic system in nonneuronal cells

In this chapter, we will explore the role of ACh and nAChRs in cells other than neurons. As seen below, ACh and nAChRs are expressed in many cells (Table 8.1). ACh has been found in many tissues and organisms. Plants, bacteria, algae, and fungi are but a few of the evolutionarily "old" organisms that use ACh. Cells have been using ACh as a chemical signal for 3 billion years and only relatively recently (500 million years) used in the nervous system (Grando et al., 2020). Thus, in addition to a large number of organisms, a plethora of nonneuronal cell types express ACh as well as ChAT, AChE, and both nAChRs and mAChRs. ACh and ChAT have been detected in immune cells, lung, skin, endothelial cells, most epithelial cells, cultured glial cells, and placenta, among others. ACh is used as a signal to control processes of adhesion, angiogenesis, migration, immune function, cell absorption, growth, and proliferation (Sharma & Vijayaraghavan, 2002; Wessler et al., 1999; Wessler & Kirkpatrick, 2008) during development and in normal and aberrant (cancer) cells in adults. Expression of nAChRs in lung cancer will be discussed in Chapter 5.

Activation of nAChRs can produce changes in gene expression in cells such as endothelial cells (Sharma & Vijayaraghavan, 2002). In nonneuronal cells, effects of activation of nAChRs in some cells may produce the opposite effect of activation in other cell types. In this chapter, we will examine the expression of nAChR subtypes regulating specific functions in nonneuronal cells. We will focus on the role of cholinergic signaling in (1) the immune system and (2) skin. The knowledge of cholinergic signaling in these areas is more advanced that in other nonneuronal cells.

Nicotinic Acetylcholine Receptors in Health and Disease
https://doi.org/10.1016/B978-0-12-819958-9.00010-4

Table 8.1 Cholinergic system in nonneuronal cells.

nAChR subunit expression in human cells
Airway epithelial cells
Skin keratinocytes
Skin melanocytes
Intestine
Endothelial cells
Multiple types of immune cells
Mesenchymal cells
Mesothelial cells
Astrocytes
ACh or ChAT expression in human cells
Eye
Kidney
Skin
Placenta
Urogenital tract
Lung
GI tract
Hair
Platelets
Breast
Sperm
Liver
Immune cells
Astrocytes
Bone cells and osteoclast
Stem cells
Vagina

Adapted from Wessler, I., Kirkpatrick, C., & Racke, K. (1999). The cholinergic "pitfall": Acetylcholine, a universal cell molecule in biological systems, including humans. *Clinical and Experimental Pharmacology & Physiology, 26*, 198–205; Beckmann, J., & Lips, K. S. (2013). The non-neuronal cholinergic system in health and disease. *Pharmacology, 92* (5–6), 286–302. https://doi.org/10.1159/000355835. Wessler, I, & Kirkpatrick, C. J. (2008). Acetylcholine beyond neurons: The non-neuronal cholinergic system in humans. British Journal of Pharmacology 154, 1558–1571; Wessler, I. K., & Kirkpatrick, C. J. (2017). Nonneuronal acetylcholine involved in reproduction in mammals and honeybees. Journal of neurochemistry 142(Suppl. 2), 144–150.

8.2 Nonneuronal cholinergic system overview

nAChR RNAs and proteins have been detected in the cells we will discuss below. Thus, a complete nonneuronal cholinergic system (ChAT, AChE, and choline transporter) is present in many cells and in some that are not part of or don't directly receive signals from the nervous system

(Table 8.1). Choline and acetyl–CoA are also ubiquitous. ACh can be synthesized by ChAT or carnitine acetyltransferase. Specific nonneuronal cell types vary in which enzyme is expressed. ACh can also function via autocrine or paracrine mechanisms in many tissues and cell types (Beckmann & Lips, 2013) and not just through synaptic signaling.

ACh is released from neurons (for example, the parasympathetic nervous system) onto nonneuronal cells such as immune cells. There is widespread communication of neurons with various targets such as the skin, glands, and many targets in the autonomic nervous system. For the most part, ACh is not stored in vesicles in nonneuronal cells but released as a local signal (Wessler et al., 1999). Organic cation transporters (OCTs) may be involved in nonneuronal release (Wessler & Kirkpatrick, 2008). This is clearly the case for release from the placenta. Airway epithelial use OCTs also. OCTs are widely expressed. Release of ACh from human skin is not blocked by botulinum toxin, which disrupts synaptic vesicle release in neurons (Wessler & Kirkpatrick, 2008). It is likely that at least in most nonneuronal cells, neuronal-type vesicular release is not used. The mediatophore protein may also be used for ACh release from nonneuronal cells (Beckmann & Lips, 2013). The high-affinity choline transporter is also used in some nonneuronal cells for choline reuptake, and in others, choline-like transport proteins are used (Beckmann & Lips, 2013). In summary, multiple elements of the cholinergic system are present in multiple cell types. ACh is synthesized, released under control, degraded, and can signal via nAChRs.

8.3 Cholinergic system in immune cells

The immune system has innate and adaptive responses. Innate responses are rapid and not specific to a particular antigen or pathogen. Adaptive responses are specific using B cells, T cells, and other immune cells, but take a week or more to develop. T cells and B cells express ChAT RNA and protein and release ACh. T cells release ACh and use autocrine pathways to regulate proliferation and cytokine release (Fujii et al., 2017).

Dendritic cells (DCs) and macrophages also express ChAT (Reardon et al., 2013) as well as α7 nAChRs. Macrophages also express toll-like receptors (TLRs), a part of the innate immune response. Activation of TLR by microbial components stimulates release of TNF-α. ChAT expressing cells are detected in the lymph nodes, spleen, and other secondary lymphoid organs and in mature CD4+ T cells and mature B cells. However, ChAT was not detected in cells in the thymus or bone marrow (Reardon et al.,

2013). This is consistent with ChAT expression being stimulated in mature T or B cells. Expression of ChAT in B cells after stimulation with various compounds (i.e., LPS, which activates TLR)) was dependent on the presence of MyD88, a molecule involved in transmitting TLR signals as part of the innate immune response (Reardon et al., 2013). Microbial colonization, expected to activate TLRs, induced a transient ChAT expression. Increased expression of ChAT can reoccur if B cells are stimulated again. B cells were not stimulated to produce ACh by norepinephrine (NE) as are T cells, thus responding to different stimuli (Reardon et al., 2013). The ChAT-expressing B cells don't suppress adaptive immune responses or change inflammatory cytokine production, but reduce neutrophil recruitment locally (Reardon et al., 2013) without vagus nerve involvement. The neutrophils that are present are active. Interestingly, van der Zanden et al. (2009) showed that ACh and vagus nerve stimulation can enhance phagocytic activity in peritoneal macrophages via an $\alpha4\beta2$ nAChR-mediated mechanism. Thus, ACh-expressing cells are stimulated by activation of the innate immune response.

Immune cells have been shown to synthesize ACh besides T and B cells. The evidence consists of ChAT detection by immune techniques, enzyme activity, and RNA (Fujii et al., 2017). Mouse mononuclear lymphocytes (MNLs) are also expressed ChAT RNA if stimulated with concanavalin A (CONA) but not otherwise. ACh is present in human blood, and most is thought to arise from T and B cells and a small amount from monocytes as well (Fujii et al., 2017). In rat, CD4+ T cells have more than CD8+ and both have more than B cells. ChAT activity is detected in human mononuclear leukocytes (MNLs) and rat lymphocytes (Fujii et al., 2017). Numerous T and B cell cancer cell lines and MNLs express ChAT RNA and activity. AChE is also detected in MNLs, DC, and macrophages. The same ChAT RNA (same gene) as in the brain is expressed in T lymphocytes (Fujii et al., 2017). While some nonneuronal cells also express carnitine acetyltransferase that can synthesize ACh, T cell activation increased ChAT activity, but not carnitine transferase (Fujii et al., 2017). It is clear that stimulation of T cells upregulates ACh. Activation of B cells also increases ChAT RNA expression (Fujii et al., 2017). DC can also express ChAT but only after activation, but it is not clear that they make ACh. ChAT expression was not detected in mouse macrophages, even after stimulation (Fujii et al., 2017). However, human alveolar macrophages express ChAT and ACh (Wessler & Kirkpatrick, 2001). Activation of T cells by antigens can increase expression of ACh and ChAT (Kawashima et al., 2007).

In neurons, ACh is packaged into synaptic vesicles for release. This is accomplished by a vesicular acetylcholine transporter (VAChT). There is some evidence that human T and B cells express VAChT (Tayebati et al., 2002), while other studies don't show that ACh is stored in vesicles (Fujii et al., 2017). Mediatophore releases ACh in the Torpedo electroplaque (Dunant et al., 2009). Mediatophore is present in some human leukemic cells and is upregulated by immune activation (Fujii et al., 2009). It is likely that mediatophore is involved at some level in ACh release from T cells as it is in other nonneuronal cells.

AChE mRNA is expressed in many immune cells including DC, macrophages, and T and B lymphocytes. AChE was detected in MNLs, DC, and macrophages. nAChRs are also regulated by endogenous ligands such as SLURP-1 and SLURP-2 (see Chapter 9). SLURP-1 and 2 act to regulate multiple nAChRs. RNAs for SLURPs are detected in many organs, as well as in DCs, macrophages, and MNLs (Fujii et al., 2017). SLURP protein has been detected in CD4+ T cells and DCs and may increase the autocrine response of T cells to ACh. There is evidence that SLURP-1 may increase T cell ACh release (Fujii et al., 2017). Thus, regulatory signals such as SLURPs also may act on nAChRs in nonneuronal cells.

Immune cells express both nAChRs and mAChRs. Multiple immune cells express nAChRs. MNLs isolated from mice expressed $\alpha2$, $\alpha3$–$\alpha10$ and $\beta4$ nAChR RNAs. Mouse immature T cells express $\alpha3$, $\alpha5$, $\alpha7$, $\beta2$, and $\beta4$ nAChR RNAs, while a subset of these ($\alpha2$, $\alpha5$, and $\alpha7$) are present in peripheral T cells (Kuo et al., 2002). Several subunits are expressed, but $\alpha7$ seems to have a prominent role with $\alpha4\beta2$ nAChRs also involved. Roles of other subtypes are less clear. Mouse DCs expressed the same complement of nAChR RNAs as MNLs except $\alpha9$ (Kawashima et al., 2007) and expressed ChAT if stimulated with LPS. Macrophages also expressed the same set as MNLs and $\alpha3$ RNA was detected as well, but no ChAT was detected even after stimulation (Kawashima et al., 2007). nAChR subunit proteins ($\alpha4$, $\alpha7$, and $\beta2$) are present in mouse B cells (Skok et al., 2005). Epibatidine binding indicated that assembled $\alpha4\beta2^*$ nAChRs were present, and α–Bgt binding indicated assembled $\alpha7$ nAChRs. Skok et al. (2005) also showed that these nAChRs may be involved in preimmune status and B cell regulation. Knockout (KO) mice for $\alpha4$, $\beta2$, and $\alpha7$ all showed lower levels of immunoglobulin G (IGG), but not IGM in serum. However, after immunization with cytochrome c, $\beta2$ and $\alpha4$ KO mice had a higher IgG response (Skok et al., 2005). Further support for a role of nAChRs in B cell function comes from work showing that mice lacking $\alpha7$ have a greater IgG response to a common test antigen, ovalbumin, than wild-type mice (Kawashima et al., 2007). Thus, α7nAChRs can regulate antibody production. T cells, B cells, DCs, and macrophages all express α7 nAChRs. The Dupα7 gene is also expressed. This gene contains exons 5–10 from the normal α7 nAChR gene and four new exons 1–4 encoded by the FAM7 gene (Fujii et al., 2017). This is not known to function as a channel, but may regulate α7 function in immune cells. The Dupα7 may act as a negative regulator since it doesn't contain binding sites for ACh, but might assemble with α7 subunits.

To summarize, all of the necessary components for the cholinergic system such as ACh, ChAT, AChE, and nAChRs are present in various immune cells. Multiple nAChR subunits are present in T cells, B cells, macrophages, DCs, and MNLs. Both assembled α7 nAChRs and α4β2* nAChRs have also been detected in some immune cells.

8.3.1 Cholinergic antiinflammatory network

Over the last 20 years or so, the existence of an "antiinflammatory cholinergic network" has been uncovered. ACh is released from the vagus nerve. ACh leads to reduction of inflammatory cytokines (Rosas-Ballina & Tracey, 2009). This forms the basis for the cholinergic antiinflammatory network. Clinical applications arise from this by using cholinergic compounds to regulate inflammation. For example, nicotine has been used to treat inflammatory conditions such as inflammatory bowel disease. Interactions and signaling between neural and immune cells are an important regulator of immune response and an important function of nonneuronal nAChRs.

Nervous system signaling through the vagus nerve inhibits inflammatory signals by reducing TNF-α release, a mediator of a number of inflammatory events (Borovikova et al., 2000; Wang et al., 2003). ACh released from the vagus nerve was shown to inhibit the release of TNF-α and other cytokines such as IL-1beta, IL-6, and IL-18 from cultured human macrophages (Borovikova et al., 2000). Stimulation of the vagus nerve in vivo reduced serum and liver TNF-α (Borovikova et al., 2000). Blockade of mAChRs with atropine didn't eliminate the effect of ACh on TNF-α release. α-Bgt did block the effects of ACh, suggesting that α7 nAChRs were involved (Borovikova et al., 2000).

Macrophages were shown to express α7 nAChRs (Wang et al., 2003). RT-PCR and immunoblotting were used to identify the α7 protein and RNA and also showed that α-Bgt binds in clusters to the surface of the human macrophages. Nicotine was shown to inhibit the release of TNF-α from macrophages, as was previously shown for ACh (Wang et al., 2003). Expression of α7 nAChRs was required for the inhibitory actions of nicotine. Wang et al. (2003) also showed that α7 KO mice expressed higher levels of TNF-α, IL-6, and IL-1beta (inflammatory cytokines) in response to a challenge with LPS. Macrophages from mice lacking α7 also didn't respond to ACh or nicotine by reducing TNF-α expression. In α7 KO mice, vagal nerve stimulation didn't reduce the amount of TNF released as would normally occur (Wang et al., 2003). Thus, α7 nAChRs on macrophages play an important role in mediating the antiinflammatory response of the cholinergic system. Modulation of α7 nAChRs may be useful in controlling the inflammatory response. Evidence also supports a role for α7 in reducing antigen presentation by DCs, reducing DC and macrophage function, promotion of T cell differentiation into Tregs, and suppression of Th1 and Th2 cell production (Fujii et al., 2017).

Smoking can alter the risk of autoimmune inflammatory disorders. Smoking (nicotine) can increase the risk of Crohn's disease and reduce the incidence of ulcerative colitis (Fujii et al., 2017). Nicotine has been shown to reduce monocyte production of TNF-α, IL-1b, and IL 12 and that α7 and α9–containing nAChRs are required for this effect (St-Pierre et al., 2016). This demonstrates that other nAChR subtypes besides α7 nAChRs participate in the cholinergic antiinflammatory response. Nicotine also lowers the level of α5 and α7 RNAs in human MNLs (Fujii et al., 2017). Nicotine in a range available in smokers' blood lowers α3, α5, α6, and α7 RNA levels in cultured leukemic T cells also (Kimura et al., 2003). This ties together smoking and effects on immune function through potential regulation of nAChR expression.

We described the results above of Wang et al. (2003) on the role of the vagus nerve in the cholinergic antiinflammatory reflex. Evidence supports the spleen as being the important player in that splenectomy in rats abolished the antiinflammatory effect of nicotine and vagal stimulation. Other immune organs also receive sympathetic innervation (Fujii et al., 2017). Clearly, the vagus nerve is key to the cholinergic antiinflammatory response (Wang et al., 2003), but the vagus nerve doesn't innervate any immune organ including the spleen (Fujii et al., 2017). Rosas–Ballina et al. (2011) propose that the vagus action is through signals to the celiac ganglion, which are then transmitted to the spleen (Fig. 8.1). Direct activation of the splenic nerve also leads to reduction in TNF-α production (Rosas–Ballina et al., 2011). Vagal nerve stimulation was shown to stimulate ACh release in

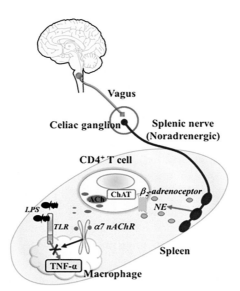

Fig. 8.1 Cholinergic antiinflammatory action via spleen T cells and macrophage nAChRs. Vagus nerve stimulation of noradrenergic neurons in the celiac ganglion leads to release of norepinephrine (NE) in the spleen. CD4+ T cells resident in the spleen release acetylcholine (ACh) in response to NE. ACh acts on α7 nAChRs on macrophages in the spleen to inhibit tumor necrosis factor-α (TNF-α) production normally stimulated by lipopolysaccharide (LPS) interactions with toll-like receptors (TLRs). Note that nAChRs are on the macrophages in this model. *(From Fujii, T., Mashimo, M., Moriwaki, Y., Misawa, H., Ono, S., Horiguchi, K., & Kawashima, K. (2017). Physiological functions of the cholinergic system in immune cells.* Journal of Pharmacological Sciences, *134 (1), 1–21. https://doi. org/10.1016/j.jphs.2017.05.002.)*

the spleen (Rosas–Ballina et al., 2011). They also showed that ACh release was increased in the presence of increasing amounts of NE in lymphocytes isolated from the spleen. In nude mice, which are lacking T cells, vagal stimulation didn't reduce TNF-α release. CD3 and CD4+ T cells were shown to express ChAT while CD8+ was not (Rosas–Ballina et al., 2011). Activation increased ACh release. ChAT + lymphocytes were close to splenic nerve fibers producing catecholamines. T cells in the spleen express adrenergic receptors. ChAT activity was important for the effects on TNF-α levels as vagal nerve stimulation didn't reduce TNF-α in CD4+ T cells in which ChAT was reduced by siRNA treatment (Rosas–Ballina et al., 2011) (Fig. 8.1). T cells may also be activated outside the spleen by the vagus nerve, migrate into the spleen, and release ACh onto sympathetic neurons

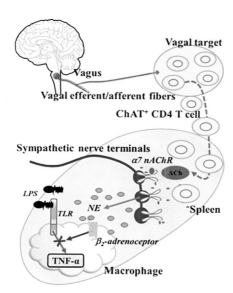

Fig. 8.2 Cholinergic antiinflammatory effects mediated by nonresident T cells and sympathetic nerve nAChRs. The vagus nerve activates T cell outside the spleen with subsequent movement of T cells into the spleen. The CD4+ T cells release ACh onto sympathetic nerve terminals containing α7 nAChRs stimulating release of NE in the spleen. NE acting through β2-adrenoreceptors blocks TNF-α production normally stimulated by LPS interactions with TLRs. Note that nAChRs are on the sympathetic neurons in this model. *(From Fujii, T., Mashimo, M., Moriwaki, Y., Misawa, H., Ono, S., Horiguchi, K., & Kawashima, K. (2017). Physiological functions of the cholinergic system in immune cells.* Journal of Pharmacological Sciences, 134 *(1), 1–21. https://doi.org/10.1016/j.jphs.2017. 05.002.)*

controlling NE release (Fig. 8.2). Vida et al. (2011) showed that vagus nerve stimulation increased production of NE in the spleen. The efferent vagal nerve stimulation increased NE release in an α7–nAChR–dependent manner. Their model proposes that ACh released into celiac ganglia activates α7-nAChRs on splenic neurons. ACh agonists applied to the celiac mesenteric ganglia produce the same effects as vagus nerve stimulation. Splenic neuron stimulation then leads to the release of NE. This model doesn't require expression of α7 nAChRs in the spleen, but on celiac ganglion neurons (Fig. 8.3).

However, since NE can also suppress TNF-α alone (Fujii et al., 2017), the role of α7 nAChRs in the spleen is not clear (Fig. 8.3). There are alternative models as to how vagal stimulation results in reduced cytokine

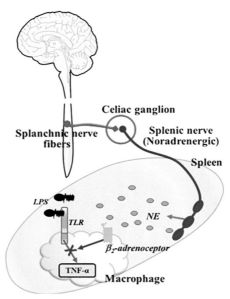

Fig. 8.3 Cholinergic antiinflammatory effects mediated by celiac ganglion nAChRs. Activation of splenic neurons in the celiac ganglion release NE into the spleen. This model doesn't require nAChRs in the spleen, T cells, or macrophages. However, ACh is released in the celiac ganglion onto nAChRs stimulating the noradrenergic splenic nerve. NE released from into the spleen inhibits TNF-α production as in Fig. 8.2. *(From Fujii, T., Mashimo, M., Moriwaki, Y., Misawa, H., Ono, S., Horiguchi, K., & Kawashima, K. (2017). Physiological functions of the cholinergic system in immune cells.* Journal of Pharmacological Sciences, 134 *(1), 1–21. https://doi.org/10.1016/j.jphs.2017.05.002.)*

production (Figs. 8.1 and 8.2). However, earlier work (Wang et al., 2003) showed TNF-α production after LPS stimulation was decreased with nicotine, and this effect was blocked in macrophages treated with α7 antisense oligonucleotides, supporting some role for signaling through α7 nAChRs in reduction of cytokine expression. The vagus nerve signals to the splenic nerve, which then releases norepinephrine (NE). NE then acts in the spleen to stimulate release of ACh from CD4+ T cells (ACh has been identified in the spleen). Immune cells such as macrophages, shown to express α7 nAChRs, are inhibited and release less TNF-α and other pro–inflammatory cytokines (Fujii et al., 2017; Rosas–Ballina et al., 2011).

In summary, immune cells express nAChRs and release ACh. This local signaling in the spleen is important for immune regulation. There are multiple mechanisms proposed for cholinergic signaling though α7 functions via the vague nerve and for a role of NE release by sympathetic nerve or splenic nerve in the antiinflammatory response. However, α7 nAChRs are clearly involved as well as ACh release from T cells. nAChRs play an important role in immune signaling. α7 nAChRs are crucial to the cholinergic antiinflammatory network.

8.3.2 Nicotinic cholinergic signaling in T cells and mast cells

Given the role of nAChRs in the cholinergic antiinflammatory response, how else are nAChRs involved in the immune system? So far there's no evidence that the nAChRs function as ion channels in immune cells. Electrophysiological recordings have been made, and no channel function has been detected. We know that modulating the function of nAChRs has effects on the immune system such as immune suppression produced by nicotine exposure from smoking. Clues to at least one mechanism are found in the paper by Razani-Boroujerdi et al. (2007). They investigated signaling in T cells, which express α7 containing nAChRs. They demonstrated that T cell lines such as Jurkat cells express epibatidine binding as well as spleen cells. Epibatidine binds to both heteromeric and α7 homomeric nAChRs. Fluorescently labeled α–Bgt also bound to spleen cells and to Jurkat cells as well (Razani-Boroujerdi et al., 2007). Even more spleen cells bound α–Bgt after nicotine treatment or exposure to immune stimulation by CONA. α–Bgt binds to muscle nicotinic receptors as well as α7 nAChRs. Antibodies to α7 demonstrated that a 58 kD band (typical of α7 nAChR proteins) was present in rat spleen T cells and Jurkat cells (Razani-Boroujerdi et al., 2007). Nicotine increased the amount of α7 nAChR protein detected in rat spleen cells. A full-length α7 nAChR RNA was expressed in Jurkat cells and human peripheral blood T cells and had the same sequence as that expressed in the brain. The 58 kD protein was also phosphorylated at tyrosine in splenic T cells, and the level of phosphorylation increased after nicotine treatment or exposure to an anti-T cell receptor antibody. Nicotine treatment of T cells was also shown to increase internal Ca^{2+} levels (Sopori et al., 1998). α7 nAChRs gate Ca^{2+} and activation can lead to an increase in the internal Ca^{2+} levels in neurons. Since no ion channel activity has been detected, α7 nAChRs in T cells may function differently. Protein tyrosine kinase (PTK) activity was required for an increase in internal Ca^{2+} after nicotine treatment in Jurkat cells and splenic T cells, while in PC12 (neuroendocrine cells expressing α7 nAChRs), it was not (Razani-Boroujerdi et al., 2007). Surprisingly, MLA and α–Bgt (both α7 nAChR antagonists in neurons or PC 12 cells) each increased internal Ca^{2+} levels in the absence of nicotine in Jurkat cells, thus acting as agonists rather than antagonists.

T cells are activated by T cell receptor (TCR) binding and depend on Lck and Fyn kinases to increase intracellular Ca^{2+} release from stores and Ca^{2+} influx from without. T cells lacking the Lck kinase were not activated by either nicotine or TCR ligation (Razani-Boroujerdi et al., 2007). Further studies showed that nicotine and TCR ligation signaling that produces increased Ca^{2+} also required an intact TCR/CD3 complex. Thus, it appears that T cells are activated by nicotine and TCR ligation by similar mechanisms. Antisense knockdown of $\alpha 7$ nAChRs was used to show that they were required for the nicotine-mediated increase in Ca^{2+} in the Jurkat cells or splenic T cells, but not for the Ca^{2+} response to TCR activation. Co-immunoprecipitation studies showed that $\alpha 7$ nAChR protein and the CD3 complex were physically associated with each other in T cells. The nicotine-induced increase in Ca^{2+} from stores doesn't depend on external Ca^{2+}, indicating that the $\alpha 7$ nAChRs are not acting as ion channels (Razani-Boroujerdi et al., 2007). This indicates that the $\alpha 7$ protein, identical to that expressed in neurons, was associated in some way with a TCR/CD3 zeta chain complex. Nicotine increased the response of the T cells in a manner that requires $\alpha 7$ protein, Lck kinase and interaction with CD3 zeta that doesn't require $\alpha 7$ nAChRs to function as an ion channel.

Smoking (nicotine) has antiinflammatory effects in some allergic disorders, but the mechanisms by which this happens in various immune cells are not clear. In other cases, immune suppression by nicotine exacerbates infections. Mast cells mediate inflammatory responses, and after IgE binding, the cells initially release histamine and later release multiple cytokines and leukotrienes. Nicotine can inhibit this response in mast cells, but the mechanism has not been identified. Some clues come from the work of Mishra et al. (2010). Exposure to low levels of nicotine (at a level lower than that in smokers) for at least 8 h reduced expression of leukotriene C4 after cell activation. The level of nicotine used was orders of magnitude lower than that needed to produce whole cell currents through $\alpha 7$ nAChRs in neurons. This effect lasted more than 24 h (Mishra et al., 2010). The reduction was greater than that produced by an antiasthma drug. The nicotine metabolite cotinine did not produce this effect. Nicotine pretreatment also reduced release of the cytokine TNF-α. However, nicotine pretreatment did not affect the release of histamine, an acute response (Mishra et al., 2010). Nicotine pretreatment for 8 h also reduced the PI3K/Akt/MAPK/NF-kB pathway. Nicotine also reduced cPLA2 activity, suggesting a mechanism for nicotine's action in reducing cytokine and leukotriene expression by mast cells (Mishra et al., 2010). A mast cell line bound α-Bgt and expressed $\alpha7$, $\alpha9$, and $\alpha10$ nAChR subunits. α-Bgt binds $\alpha7$ and $\alpha9$ nAChR homomers and $\alpha9/\alpha10$ heteromers. PCR demonstrated that the mast cells expressed full length $\alpha7$, $\alpha9$, and $\alpha10$ RNAs of the same sequence as those present in neurons. These effects of nicotine and the delayed mast cell response were blocked by $\alpha7/\alpha9$ antagonists MLA and α-Bgt. This was confirmed by using siRNA to knock down $\alpha7$, $\alpha9$, and $\alpha10$ nAChR expression. This receptor combination mediating these effects in mast cells appears to be distinct from that seen in T cells (Razani–Boroujerdi et al., 2007).

8.3.3 Therapeutic modulation of $\alpha7$ nAChRs in immune function

Nicotine reduces the effects of various inflammatory diseases such as inflammatory pain and Crohn's disease. However, due to the many harmful effects of nicotine, it can't be used to treat immune disorders. Drugs that target nAChRs in immune function may be useful. The most prominent subtype involved in immune modulation is the $\alpha7$ nAChR. How $\alpha7$ nAChRs function in the antiinflammatory response is not clear. $\alpha7$ nAChRs in neurons form channels that open quickly and desensitize rapidly. They also are permeable to Ca^{2+} in addition to Na^+, 10–20× more so than for other nAChR subtypes (Kabbani & Nichols, 2018). Since $\alpha7$ nAChRs desensitize quickly, it was postulated that they may signal by other mechanisms. However, $\alpha7$ nAChRs haven't been shown to form ion channels in T cells but are important for signaling. It has recently been shown that after opening, Na^+ and Ca^{2+} enter $\alpha7$ nAChRs, but the channels quickly shut and transition to a ligand bound and desensitized state wherein they couple to a G protein, Gaq. This activates PLC and IP3-mediated Ca^{2+} release from intracellular stores (Kabbani & Nichols, 2018). Thus, $\alpha7$ nAChR signaling in immune cells may be due to coupling of the ionotropic receptors to G-protein signaling mechanisms.

Recently, $\alpha7$ silent agonists, compounds that activate the channel poorly but produce desensitization, were shown to be effective in reducing pain (Godin et al., 2020). A specific silent agonist, 1-ethyl-4-(3(bromo)phenyl) piperazine (*m*-bromo PEP), reduced mouse bone marrow-derived monocyte/macrophage counts and reduced cytokine production from these cells (Godin et al., 2020). M-bromo PEP also reduced disease in experimental autoimmune encephalitis and chronic pain in mice (Godin et al., 2020).

It is possible that this effect and α7 signaling in others cells involved in the cholinergic antiinflammatory response may be mediated by metabotropic effects of the receptor.

8.3.4 Immune summary

α7 nAChRs are the most prominent subtype involved in the immune system, but there is evidence also for α4β2* and α3-containing nAChRs being involved in the immune response. α7 nAChRs may be acting not as ion channels but coupling with other molecules such as the TCR to influence signaling. Mast cells may use receptors containing α9 and or α9/α10 nAChR subunits to also mediate responses. Signaling may also occur through α7 nAChR signaling via a nonconducting state through metabotropic mechanisms. Due to the role of nAChRs in immune responses of diverse kinds, more research is needed to develop drugs (other than nicotine) to regulate immune responses in therapeutic ways.

8.4 Cholinergic system in skin cells

8.4.1 Nicotine effects on skin

Smoking has been long known to have visible effects on skin, such as yellowing, wrinkles, and rapid aging (Ortiz & Grando, 2012). Smokers can develop a keratosis of the tongue. Smoking can lead to free radical formation, that may account for some effects on skin aging as well as disrupting the extracellular matrix (Ortiz & Grando, 2012). Nicotine can bind melanin and remain in the skin. Nicotine alone, separate from other components of smoke, can alter apoptosis. Smoking has been shown to increase squamous cell carcinoma (De Hertog et al., 2001), but the association of smoking with basal cell carcinoma and melanoma is not clear (Ortiz & Grando, 2012). Nicotine is a mitogen, and this activity is influenced by activity through α7 nAChRs. The role of nAChRs in cancer is presented in Chapter 5. Smoking inhibits wound healing in several ways (Ortiz & Grando, 2012). However, nicotine in low doses can enhance healing and only inhibits in large doses (Ortiz & Grando, 2012). Nicotine patches can produce contact dermatitis, and smoking is associated with numerous chronic dermatoses (Ortiz & Grando, 2012). In some cases, nicotine mitigates symptoms of skin diseases, even in cases where smoking makes them worse (due to other components of smoke). Some of the therapeutic effects of nicotine in skin diseases may be due to nicotine reducing the levels of cytokines from keratinocytes and from epithelial cells from the dermis (Ortiz & Grando, 2012). Nicotine treatment alone reduces inflammation in ulcerative colitis. α7 nAChR agonists reduce damage in ulcerative colitis models (Ortiz & Grando, 2012). Some of these effects may occur independently from interactions with nAChRs, but some are clearly mediated by α7 nAChRs and potentially other subtypes including those containing α3 nAChR subunits. Given some of the effects due to nAChRs in skin, many of the skin problems described above can be accounted for by aberrant nAChR signaling due to nicotine exposure.

8.4.2 nAChR functions in skin

Nicotine exposure to the skin, as in other nonneuronal cells, usurps and potentially interferes with the normal function of ACh and nAChRs in the skin. ACh is released from several cell types in and near the skin including from keratinocytes, dermal fibroblasts, epithelial cells, melanocytes, immune cells, and local nerves (Kindt et al., 2008). ACh levels in human skin are around 1000 pmol/gram, compared with the levels in airways of 33 pmol, and in oral mucosa 8 pmol (Arredondo, Hall, et al., 2003). nAChRs are present in multiple cell types in the skin. mAChRs are present as well, but the actions of nAChRs will be emphasized here. The cholinergic system in the skin is involved in numerous processes including cell migration, proliferation, and differentiation (Kurzen et al., 2006) and growth of keratinocytes.

Healthy human skin expresses the nAChR subunit proteins α3, α5, α7, α9, and α10 (Kindt et al., 2008) and mRNAs. β2 and β4 subunits have also been detected in the skin (Nguyen et al., 2004). Expression levels varied, but nAChRs are expressed in several layers of the epidermis including, spinous, granule, and basal, and in keratinocytes and mast cells in skin (Table 8.2). The highest expression occurred for α3 and α9 subunits, and α5 was the lowest. α7 nAChR and α9 nAChR subunit proteins were detected in keratinocytes; α10 was detected in all layers. Changes in nAChR expression occurred in human patients with atopic dermatitis (AD), and these varied between areas of skin with lesions and those without (Kindt et al., 2008). α3, α7, α9, and α10 proteins were reduced in both lesioned and nonlesioned AD skin. α5 expression was stable. α10 reduction was greater in lesioned, and α3 reduction was greater in nonlesioned skin (Kindt et al., 2008). High levels of ACh in the skin have been associated with AD. Mast cells present in the skin expressed α10 nAChR subunits; α3 and α5 nAChR subunits also were present in AD lesioned skin. ACh released from skin could stimulate mast cell degranulation and lead to skin inflammation seen in AD.

Table 8.2 nAChR expression in skin.

Subunit	Layers of epidermis
α3	Stratum basale, stratum granulosum
α5	Stratum basale, stratum granulosum
β2	Stratum basale, stratum granulosum
β4	Stratum basale, stratum granulosum
α7	Stratum spinosum, stratum granulosum
α9	Stratum basale, lowest suprabasal
α10	Stratum spinosum, stratum granulosum

Subunit	Pilosebasceous unit
α3	
α5	
α7	
α9	
α10	
β2	
β4	

Melanocytes
α3
α5
α7
β2

Dermis
α3
α5
α7
β2
β4

From Kurzen, H., Wessler, I., Kirkpatrick, C. J., Kawashima, K., & Grando, S. A. (2007). The non-neuronal cholinergic system of human skin. *Hormone and Metabolic Research, 39* (2), 125–135. https://doi.org/10.1055/s-2007-961816.

nAChRs control aspects of keratinocyte function such as adhesion, differentiation, migration, and proliferation. Keratinocytes are important functional and structural elements of skin. nAChRs and mAChRs have roles in keratinocyte cell adhesion (Nguyen et al., 2004). α3, α5, α7, α10, β2, and β4 subunits are present in keratinocytes (Nguyen et al., 2004). Keratinocytes also express the α9-containing nAChRs, which possess pharmacological properties of both nAChRs and mAChRs. As described previously, each nAChR subtype has a unique pharmacology.

α9/α10 heteromers are even more distinct in that neither muscarine (classic mAChR agonist) nor nicotine (class nAChR agonist) activates this receptor subtype (Kurzen et al., 2007). ACh does, but this activation is reduced by nicotine or muscarine. α9α10 heteromeric nAChRs are blocked by atropine and α-Btx. Some nAChRs such as α9 and α7 are also inhibited by atropine (Kurzen et al., 2007). Strychnine (classic GABA antagonist) also inhibits α9 nAChRs. The role of ACh in various functions in the skin is complex and mediated by mAChRs, classic nAChRs, and nAChRs with a mixed pharmacology. Choline also can act as a cholinergic agonist and can attract keratinocytes via actions through α7 nAChRs (Kurzen et al., 2007).

Human keratinocytes in addition to expressing a number of nAChR subunits also release ACh. It was shown that the expressed subunits form functional receptors. Human mature keratinocytes exhibit single channel currents in response to acetylcholine application, which are blocked by nACh antagonists mecamylamine or κ-bungarotoxin (Grando et al., 1995). Keratinocytes also release cytokines that can stimulate lymphocyte movement into the skin. Keratinocytes and T cells interact and lead to ACh release from both cell types and subsequent signaling through mAChRs and nAChRs. These signals can effect differentiation, growth, adhesion, and apoptosis (Kurzen et al., 2007).

Few immature keratinocytes demonstrate channel activity. κ-bungarotoxin can block α3 or α4 subunit containing nAChRs (Boulter et al., 1987). Binding studies demonstrated surface expression of a large number of κ-bungarotoxin sensitive receptors in mature keratinocytes (comparable with the number seen on the surface of neurons), and a much smaller but still detectable number in undifferentiated cells (Grando et al., 1995). Keratinocytes expressed multiple nAChR subunit RNAs (detected by PCR), and α3 and β2 receptor proteins were also detected in cultured keratinocytes and the skin (Grando et al., 1995). Mecamylamine and κ- bungarotoxin applied to cultured keratinocytes caused cell detachment. ACh and nicotine promoted attachment of cultured keratinocytes to a surface. These also promote migration that was abolished with mecamylamine treatment.

Pharmacological blockade of α3, α7, or α9 containing nAChRs also produces keratinocyte cell-cell detachment (Nguyen et al., 2004). Antisense knockdown targeting α3 or α9 (and M3 mAChRs) also produced detachment. A combination of α3 and α9 knockdown produced a greater increase in detachment than either α3 or α9 alone (Nguyen et al., 2004). Antisense reduction of α3 and α9 subunits expression reduced levels of adhesion molecules E cadherin, β-catenin, and γ- catenin. In α3 KO mice, abnormalities in epithelium and loss of keratinocyte attachment were observed, but not in α9 KO mice (Nguyen et al., 2004). α3 KO mice showed reduced E-cadherin and γ-catenin protein levels, but increased β-catenin expression (Nguyen et al., 2004). α9 KO mice, in contrast, had increased E cadherin and β-catenin protein expression and reduced γ-catenin. Increased adhesion and movement are produced by nicotinic agonists, and both mAChRs and nAChRs are involved (Nguyen et al., 2004). α7 nAChR KO mice have reduced levels of pro-apoptotic regulators Bad and Bax indicating a role for nAChRs in control of keratinocyte cell apoptosis (Arredondo, Nguyen, et al., 2003). Cholinergic signaling affected cell adhesion molecule expression and phosphorylation and plays multiple roles in mature and immature keratinocytes.

Pharmacological blockade of multiple AChR subtypes in an in vitro skin system prevented epidermal proliferation and differentiation (Kurzen et al., 2007). Blockage of only mecamylamine-sensitive nAChRs (which would not include α9 or mAChRs) was not as effective. Reduced cell adhesion produced death (Kurzen et al., 2007). Activation of nAChRs increased the thickness of the epithelium. ACh and choline function in development of epithelial layers and functioning of keratinocytes, with choline acting as a chemoattractant for human keratinocytes (Kurzen et al., 2007).

AChRs are also present in the pilosebasceous unit (hair follicle, sebaceous glands, and piloerector muscle). Subunit expression varies in different parts, such as the infidibulum or the root sheath, but overall α3, α5, α7, α9, α10, β2, and β4 nAChR subunits are detected. Nicotine exposure increases sebum production. Sebocytes express multiple nAChR subunits including α3, α9, β4, α7, β2, and α5 (Kurzen et al., 2007).

Human melanocytes also express multiple nAChR subunits as well as mAChRs. α3, α5, α7, and β2 neuronal nAChR subunits have been detected (Kurzen et al., 2007). AChR activation increases skin pigmentation.

The cholinergic system is present in the dermis as well as epidermis. Fibroblast express α3β2(β4)+/−α5 nAChRs, as well as α7 and α9-containing subtypes detected by PCR, antibodies, and immunofluorescence (Kurzen et al., 2007). AChE is also present, and mAChRs are present as well. Low levels of nicotine stimulate fibroblast proliferation (Peacock et al., 1993). Dermal fibroblasts are important for wound healing (Arredondo, Hall, et al., 2003). Dermal fibroblasts express high-affinity epibatidine-binding sites, indicating that assembled nAChRs are present. α3, α5, α7, β2, and β4 proteins can be detected in cultured dermal fibroblasts and dermis. Treatment with nicotine for 24 h at 10 μM doubled the number of binding sites (Arredondo, Hall, et al., 2003). α3, α5, α7, β2, and β4 RNA and protein levels were increased by nicotine treatment in dermal fibroblasts. α5 expression was increased more than α3 and could lead to an increase in α5-containing nAChRs (Arredondo, Hall, et al., 2003).

Smoking produces numerous negative effects on the skin, including on fibroblast growth, viability and development. Since nAChRs are expressed in dermal fibroblasts, it was not surprising that nicotine exposure reduces fibroblast proliferation (Arredondo, Hall, et al., 2003). Twenty-four-hour treatment with 50 μM nicotine increased the RNA and protein levels of cell cycle regulators p21, cyclin D1, PCNA, Ki-67, caspase3, and Bcl-2 (Arredondo, Hall, et al., 2003). nAChRs were involved because mecamylamine (a broad-spectrum nAChR antagonist) blocked the increases in RNA expression for all of the genes tested. α3-containing nAChRs were implicated since α3 KO mice expressed lower levels of PCNA, p21, cyclin D1 Ki-67 and bcl-2 RNAs, and higher levels of RNA for p53, bax, and caspase 3 (Arredondo, Hall, et al., 2003). Protein changes were similar except that the cyclin D1 protein level didn't change (Arredondo, Hall, et al., 2003). Antisense α3 oligonucleotide treatment also blocked the increase in PCNA, cyclin D-1, p21, and Ki-67 proteins seen with nicotine treatment in dermal fibroblasts.

Tissue remodeling can be effected by smoking. MMP-1, collagen type Ia1, and elastin are involved in this process. Twenty-four-hour treatment with 10 μM nicotine increased the protein and RNA expression of all three proteins (Arredondo, Hall, et al., 2003). Fibroblasts from α3 KO mice exhibited elastin protein and RNA expression increased. The collagen protein level increased while the RNA level didn't change. Both RNA and protein for MMP-1 were increased (Arredondo, Hall, et al., 2003). RNA levels of all three of these proteins were decreased in α7 KO mice (Arredondo, Hall, et al., 2003). These results suggest that multiple nAChR subtypes mediate regulation of proteins involved in tissue remodeling.

ACh is also present in endothelial cells of the skin as well as ChAT, the choline uptake system, nAChRs and mAChRs. Endothelial cells in blood vessels contribute to the control of blood flow. Nicotine acting through nAChRs blocked by α-Bgt (α7) increases movement of bacteria into cultured human vascular endothelial cells (Kurzen et al., 2007).

8.4.3 Potential mechanisms of action in skin cells

Except in some keratinocytes, nAChRs don't function as ion channels in the skin as they do in neurons. It is possible that the vast array of nAChR subtypes expressed in the skin function in some cells as ion channels, but other mechanisms are used as well. A mechanism of action in nonneuronal (as well as neuronal cells) may be due to influx of Ca^{2+} through nAChRs. Nicotine induces an increase in intracellular Ca^{2+} that is blocked by the antagonist mecamylamine in keratinocytes (Zia et al., 2000). This increase in Ca^{2+} coincides with a decrease in crawling locomotion of cultured keratinocytes. $\alpha7$ nAChRs are also involved in a similar rise in Ca^{2+} in nasal epithelial cells (Sharma & Vijayaraghavan, 2002). Nicotine also induced a calcium oscillation in oligodendrocyte precursor cells (Rogers et al., 2001) and an increase in intracellular Ca^{2+} in cultured hippocampal astrocytes. $\alpha4\beta2$ * nAChRs were implicated in the former, while $\alpha7$-containing nAChRs were implicated in the latter (Sharma & Vijayaraghavan, 2002). Similar changes in Ca^{2+} levels due to nAChRs may also be involved in nChR effects in the skin.

Increased intracellular Ca^{2+} and the signaling events that follow can be produced by nAChRs in multiple ways. nAChRs have a PCa/PNa (ration of channel permeability of Ca^{2+} compared with Na^+) of 1 for some types and up to 20 for $\alpha7$ nAChRs (Sharma & Vijayaraghavan, 2002). Depolarization via activation of nAChRs can activate voltage-gated Ca^{2+} channels (VGCC) and thus also increase intracellular Ca^{2+}. However, few nonneuronal cells have significant amounts of VGCC, and this mechanism may not be the major mode of action of nAChRs in nonneuronal cells (Sharma & Vijayaraghavan, 2002). There is evidence in cultured hippocampal astrocytes for activation of nAChRs leading to release of Ca^{2+} from internal stores. This could be a major effector of nAChRs in nonneuronal cells. nAChRs also may be signaling through a metabotropic mechanism (Kabbani & Nichols, 2018) in some skin cells to mediate signaling via release of Ca^{2+} from stores. Given the important roles of nAChRs in skin cells, the mechanisms involved need to be explored further.

8.4.4 Skin summary

The cholinergic system is highly expressed in several layers and cell types of the skin. Multiple nAChR subunit RNAs and proteins have been detected as well as assembled receptors. nAChRs are involved in control of multiple events in the skin including adhesion, migration, proliferation, apoptosis, and inflammation. While the mechanisms of action are not completely understood, more research on skin nAChRs would possibly benefit treatment of multiple skin disorders.

References

Arredondo, J., Hall, L. L., Ndoye, A., Nguyen, V. T., Chernyavsky, A. I., Bercovich, D., Orr-Urtreger, A., Beaudet, A. L., & Grando, S. A. (2003). Central role of fibroblast $\alpha3$

nicotinic acetylcholine receptor in mediating cutaneous effects of nicotine. *Laboratory Investigation*, *83*(2), 207–225. https://doi.org/10.1097/01.LAB.0000053917.46614.12.

Arredondo, J., Nguyen, V. T., Chernyavsky, A. I., Bercovich, D., Orr-Urtreger, A., Vetter, D. E., & Grando, S. A. (2003). Functional role of α7 nicotinic receptor in physiological control of cutaneous homeostasis. *Life Sciences*, *72*(18–19), 2063–2067. https://doi.org/10.1016/S0024-3205(03)00084-5.

Beckmann, J., & Lips, K. S. (2013). The non-neuronal cholinergic system in health and disease. *Pharmacology*, *92*(5–6), 286–302. https://doi.org/10.1159/000355835.

Borovikova, L. V., Ivanova, S., Zhang, M., Yang, H., Botchkina, G. I., Watkins, L. R., Wang, H., Abumrad, N., Eaton, J. W., & Tracey, K. J. (2000). Vagus nerve stimulation attenuates the systemic inflammatory response to endotoxin. *Nature*, *405*(6785), 458–462. https://doi.org/10.1038/35013070.

Boulter, J, Connolly, J, Deneris, E, Goldman, D, Heinemann, S, & Patrick, J. (1987). Functional expression of two neuronal nicotinic acetylcholine receptors from cDNA clones identifies a gene family. *Proceedings of the National Academy of Sciences of the United States of America*, *84*, 7763–7767.

De Hertog, SA, Wensveen, CA, Bastiaens, MT, Kielich, CJ, Berkhout, MJ, Westendorp, RG, Bavinck, JNB, & Leiden Skin Cancer Study. (2001). Relation between smoking and skin cancer. *Journal of Clinical Oncology*, *19*(1), 231–238.

Dunant, Y., Cordeiro, J. M., & Gonçalves, P. P. (2009). Exocytosis, mediatophore, and vesicular Ca^{2+}/H^+ antiport in rapid neurotransmission. *Annals of the New York Academy of Sciences*, *1152*, 100–112. https://doi.org/10.1111/j.1749-6632.2008.04000.x.

Fujii, T., Masai, M., Misawa, H., Okuda, T., Takada-Takatori, Y., Moriwaki, Y., Haga, T., & Kawashima, K. (2009). Acetylcholine synthesis and release in NIH3T3 cells coexpressing the high-affinity choline transporter and choline acetyltransferase. *Journal of Neuroscience Research*, *87*(13), 3024–3032. https://doi.org/10.1002/jnr.22117.

Fujii, T., Mashimo, M., Moriwaki, Y., Misawa, H., Ono, S., Horiguchi, K., & Kawashima, K. (2017). Physiological functions of the cholinergic system in immune cells. *Journal of Pharmacological Sciences*, *134*(1), 1–21. https://doi.org/10.1016/j.jphs.2017.05.002.

Godin, J. R., Roy, P., Quadri, M., Bagdas, D., Toma, W., Narendrula-Kotha, R., Kishta, O. A., Damaj, M. I., Horenstein, N. A., Papke, R. L., & Simard, A. R. (2020). A silent agonist of α7 nicotinic acetylcholine receptors modulates inflammation ex vivo and attenuates EAE. *Brain, Behavior, and Immunity*, *87*, 286–300. https://doi.org/10.1016/j.bbi.2019.12.014.

Grando, S. A., Horton, R. M., Pereira, E. F. R., Diethelm-Okita, B. M., George, P. M., Albuquerque, E. X., & Conti-Fine, B. M. (1995). A nicotinic acetylcholine receptor regulating cell adhesion and motility is expressed in human keratinocytes. *Journal of Investigative Dermatology*, *105*(6), 774–781. https://doi.org/10.1111/1523-1747.ep12325606.

Grando, S. A., Kawashima, K., & Wessler, I. (2020). A historic perspective on the current progress in elucidation of the biologic significance of non-neuronal acetylcholine. *International Immunopharmacology*, *81*, 106289. https://doi.org/10.1016/j.intimp.2020.106289.

Kabbani, N., & Nichols, R. A. (2018). Beyond the channel: Metabotropic signaling by nicotinic receptors. *Trends in Pharmacological Sciences*, *39*(4), 354–366. https://doi.org/10.1016/j.tips.2018.01.002.

Kawashima, K., Yoshikawa, K., Fujii, Y. X., Moriwaki, Y., & Misawa, H. (2007). Expression and function of genes encoding cholinergic components in murine immune cells. *Life Sciences*, *80*(24–25), 2314–2319. https://doi.org/10.1016/j.lfs.2007.02.036.

Kimura, R., Ushiyama, N., Fujii, T., & Kawashima, K. (2003). Nicotine-induced Ca^{2+} signaling and down-regulation of nicotinic acetylcholine receptor subunit expression in the CEM human leukemic T-cell line. *Life Sciences*, *72*, 77–78. https://doi.org/10.1016/S0024-3205(03)00077-8.

Kindt, F., Wiegand, S., Niemeier, V., Kupfer, J., Löser, C., Nilles, M., Kurzen, H., Kummer, W., Gieler, U., & Haberberger, R. V. (2008). Reduced expression of nicotinic α subunits 3, 7, 9 and 10 in lesional and nonlesional atopic dermatitis skin but enhanced expression of α subunits 3 and 5 in mast cells. British Journal of Dermatology, 159(4), 847–857. https://doi.org/10.1111/j.1365-2133.2008.08774.x.

Kuo, Y. P., Lucero, L., Michaels, J., DeLuca, D., & Lukas, R. J. (2002). Differential expression of nicotinic acetylcholine receptor subunits in fetal and neonatal mouse thymus. Journal of Neuroimmunology, 130(1–2), 140–154. https://doi.org/10.1016/S0165-5728 (02)00220-5.

Kurzen, H., Henrich, C., Booken, D., Poenitz, N., Gratchev, A., Klemke, C. D., Engstner, M., Goerdt, S., & Maas-Szabowski, N. (2006). Functional characterization of the epidermal cholinergic system in vitro. Journal of Investigative Dermatology, 126(11), 2458–2472. https://doi.org/10.1038/sj.jid.5700443.

Kurzen, H., Wessler, I., Kirkpatrick, C. J., Kawashima, K., & Grando, S. A. (2007). The non-neuronal cholinergic system of human skin. Hormone and Metabolic Research, 39 (2), 125–135. https://doi.org/10.1055/s-2007-961816.

Mishra, N. C., Rir-sima-ah, J., Boyd, R. T., Singh, S. P., Gundavarapu, S., Langley, R. J., Razani-Boroujerdi, S., & Sopori, M. L. (2010). Nicotine inhibits FcεRI-induced cysteinyl leukotrienes and cytokine production without affecting mast cell degranulation through α7/α9/α10-nicotinic receptors. Journal of Immunology, 185(1), 588–596. https://doi.org/10.4049/jimmunol.0902227.

Nguyen, V. T., Chernyavsky, A. I., Arredondo, J., Bercovich, D., Orr-Urtreger, A., Vetter, D. E., Wess, J., Beaudet, A. L., Kitajima, Y., & Grando, S. A. (2004). Synergistic control of keratinocyte adhesion through muscarinic and nicotinic acetylcholine receptor subtypes. Experimental Cell Research, 294(2), 534–549. https://doi.org/10.1016/j. yexcr.2003.12.010.

Ortiz, A., & Grando, S. A. (2012). Smoking and the skin. International Journal of Dermatology, 51(3), 250–262. https://doi.org/10.1111/j.1365-4632.2011.05205.x.

Peacock, ME, Sutherland, DE, Schuster, GS, Brennan, WA, O'Neal, RB, Strong, SL, & Van Dyke, TE. (1993). The effect of nicotine on reproduction and attachment of human gingival fibroblasts in vitro. Journal of Periodontology, 7, 658–665.

Razani-Boroujerdi, S., Boyd, R. T., Dávila-García, M. I., Nandi, J. S., Mishra, N. C., Singh, S. P., Pena-Philippides, J. C., Langley, R., & Sopori, M. L. (2007). T cells express α7-nicotinic acetylcholine receptor subunits that require a functional TCR and leukocyte-specific protein tyrosine kinase for nicotine-induced Ca^{2+} response. Journal of Immunology, 179(5), 2889–2898. https://doi.org/10.4049/jimmunol.179.5.2889.

Reardon, C., Duncan, G. S., Brüstle, A., Brenner, D., Tusche, M. W., Olofsson, P., Rosas-Ballina, M., Tracey, K. J., & Mak, T. W. (2013). Lymphocyte-derived ACh regulates local innate but not adaptive immunity. Proceedings of the National Academy of Sciences of the United States of America, 110(4), 1410–1415. https://doi.org/10.1073/ pnas.1221655110.

Rogers, S. W., Gregori, N. Z., Carlson, N., Gahring, L. C., & Noble, M. (2001). Neuronal nicotinic acetylcholine receptor expression by O2A/oligodendrocyte progenitor cells. Glia, 33(4), 306–313. https://doi.org/10.1002/1098-1136(20010315)33:4<306:: AID-GLIA1029>3.0.CO;2-W.

Rosas-Ballina, M., Olofsson, P., Ochani, M., Valdes-Ferrer, S., Levine, Y., Reardon, C., Tusche, M., Pavlov, A., & U., Chavan, S., Mak, T., & Tracey, KJ. (2011). Acetylcholine-synthesizing T cells relay neural signals in a vagus nerve cicuit. Science, 334, 98–101.

Rosas-Ballina, M., & Tracey, K. J. (2009). Cholinergic control of inflammation. Journal of Internal Medicine, 265(6), 663–679. https://doi.org/10.1111/j.1365-2796.2009. 02098.x.

Sharma, G., & Vijayaraghavan, S. (2002). Nicotinic receptor signaling in nonexcitable cells. *Journal of Neurobiology*, *53*(4), 524–534. https://doi.org/10.1002/neu.10114.

Skok, M., Grailhe, R., & Changeux, J. P. (2005). Nicotinic receptors regulate B lymphocyte activation and immune response. *European Journal of Pharmacology*, *517*(3), 246–251. https://doi.org/10.1016/j.ejphar.2005.05.011.

Sopori, M. L., Kozak, W., Savage, S. M., Geng, Y., & Kluger, M. J. (1998). Nicotine-induced modulation of T cell function: Implications for inflammation and infection. *Advances in Experimental Medicine and Biology*, *437*, 279–289. https://doi.org/10.1007/978-1-4615-5347-2_31.

St-Pierre, S., Jiang, W., Roy, P., Champigny, C., LeBlanc, É., Morley, B. J., Hao, J., Simard, A. R., & Chellappan, S. P. (2016). Nicotinic acetylcholine receptors modulate bone marrow-derived pro-inflammatory monocyte production and survival. *PLoS ONE*, *11*(2), e0150230. https://doi.org/10.1371/journal.pone.0150230.

Tayebati, S. K., El-Assouad, D., Ricci, A., & Amenta, F. (2002). Immunochemical and immunocytochemical characterization of cholinergic markers in human peripheral blood lymphocytes. *Journal of Neuroimmunology*, *132*(1–2), 147–155. https://doi.org/10.1016/S0165-5728(02)00325-9.

van der Zanden, E. P., Snoek, S. A., Heinsbroek, S. E., Stanisor, O. I., Verseijden, C., Boeckxstaens, G. E., Peppelenbosch, M. P., Greaves, D. R., Gordon, S., & De Jonge, W. J. (2009). Vagus nerve activity augments intestinal macrophage phagocytosis via nicotinic acetylcholine receptor α4β2. *Gastroenterology*, *137*(3), 1029–e4. https://doi.org/10.1053/j.gastro.2009.04.057.

Vida, G., Pena, G., Deitch, E. A., & Ulloa, L. (2011). α7-cholinergic receptor mediates vagal induction of splenic norepinephrine. *Journal of Immunology*, *186*(7), 4340–4346. https://doi.org/10.4049/jimmunol.1003722.

Wang, H., Yu, M., Ochani, M., Amelia, C. A., Tanovic, M., Susarla, S., Li, J. H., Wang, H., Yang, N., Ulloa, L., Al-Abed, Y., Czura, C. J., & Tracey, K. J. (2003). Nicotinic acetylcholine receptor α7 subunit is an essential regulator of inflammation. *Nature*, *421* (6921), 384–388. https://doi.org/10.1038/nature01339.

Wessler, I. K., & Kirkpatrick, C. J. (2001). The non-neuronal cholinergic system: An emerging drug target in the airways. *Pulmonary Pharmacology and Therapeutics*, *14*(6), 423–434. https://doi.org/10.1006/pupt.2001.0313.

Wessler, I., & Kirkpatrick, C. J. (2008). Acetylcholine beyond neurons: The non-neuronal cholinergic system in humans. *British Journal of Pharmacology*, *154*(8), 1558–1571. https://doi.org/10.1038/bjp.2008.185.

Wessler, I., Kirkpatrick, C., & Racke, K. (1999). The cholinergic "pitfall": Acetylcholine, a universal cell molecule in biological systems, including humans. *Clinical and Experimental Pharmacology & Physiology*, *26*, 198–205.

Zia, S., Ndoye, A., Lee, T. X., Webber, R. J., & Grando, S. A. (2000). Receptor-mediated inhibition of keratinocyte migration by nicotine involves modulations of calcium influx and intracellular concentration. *Journal of Pharmacology and Experimental Therapeutics*, *293* (3), 973–981.

CHAPTER NINE

Regulation of nAChR expression: Posttranscriptional regulation of nAChRs

While transcriptional control is important for regional expression, posttranscriptional events are critical for determining nAChR function. In the case of neuronal nAChRs, these events are paramount as in many cases changes in receptor expression take place with no changes in RNA levels. In the case of nAChRs, posttranscriptional regulation appears to be more important than transcriptional mechanisms in controlling nAChR expression and function. A number of transcription factors involved in neuronal nAChR expression and promoter sequences have been identified (Albuquerque, Pereira, Alkondon, & Rogers, 2009; Boyd, 1994, 1996; Danthi & Boyd, 2006; Deneris et al., 2000; Du et al., 1997; Fornasari et al., 1997; Nagavarapu et al., 2001). These are important for controlling what nAChR subunits can be expressed in a specific cell type but this chapter will focus on posttranscriptional events. nAChR number and function can be upregulated or downregulated by various posttranscriptional mechanisms. Difficulties in expressing specific subtypes in specific cells or cell lines led to studies investigating the mechanisms required for assembly and surface expression. This led to the discovery of multiple mechanisms used in the brain to regulate nAChR expression. nAChRs must fold properly, assemble into pentamers, and traffic to the surface. These events are some of the most important for controlling the function of nAChRs.

9.1 Assembly of NMJ nAChRs

The first studies in nAChR assembly used NMJ nAChRs. The receptor is composed of five subunits polypeptides (2α, β, γ, and δ) with cleavable signal peptides. After cleavage, the subunits are assembled in a specific order (Blount et al., 1990; Merlie & Smith, 1986; Wang et al., 1996). The α subunit is folded first (and becomes able to bind α-Bgt) and combines with the nascent δ and γ subunits to form two heterotrimers. This process takes about

30 min (Wang et al., 1996). The heterodimers each form one ligand-binding site. Interaction between the N-terminal domains of the subunits in the ER is crucial to this process (Wang et al., 1996). After this, these heterodimers associate into the pentameric form by combination with the β subunit. Unassembled subunits degrade (Merlie & Lindstrom, 1983). The fully assembled receptor is then trafficked through the Golgi apparatus where it is modified by glycosylation and acylation and then placed in the surface plasma membrane (PM). This process takes about 90 min (Merlie & Lindstrom, 1983; Wang et al., 1996). N-linked glycosylation of muscle nAChR subunits is required for assembly (Wanamaker & Green, 2005).

9.2 Assembly and trafficking of neuronal nAChRs

Neuronal nAChRs are composed of a large family of subunits assembled into many subtypes. Because of this it was reasonable to assume that the assembly and transport process would be more complicated than for NMJ nAChRs. Since there are multiple subtypes of neuronal nAChRs, some of the molecules are more involved in the processing of some subtypes over others. It was shown that Rapsyn was required for NMJ receptor clustering (Jones et al., 2010), and this opened the way for studies on many other molecules that interact with nAChRs. The number of molecules involved in neuronal nAChR assembly, trafficking, and function seems to be growing every year. This is an active area of research due the potential application of modifiers of these molecules as therapies for disorders involving nAChR function (Matta et al., 2021).

Some of the prominent signals are shown in Table 9.1. NACHO has an important role in α7 nAChR assembly and promotes assembly of other subtypes as well. NACHO in the ER interacts with RPN1/1 (oligosaccharyltransferases, OST) and calnexin to promote folding of α4β2 and α7 nAChRs (Matta et al., 2021). RIC-3 and BCL-2 in conjunction with NACHO promote α7 nAChR assembly and trafficking (Matta et al., 2021). Nicotine also interacts with α4β2 nAChRs at the orthosteric site to promote assembly of α4β2 nAChRs (see below for more on nicotine). Prototoxins such as Lynx1 and 2 interact with α4β2 and α7 nAChRs to modulate trafficking and function (Matta et al., 2021). nAChRs are further processed in the Golgi on the way to the surface. Additional processing modulates nAChR function when at the surface as well as turnover of surface receptors. nAChR assembly is not static but is modulated by these molecules and others. For example, Fig. 9.1 depicts additional molecules

Table 9.1 Some molecules that affect nAChR trafficking and assembly.

nAChR	α7	α4β2	α6β2β3	α3β2	α6β4
Protein chaperones	NACHO, Bcl-2, RIC-3	NACHO	NACHO, SULT2B1	NACHO	IRE1a, SULT2B1
Chemical chaperones	Nicotine, phenylbutyrate, valproate, polyamines	Nicotine, menthol, butryate, polyamines	Nicotine, menthol	Nicotine	Nicotine
Auxiliary subunits	Prototoxin	Prototoxin	BARP	BARP	BARP

From Matta, J. A., Gu, S., Davini, W. B., & Bredt, D. S. (2021). Nicotinic acetylcholine receptor redux: Discovery of accessories opens therapeutic vistas. Science, 373 (6556). https://doi.org/10.1126/science.abg6539.

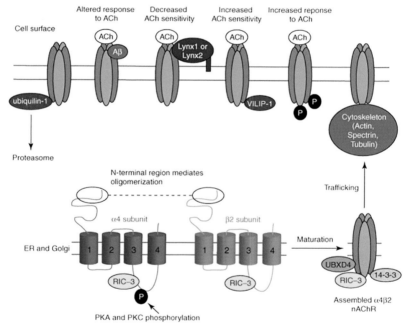

Fig. 9.1 Multiple proteins are involved in the maturations, assembly, and trafficking of nAChRs. nAChR subunits are processed through the ER and Golgi, and sequences in the N terminus of each subunit mediate interactions during assembly. Proteins such as resistance to inhibitors of cholinesterase (RIC-3), 14-3-3, visinin-like protein (VILIP-1), and UBXD4 are involved in tracking to the surface. nAChRs interact with cytoskeletal elements to regulate the distribution and stability. Other proteins such as Aβ and Lynx act to influence the function of the nAChRs at the surface while others influence turnover (Ubiquilin-1). *(From Jones, A. K., Buckingham, S. D., & Sattelle, D. B. (2010). Proteins interacting with nicotinic acetylcholine receptors: Expanding functional and therapeutic horizons. Trends in Pharmacological Sciences, 31 (10), 455–462. https://doi.org/10.1016/j.tips.2010.07.001.)*

interacting with α4β2 nAChRs that modulate function, stability, and trafficking (Jones et al., 2010). These include RIC-3, 14-3-3, Lynx 1 and 2. As for NMJ nAChRs, the N-terminal regions of neuronal nAChRs mediate oligomerization. Other nAChR subtypes have similar but in some cases subtype-specific modulators. This topic can't be covered exhaustively here, but we will highlight some of the major molecules involved in the process. RIC-3 is described in Chapter 10. As can be seen, multiple regulatory molecules are required for proper nAChR assembly and surface expression.

9.3 NACHO

The actions of NACHO (novel nAChR regulator) are specific to nAChRs. NACHO was identified in a screen for molecules that enhanced α7 responses (Gu et al., 2016). NACHO increases α7 folding, maturation through the Golgi and cell surface expression (Gu et al., 2016). NACHO is resident in the ER membrane (Crespi et al., 2018). NACHO and α7 protein don't physically interact indicating that the NACHO acts as a chaperone through interaction with other proteins such as calnexin (Crespi et al., 2018); however, the first two transmembrane (TM) domains of α7 are required for NACHO function (Matta et al., 2021). KO mice for NACHO eliminate α7 function and assembly, and KO mice for NACHO have deficits in cognitive function and increased motor activity. KO mice demonstrate a loss of both presynaptic and postsynaptic nAChRs (Matta et al., 2017). The functions of multiple other receptors such as TRPV1, AMPA, and GABA are not impaired in the KO mice (Matta et al., 2021). NACHO also increases surface levels of α4β2 nAChRs (Gu et al., 2016) and expression of α3β2 and α3β4 nAChRs (Matta et al., 2017; Mazzaferro et al., 2021). However, NACHO only upregulates the high-affinity (α4)2(β2)3 form of the receptor (Mazzaferro et al., 2021). NACHO knockdown doesn't eliminate α4β2 function, but reduces it (Gu et al., 2016). NACHO and RIC-3 can act synergistically on α7 nAChRs, and some areas of the brain with high α7 expression don't express RIC-3 (Gu et al., 2016). This indicates that NACHO serves to increase assembly and surface expression in these brain areas without RIC-3. NACHO also doesn't change the kinetics of deactivation or desensitization of the α4β2 or α7 nAChRs (Gu et al., 2016). NACHO has a specific patter of expression (hippocampus, cortex, and olfactory), but is not ubiquitously expressed (Crespi et al., 2018; Gu et al., 2016). NACHO alone doesn't enhance α6β2* nAChR expression (Gu et al., 2016). However, in combination with additional regulators such as BARP (β-anchoring and regulatory protein), LAMP5 (lysosomal-associated membrane protein 5), and SULT2B1 (sulfotransferase), NACHO can promote increased surface expression of α6β2β3-containing nAChRs (Matta et al., 2021).

9.4 14-3-3η and adenomatous polyposis coli (APC)

14-3-3 family is highly expressed in the nervous system and was found to be the first cytosolic protein to bind to nACRs (Jeanclos et al., 2001). 14-3-3η is one of a family of chaperone proteins. 14-3-3η has been shown in vitro to increase expression of α4β2 nAChRs. However, it was shown that the upregulated nAChRs had lower sensitivity. This was due to a selective upregulation of the low sensitivity form of α4β2 nAChRs, (α4)3(β2)2 (Mazzaferro et al., 2021). The interaction was with the TM3-TM4 cytoplasmic domain of the α4 subunit. Mutation in the α4 cytoplasmic domain reduced surface expression of α4β2 nAChRs (Jeanclos et al., 2001). 14-3-3η didn't alter the function of the α4β2 nAChR. Interactions with the α4β2 nAChRs took place within the ER/Golgi compartment (Jeanclos et al., 2001). 14-3-3η was also shown to interact with α4β2 nAChRs in vivo (Jeanclos et al., 2001). 14-3-3η interactions with the TM3-TM4 cytoplasmic also increased α3* nAChR surface expression (Jones et al., 2010). Adenomatous polyposis coli (APC) is required for localizing nAChRs to postsynaptic sites with 14-3-3η linking APC to the α3* nAChR (Rosenberg et al., 2008).

9.5 ly6/uPAR prototoxins

A large family of proteins, the ly6 prototoxins, are expressed in the mammalian brain, immune system, and elsewhere (Miwa et al., 2019). These proteins have structural homology to α-Bgt, the snake venom toxin that binds tightly to NMJ and α7 homomeric nAChRs (Matta et al., 2021; Miwa et al., 2019) and has been described elsewhere in this book. While α-Bgt binding to nAChRs is irreversible, ly6 protoxins act at nAChRs with lower affinity in a reversible manner. Members of the family interact with multiple nAChRs and alter the function and processing of neuronal nAChRs. Some prototoxins have preferential interactions with specific nAChR subtypes (Miwa et al., 2019). The prototoxins in the brain have unique, mostly nonoverlapping expression patterns (Miwa et al., 2019). By interacting with nAChRs, these prototoxins alter expression and function that in turn have effects on multiple cholinergic functions including memory, behavior, and development. Prototoxins often act as allosteric modulators, although they can interact with nAChRs at the orthosteric site as well. The functions of several of these family members will be described here.

9.5.1 Lynx1(ly6/ neurotoxin 1)

Lynx 1 is highly expressed in the cerebellum, hippocampus, retina, spinal cord, and cortex (Miwa et al., 2019). Lynx 1 is expressed at synapses on neurons (Crespi et al., 2018). Lynx 1 is also expressed in the habenulopeduncular tract, which plays an important role in nicotine aversion and that expresses α3β4α5* nAChRs. Lynx 1 is a glycosylphosphatidylinositol (GPI)-anchored protein expressed in the ER, and it is enriched on the plasma membrane. One role of lynx 1 is to regulate nAChR assembly. Lynx1 interacts with nAChR dimers in the ER (Miwa et al., 2019; Nichols et al., 2014). Interestingly, lynx 1 acts within the ER and promotes assembly of the low-nicotine sensitivity form of α4β2 nAChRs (α4)3(β2)2 (Nichols et al., 2014). Lynx1 did this at least in part by stabilizing α4/α4 dimers (Nichols et al., 2014). Cleaving of lynx 1 from the membrane removes its effect on nAChRs. This is consistent with in vitro studies in oocytes showing that α4β2 in the presence of lynx 1 had reduced agonist sensitivity to acetylcholine (Miwa et al., 2019). Soluble forms without the anchor have function as well (Miwa et al., 2019).

Lynx1 also binds to (α3)2(β4)3 and (α3)3(β4)2 nAChRs (Miwa et al., 2019). In vitro lynx 1 reduces cell surface expression of the (α3)2(β4)3 subtype and reduced function by altering the single channel properties of the (α3)3(β4)2 nAChRs (reduced conductance, lowering number of long bursts, and increased time in the desensitized/ inactivated state) (George et al., 2017). The functional effects are due to interactions at the α3(−)/(+)α3 interface (George et al., 2017). Lynx 1 in vitro also reduced the function of (α3β4)2α5 nAChRs by acting at the α3(−)/(+)α5 interface. Lynx 1 produced increased closed dwell times and reduced surface expression that is dependent on linkage of lynx 1 to the surface (George et al., 2017).

Lynx 1 in soluble form was shown to bind to α3, α4, α5, α6, α7, β2, and β4 nAChR subunits in rat brain and human cortical extracts (Thomsen et al., 2016). Various forms of amyloid-β interact with nAChRs (see Chapter 5). In this assay, amyloid-β interfered with lynx1/nAChR binding to the α3, α4, α5, and α7 subunits (Thomsen et al., 2016). Soluble lynx 1 in vitro inhibited α7, α4β2, α3β4, and α3β2 nAChR function (Miwa et al., 2019; Thomsen et al., 2016).

The expression of lynx 1 is developmentally regulated and can function to alter plasticity in the primary visual cortex perhaps by altering dendritic spine turnover (Miwa et al., 2019). As lynx 1 expression increases in primary auditory cells, sensitivity to nicotine decreases. Lynx 1 KO mice have a higher sensitivity to nicotine in primary auditory cells that can be blocked by DHβE. This is consistent with a negative effect on α4β2* nAChRs by lynx 1 that is a normal developmental event (Miwa et al., 2019), but was disrupted by lynx 1 KO. Lynx 1 KO mice also have enhanced memory and associative learning, consistent with relief of lynx 1 inhibition of α4β2 nAChRs (Miwa et al., 2019). Other types of memories such as contextual are not altered in the KO mice. Lynx 1 KO mice have normal nociceptive signaling in the hot plate test, but in these mice nicotine reduces nociceptive behavior (Miwa et al., 2019).

9.5.2 Lynx 2 (ly6/neurotoxin 2)

Lynx2 is not expressed in the same brain regions as lynx 1. For example, lynx 2 is expressed in the CA1 regions of the hippocampus while lynx 1 is in CA3 neurons. Lynx 2 is also present in prefrontal cortex and amygdala (Miwa et al., 2019). Lynx 2 also represses nAChR activity by binding to them as does lynx 1. Lynx 2 immunoprecipitates with α7, α4β2, and α4β4 nAChRs (Tekinay et al., 2009; Wu, Puddifoot, Taylor, & Joiner, 2015). Lynx 2 causes α4β2 nAChRs to desensitize faster when exposed to ACh and changes the EC_{50} for ACh, nicotine, and epibatidine (Tekinay et al., 2009; Wu et al., 2015). Surface expression of α4β2 nAChRs is reduced by lynx 2 (Wu et al., 2015). Lynx2 also dramatically reduced the chaperoning activity of nicotine (see below) on α4β2 nAChRs expressed in cell lines (Wu et al., 2015). Lynx 2 also regulates the surface expression of α7 nAChRs (Puddifoot et al., 2015). Lynx 2 KO mice show enhanced anxiety and fear (Matta et al., 2021; Tekinay et al., 2009). Deletion of lynx 2 led to increased glutamate release in response to nicotine (Tekinay et al., 2009).

Most of the work testing of lynx1/lynx 2 binding to nAChRs has been done in cell extracts or heterologous expression systems (Miwa et al., 2019). However, work with KO animals demonstrates an in vivo function of lynx 1 and lynx 2 in regulation of cholinergic function. The functional and behavioral effects seen in lynx KO mice also support the in vivo presence of binding interactions between the lynx proteins and nAChRs that have functional consequences.

9.5.3 Secreted Ly-6/uPAR-related proteins (SLURPs)

In addition to membrane-linked members of the ly6 prototoxin family, some such as SLURP-1 and SLURP-2 are secreted. They are expressed in a number of nonneuronal cells, as are nAChRs. SLURP-1 functions to control growth and inflammation in some cells such as keratinocytes. SLURP-1 is also in immune cells and intestinal epithelial cells (Lyukmanova, Shulepko, Kudryavtsev, et al., 2016) and is expressed in spinal cord sensory neurons (Crespi et al., 2018). Recombinant SLURP-1 reduces ACh-induced currents of $\alpha7$ nAChRs expressed in Xenopus oocytes (Lyukmanova, Shulepko, Kudryavtsev, et al., 2016). Human SLURP-1 binds $\alpha7$ nAChRs in human cortical extracts and cell lines (Lyukmanova, Shulepko, Kudryavtsev, et al., 2016). Binding to other subunits in the human brain was not detected ($\alpha3$, $\alpha4$, $\alpha5$, $\alpha6$, $\beta2$, and $\beta4$) (Lyukmanova, Shulepko, Kudryavtsev, et al., 2016), but subsequent studies show that it also can effect function of other nAChRs besides $\alpha7$ nAChRs. Various recombinant forms of SLURP-1 have been used that include extra N- or C-terminal tags that may interfere with function or give different results (Durek et al., 2017). Studies using a synthetic SLURP-1 identical to the human sequence showed that in Xenopus oocytes, it inhibited ACh-induced currents of human $\alpha3\beta2$, $\alpha3\beta4$, $\alpha4\beta4$, and $\alpha9\alpha10$ nAChRs, but not $\alpha4\beta2$ nAChRs (Durek et al., 2017). The synthetic SLURP-1 also didn't interfere with α-Bgt binding to muscle-type nAChRs or human $\alpha7$ nAChRs. This is consistent with previous results that showed a recombinant SLURP-1 didn't compete for binding to $\alpha7$ nAChRs expressed in GH4C1 cells (Lyukmanova, Shulepko, Kudryavtsev, et al., 2016). Synthetic SLURP-1 blocked human $\alpha7$ nAChR function but only if PNU120596 (allosteric modulator) was present (Durek et al., 2017). This is different from what was observed previously in that no modulator was needed for inhibition by the recombinant SLURP-1, see above (Lyukmanova, Shulepko, Kudryavtsev, et al., 2016). SLURP-1 may act as a silent negative allosteric modulator, requiring the channel to be in an open state (Durek et al., 2017). More work needs to be done, but clearly SLURP-1 has the ability to modulate nAChR function.

SLURP-2 also has a role in regulating epithelial cell growth, but interacts with nAChRs (Crespi et al., 2018). In human cortical extracts, SLURP-2 immunoprecipitates $\alpha3$, $\alpha4$, $\alpha5$, $\alpha7$, $\beta2$, and $\beta4$ subunits (Lyukmanova, Shulepko, Shenkarev, et al., 2016). In keratinocytes, SLURP2 competes with ACh binding to $\alpha3$-containing nAChRs and alters keratinocyte differentiation (Crespi et al., 2018). SLURP-2 reduces ACh-induced currents for $\alpha3\beta2$ and $\alpha4\beta2$ nAChRs expressed in oocytes (Lyukmanova, Shulepko, Shenkarev, et al., 2016). In oocytes, SLURP-2 potentiates $\alpha7$ nAChR activity at low concentrations ($<1\,\mu M$, best at $30\,nM$), but inhibits at high (Lyukmanova, Shulepko, Shenkarev, et al., 2016). It has been proposed that this potentiation maybe due to SLURP-2 binding at orthosteric sites of $\alpha7$ nAChRs and promoting opening of receptors with both SLURP-2 and ACh bound. Higher levels of SLURP-2 inhibit due to blocking access to the ligand-binding site. SLURP-2 alone doesn't produce current (Lyukmanova, Shulepko, Shenkarev, et al., 2016).

9.6 Menthol

Menthols acts posttranslationally as a chemical chaperone and as an allosteric modulator. Menthol is a component of many cigarettes and enhances the addictive properties of nicotine (see Chapter 7). Smoking leads to upregulation of $\beta2^*$ nAChRs in multiple brain regions including prefrontal cortex and brainstem (Brody et al., 2013) while PET scanning using an $\alpha4\beta2^*$ nAChR ligand demonstrated an even greater density of $\alpha4\beta2^*$ nAChRs in smokers of menthol cigarettes (Brody et al., 2013). Menthol enhances the intensity of withdrawal from nicotine by mecamylamine treatment and increases rat self-administration (Henderson et al., 2017). Menthol enhances the reinforcing properties of nicotine (Cooper & Henderson, 2020). Nicotine treatment induces a conditioned place preference (CPP) response in mice, i.e., a preference for the nicotine-paired chamber. Menthol alone didn't increase CPP, but increased CPP when paired with nicotine (Henderson et al., 2017). Nicotine mediates it effects on addiction in large part by influencing signaling in midbrain dopaminergic neurons (see Chapter 4). How might the effects of menthol be mediated?

Menthol effects the expression of nAChRs. Nicotine treatment upregulated $\alpha4^*$ nAChRs in the VTA neurons and substantia nigra pars reticulate (SNr) GABergic neurons and menthol in combination produced an additional upregulation (Henderson et al., 2017). Nicotine increased the number of $\alpha6^*$nAChRs in the VTA and substantia nigra pars compacta (SNc). In VTA neurons, menthol didn't augment the nicotine produced upregulation of $\alpha6^*$ nAChRs and in SNc dopaminergic neurons menthol + nicotine didn't increase the $\alpha6^*$ levels (Henderson et al., 2017). Menthol upregulated $\alpha4^*$ nAChRs in VTA DA neurons and SNr GABAergic neurons over that produced by nicotine alone, but not $\alpha6^*$ nAChRs. In the VTA, $\alpha4\alpha6^*$ nAChRs were not increased by nicotine alone, but were by nicotine and menthol. In SNc, both increased $\alpha4\alpha6^*$ nAChR expression (Henderson et al., 2017). Cell lines expressing $\alpha4\beta2$ nAChRs treated with 500 nM menthol alone increased expression in the plasma membrane by twofold, and $\alpha6\beta2\beta3$ nAChRs expressed in cell lines were upregulated by threefold after the same menthol treatment (Henderson et al., 2016). The chaperoning effects of menthol, like nicotine, depend on nAChR movement between the Golgi and ER. Nicotine increases the number of endoplasmic reticulum exit sites (ERES) and menthol does as well in cells expressing either $\alpha4\beta2$ nAChRs or $\alpha6\beta2\beta3$ nAChRs (Henderson et al., 2016).

In transfected Neuro-2a cells, chronic menthol treatment increased $\alpha4\beta2^*$ nAChR numbers by increasing the formation of the low-sensitivity form $(\alpha4)3(\beta2)2$. This is contrary to the effect of nicotine (see below), which upregulates the high-sensitivity form, $(\alpha4)2(\beta2)3$ (Henderson et al., 2016). Menthol and nicotine together produce an upregulation of the higher-sensitivity form as occurs for nicotine alone (Henderson et al., 2017).

Menthol also effects the activity of neurons. Nicotine reduced baseline firing frequency in cultured TH+ (DA) midbrain neurons (Henderson et al., 2017). In midbrain, TH+ neurons (DA) treated with menthol for 10 days+ nicotine baseline firing was reduced even more than the level of reduction induced by nicotine alone. However, baseline firing frequency was enhanced by menthol in (TH- non DA) GABAergic cells augmenting the modest nicotine-induced enhancement (Henderson et al., 2017). In control TH+/DA midbrain neurons, ACh induced an increase in firing frequency of twofold, but after 10 days of nicotine treatment, ACh-induced a fivefold increase in frequency (Henderson et al., 2017). In nicotine + menthol treated TH+/DA neurons, ACh induced an increase in firing frequency of eightfold (Henderson et al., 2017). Thus, menthol enhanced the effects of nicotine on ACh-induced firing frequency. Menthol can have effects on nicotine-induced signaling as well since 10-day treatment with menthol + nicotine increased the firing rate of DA neurons in response to a 10-s pulse of 500 nM nicotine (Henderson et al., 2017). Menthol alone doesn't potentiate nAChR currents or activate receptors (Henderson et al., 2016). However, in Neuro-2a cells, $\alpha4\beta2$ nAChRs had a lower level of nicotine-induced desensitization and recovered 80% of their function (compared with 20% without menthol) after a 24-h menthol treatment (Henderson et al., 2016). $\alpha6\beta2\beta3$ nAChRs also expressed in Neuro-2a cells and treated with menthol recovered 60% of function after desensitization vs 40% without menthol (Henderson et al., 2016).

How does menthol produce functional effects on nAChRs? Menthol has also been shown to act as a noncompetitive antagonist for $\alpha 3\beta 4^*$ and $\alpha 7$ nAChRs (Cooper & Henderson, 2020). Computational docking studies indicate that menthol may bind $\alpha 4\beta 2$ nAChRs in the TM2 region (Henderson et al., 2018). Given the widespread use of menthol in cigarettes and vaping products, more studies need to be done to determine the mechanism of menthol action on nAChR function and regulation.

9.7 Upregulation by nicotine: An important role in addiction

It was recognized early that treatment of animals, neurons, or cell lines with nicotine led to increased binding of labeled nicotine. This is in contrast to regulation of many other receptors in that chronic exposure to agonist downregulates expression and chronic exposure to an antagonist upregulates (Wonnacott, 1990). This increased nicotine binding was shown as a higher B_{Max} (density) and not due to changes in the K_d (affinity) (Wonnacott, 1990). This occurred in multiple brain regions in mice (Marks et al., 1983). α-Bgt binding (now known to be $\alpha 7$ nAChRs) was also slightly upregulated by chronic nicotine (Marks et al., 1983; Wonnacott, 1990). Nicotine-induced upregulation was also observed in rats (Schwartz & Kellar, 1983).

This upregulation can also be monitored by ligand binding of epibatidine, cytisine, acetylcholine, α-Bgt, and other cholinergic compounds. Upregulation varies by brain region and mode of nicotine administration (osmotic minipump vs. self-administration) (Moretti et al., 2010). Higher numbers of nicotine-binding sites were also found in multiple brain regions in postmortem brains of human smokers, obviously exposed to chronic nicotine (Benwell et al., 1988; Perry et al., 1999). These upregulations occurred also in rat without changes in the levels of $\alpha 2$, $\alpha 3$, $\alpha 4$, $\alpha 5$, and $\beta 2$ nAChR RNAs (Flores et al., 1992; Marks et al., 1992). Nicotine failed to upregulate $\alpha 4$ and $\beta 2$ RNAs in cell lines as well (Peng et al., 1994) supporting a posttranscriptional mechanism of action.

The mechanism was not clear since it was also recognized that desensitized nAChRs bind nicotine with higher affinity than those in the open or closed, ligand-unbound states (see Chapter 3). Chronic exposure to nicotine promotes desensitization of some subtypes of nAChRs. Acetylcholine is removed from the synaptic cleft rapidly (<1 ms), but nicotine remains longer and being cell-permeable enters neurons (Lester et al., 2009). Nicotine also can be concentrated in acidic organelles such as synaptic vesicles. This would predict that intracellular nAChRs are in a desensitized state. Acetylcholine also can enter cells, but not as readily. Other cholinergic compounds have membrane permeability such as mecamylamine and epibatidine (Lester et al., 2009). Multiple subtypes can be upregulated, and differential effects are seen in various brain regions. More nicotine is required to upregulate $\alpha7$ nAChRs than $\alpha4\beta2$ nAChRs in mice (Marks et al., 2004). Most of the high- affinity nicotine-binding sites that are upregulated are composed of $\alpha4\beta2^*$ nAChRs (Flores et al., 1992). Several studies suggest that the upregulated nAChRs are still functional, and upregulation is due to more assembled or activatable receptors being present (Nashmi et al., 2007). Nicotine can have differential effects on nAChRs of different stoichiometries (Lester et al., 2009; Moretti et al., 2010). Nicotine has recently been shown to act as a chaperone. The effects on expression of nAChRs are not due to transcriptional changes and are cell-autonomous (Lester et al., 2009; Pauly et al., 1996).

Studies focusing on the $\alpha4\beta2$ nAChRs have shown that increases in binding sites after nicotine treatment is due to increased levels of nAChR proteins (Marks et al., 2011). This work used a $\beta2$ subunit selective ligand A85380 and antibodies to $\alpha4$ (mAb299) and $\beta2$ (mAb270). Mice were treated for 10 days with nicotine by cannula. They showed that nicotine increased binding of all three, and in regions that express mostly $\alpha4\beta2$ nAChRs, the binding was increased in a coordinated fashion (Marks et al., 2011). Quantification using autoradiography and those using immunoprecipitation gave similar results, showing that increased binding was due to increased protein expression (Marks et al., 2011) The high-affinity binding sites, $\alpha4\beta2^*$ nAChRs, can be precipitated by anti-$\beta2$ or anti-$\alpha4$ antibodies (Marks et al., 2011; Moretti et al., 2010). Upregulation of epibatidine-binding sites was correlated with increased $\alpha4$ and $\beta2$ nAChR proteins.

Nicotine binding to $\alpha4^*$ nAChRs in brain cultures and nAChRs expressed in cell lines (Lester et al., 2009) promotes upregulation. The EC_{50} for upregulation of $\alpha4\beta2$ nAChRs in cell lines is less than 100 nM (Lester et al., 2009). $\alpha4\beta2$ nAChR exists in two forms; a high-affinity form made up of $(\alpha4)2(\beta2)3$ and a low-affinity form $(\alpha4)3(\beta2)2$. Nicotine upregulates $\alpha4\beta2^*$ nAChRs by stabilizing the higher-affinity form (Lester et al., 2009). $\alpha4\beta2$

nAChRs expressed in a cell line and treated with nicotine showed a dramatic upregulation due to an increase in assembly from pools of subunits and a fivefold increase of lifetime on the surface (Kuryatov et al., 2005). Decreased turnover after nicotine treatment was also the case for α3β2 nAChRs (Wang et al., 1998). α4β2 upregulation didn't require function of the surface receptors (Kuryatov et al., 2005).

After treatment, most of the surface receptors are desensitized and the assembly intermediates also (Kuryatov et al., 2005). After nicotine removal, the receptors regain activity and thus a net increase in function (Kuryatov et al., 2005). The increase in function is in part related to the number of nAChRs on the surface, although some surface receptors may remain desensitized. The nicotine concentration used to produce this upregulation ranged from 12.5 to 88 nM, levels that can be achieved by smokers. ACh and cytisine can also upregulate α4β2 receptors in a cell line, and function is also not required for this. The ability of compounds to induce conformation changes compatible with desensitization or activation is important to the ability to upregulate nAChRs (Kuryatov et al., 2005).

Mice expressing α4* nAChRs labeled with yellow fluorescent protein (YFP) inserted into the TM3-TM4 loop were used to measure changes in nAChR proteins levels after chronic nicotine administration (Nashmi et al., 2007). After 10 days of chronic nicotine treatment, α4 proteins levels were upregulated in the SNr and the VTA. The chronic nicotine treatment increased baseline firing of the GABAergic SNr neurons as well as increased firing response to acute 1 μM nicotine (Nashmi et al., 2007), indicating that the upregulated α4* nAChRs were functional. Nicotine also increased α4* nAChRs in the medial perforant path on glutamatergic neurons with enhanced functional expression (Nashmi et al., 2007).

Cell culture studies were used to show that the upregulation was intrinsic to the receptor and not specific for neurons (Peng et al., 1994). α4β2 nAChRs were expressed in M10 cells and Xenopus oocytes. In total, 5 μM nicotine increased nicotine binding (not affinity for nicotine) and increased the amount of surface receptors detected by binding of an α4 subunit mAb. This assay also showed that the increase measured was not due to changes in the intracellular pool because surface labeling of nAChRs was done before immunochemistry (Peng et al., 1994) indicating that new protein synthesis was not required. The receptors expressed in oocytes were also upregulated by nicotine treatment.

α3β2, α3β2α5, α3β4, and α3β4α5 nAChRs were expressed in cell lines, and the effects of nicotine on upregulation were examined (Wang et al.,

1998). $\alpha3\beta2$ and $\alpha3\beta2\alpha5$ nAChRs were highly upregulated by chronic nicotine exposure, but $\alpha3\beta4$ or $\alpha3\beta4\alpha5$ were not. However, the $\alpha3\beta4$ nAChRs had a higher baseline expression and perhaps were already maximally upregulated, and $\alpha3\beta2$ were expressed at a lower level that was only increased by nicotine (Wang et al., 1998). The $\beta2$ and $\beta4$ subunits differ in ER motifs that could explain the differential effects of nicotine (Henderson & Lester, 2015; Sallette et al., 2004). $\beta4$-containing nAChRs with high PM density may already be efficient in ER transport and don't benefit from nicotine chaperoning. Nicotine may act differently on receptors containing a $\beta2$ subunit versus those with a $\beta4$ subunit.

Nicotine upregulation also occurred in SH-SY5Y cells, which express $\alpha3^*$ nAChRs (Wang et al., 1998). Both total and surface were upregulated, and protein synthesis was not required for the $\alpha3\beta2^*$ nAChR upregulation, the same as was seen above for $\alpha4\beta2$ nAChRs (Peng et al., 1994) indicating that increased assembly was involved, perhaps in addition to other mechanisms. When a protein synthesis inhibitor (cyclohexamide) was applied to the $\alpha3\beta2$ cells and expression monitored by epibatidine binding (binds to assembled receptors only), a slow loss of loss of receptors occurred that was reversed by the presence of nicotine during the cyclohexamide treatment (Wang et al., 1998). Initially, there is an increase in receptors, due to increased assembly. After that, the amount of receptors detected was very stable indicating dramatically decreased turnover in the presence of nicotine. In addition to nicotine (which is cell permeable), carbamylcholine (surface acting only) also increased expressed of $\alpha3\beta2$ nAChRs in a cell line (Wang et al., 1998). Thus, nicotine affects turnover as well as assembly.

Nicotine doesn't lead to the upregulation of all subtypes, however. $\alpha4\beta2^*\alpha5$ nAChRs are also widely present in the brain. However, in rats, chronic treatment with nicotine using osmotic minipumps doesn't increase the level of this nAChR population (Mao, Perry, Yasuda, Wolfe, & Kellar, 2008). Upregulation was not seen in multiple brains regions (Mao et al., 2008). This indicates that subtype specificity exists for nAChR upregulation by nicotine.

Many studies have been done in cell lines, but in vivo upregulation varies by subtype, brain region, stoichiometry, and location on the neuron. For example, $\alpha4^*$ nAChRs are upregulated in GABAergic neurons of SNr and VTA but not in dopaminergic neurons of the VTA. Brain-region-specific upregulation could reflect the subtypes and subunits present. GABAergic neuron upregulation may be due to the presence of $\alpha4^*$ (non-$\alpha5$)$\beta2$ nAChRs and lack of upregulation in dopaminergic neurons due to presence of α4α6β2 nAChRs that are not shown to upregulate (Henderson & Lester, 2015). α6β2β3 nAChR is also upregulated.

It was thought that chronic activation with subsequent entry of Na^+ and Ca^{2+} into cells was the basis for upregulation. However, it has been shown that channel activity is not necessary for upregulation (Kuryatov et al., 2005; Peng et al., 1994). Antagonists such as DHβE and mecamylamine can upregulate (Henderson & Lester, 2015). Binding may be necessary, but activation is not. Desensitization may be critical as concentrations of nicotine (50–100 nM) that produce upregulation of α4β2 or α6β2β3 nAChRs only activate a small number of receptors, but produce desensitization (Henderson & Lester, 2015; Henderson et al., 2014).

In summary, nicotine is thought to function as a pharmacological chaperone in an "inside out" manner (Henderson & Lester, 2015). Nicotine can penetrate the cell, and most of the nicotine binding (85%) is within the cell (Henderson and Lester, 2015). Once inside cells, nicotine binds to and increases maturation of pentameric nAChRs, increases assembly in the ER, and PM insertion (Henderson & Lester, 2015). When surface expression of nAChRs occurs, there is an increase of intracellular receptors. For nicotine to have an effect, cycling of receptors between Golgi and ER is required, and nicotine increases the number of ERES. Upregulation occurs at low levels of nicotine that don't activate most nAChRs. Coat protein complex I (COPI) regulates traffic from Golgi to ER and is required for nicotine's effects (Henderson et al., 2014). Nicotine also reduced turnover. Multiple nAChR subtypes are upregulated after nicotine treatment. This may explain the cognitive enhancements that occur after nicotine exposure in humans and rodents (Henderson & Lester, 2015). Upregulation can occur with other cholinergic ligands as well including antagonists (Pauly et al., 1996; Peng et al., 1994), but the focus has been on nicotine as the addictive component of smoke. Upregulation can occur in the brain or in cell lines. The degree of upregulation varies among brain regions with some not being upregulated after nicotine treatment at all. Timing of nicotine exposure may also be a factor. Upregulation of binding sites for various ligands such as nicotine or epibatidine correlates with increases in nAChR protein at the surface. For the most part, this upregulation is posttranscriptional due to increased expression of nAChR subunit proteins, without large changes in nAChR subunit RNA. The upregulation is cell-autonomous and intrinsic to the receptor. Nicotine interacts differently with ER-binding motifs in various nAChR subunits. Largely, these upregulated nAChRs are functional, and this functional upregulation may be important to development of tolerance and sensitization (Lester et al., 2009).

9.8 MicroRNA (miRNA)

miRNAs are small RNAs that are present in most cells including having a high level of expression in neurons. miRNAs are 21–24 nucleotides long and regulate the expression of many protein coding genes including those for nAChRs (Hogan et al., 2014). miRNAs interact with miRNA recognition elements (MREs), often in the 3' untranslated regions 3'(UTR) of RNAs. This causes an interaction with the RNA-induced silencing complex (RISC) that leads to mRNA degradation or translational silencing (Hogan et al., 2014).

miRNAs may be involved in the posttranscriptional control of nAChRs. Nicotine at doses that are achieved in smokers was shown to increase multiple miRNAs in cortical-derived neurospheres (Balaraman et al., 2012). miRNA levels were also shown to be regulated by chronic nicotine exposure in the adult mouse brain (Lippi et al., 2011). Chronic treatment for 5 days upregulated the expression of multiple miRNAs in the hippocampus, prefrontal cortex, midbrain, and subcortical limbic forebrain (Lippi et al., 2011).

Many MREs were detected in the 3'UTRs of neuronal nAChR genes using an miRNA library screen (Hogan et al., 2014). Sixteen mouse miRNAs were identified that reduced expression of constructs containing the 3'UTs of multiple nAChRs ($\alpha2$, $\alpha4$, $\alpha5$, $\alpha6$, $\alpha7$, $\alpha10$, $\beta2$, and $\beta3$) fused to a luciferase construct (to monitor RNA expression and translation). The inhibition was due to interactions with MREs and reduced protein expression. Ten were shown to have human homologs (Hogan et al., 2014), and several have MREs in human nAChR 3'UTRs. Many of the miRNAs that interact with nAChRs were also found in mouse brain. Interestingly, in the VTA, one miRNA targeting $\beta2$ was reduced while $\beta2$ RNA levels increased. Nicotine treatment reduced the expression of multiple miRNAs, which could lead to more translation due to the nAChRs remaining intact longer (Hogan et al., 2014). Another miRNA (miR-138) downregulated $\beta4$ subunit expression in constructs containing the $\beta4$ 3'UTR (Gallego et al., 2013). $\alpha7$ nAChR expression may also regulated by miRNAs. $\alpha7$ nAChR protein expression is low in the brains of APP-SI mice (model for AD). miR-98-5p was upregulated in these mice and in human AD patients. The binding site for miR-98-5p was conserved in human and mice $\alpha7$ nAChR 3'UTRs (Song et al., 2021). $\alpha7$ expression was reduced at the translational level as the miRNA didn't change $\alpha7$ nAChR subunit RNA levels

(Song et al., 2021). APP/SI mice improve in cognitive tests after miR-98-5p is knocked down, demonstrating a functional role of a miRNA in regulating nAChR expression and linking this to changes in brain function (Song et al., 2021). miRNA regulation of nAChRs also produces functional consequences in *Caenorhabditis elegans* (see Chapter 10). Nicotine upregulated acr-15 and acr-19 nAChR gene RNAs with the acr-19 gene being required for a withdrawal response to nicotine. miR-238 targeted the acr-19 3′UTR and thus regulated acr-19 RNA levels. Acr-15 acted to downregulate the effectiveness of miR-238 by interfering with the RISC and thus block repression of acr-19 expression (Rauthan et al., 2017). Other *C. elegans* nAChR genes UNC-29 and UNC-63 are also regulated by miRNAs acting at MREs in the 3′UTRs (Simon et al., 2008).

Several of the homologous miRNAs also have conserved MREs in the 3′UTR of acetylcholinesterases, choline transporters, and chaperones (Hogan et al., 2014). miRNA may regulate cholinergic function at a broader level than just affecting nAChR expression.

References

Albuquerque, E. X., Pereira, E. F. R., Alkondon, M., & Rogers, S. W. (2009). Mammalian nicotinic acetylcholine receptors: from structure to function. *Physiological Reviews*, *89*(1), 73–120.

Balaraman, S., Winzer-Serhan, U. H., & Miranda, R. C. (2012). Opposing actions of ethanol and nicotine on microRNAs are mediated by nicotinic acetylcholine receptors in fetal cerebral cortical-derived neural progenitor cells. *Alcoholism: Clinical and Experimental Research*, *36*(10), 1669–1677. https://doi.org/10.1111/j.1530-0277.2012.01793.x.

Benwell, M. E. M., Balfour, D. J. K., & Anderson, J. M. (1988). Evidence that tobacco smoking increases the density of (−)-[3*H*]nicotine binding sites in human brain. *Journal of Neurochemistry*, *50*(4), 1243–1247. https://doi.org/10.1111/j.1471-4159.1988.tb10600.x.

Blount, P., Smith, M. M., & Merlie, J. P. (1990). Assembly intermediates of the mouse muscle nicotinic acetylcholine receptor in stably transfected fibroblasts. *Journal of Cell Biology*, *111*(6 I), 2601–2611. https://doi.org/10.1083/jcb.111.6.2601.

Boyd, R. T. (1994). Sequencing and promoter analysis of the genomic region between the rat neuronal nicotinic acetylcholine receptor β4 and α3 genes. *Journal of Neurobiology*, *25*(8), 960–973. https://doi.org/10.1002/neu.480250806.

Boyd, R. T. (1996). Transcriptional regulation and cell specificity determinants of the rat nicotinic acetylcholine receptor α3 gene. *Neuroscience Letters*, *208*(2), 73–76. https://doi.org/10.1016/0304-3940(96)12561-1.

Brody, A. L., Mukhin, A. G., La Charite, J., Ta, K., Farahi, J., Sugar, C. A., Mamoun, M. S., Vellios, E., Archie, M., Kozman, M., Phuong, J., Arlorio, F., & Mandelkern, M. A. (2013). Up-regulation of nicotinic acetylcholine receptors in menthol cigarette smokers. *International Journal of Neuropsychopharmacology*, *16*(5), 957–966. https://doi.org/10.1017/S1461145712001022.

Cooper, S., & Henderson, B. J. (2020). The impact of nicotine delivery system (ENDS) flavors on nicotinic acetylcholine receptors and nicotine addiction related behaviors. *Molecules*, *25*.

Crespi, A., Colombo, S. F., & Gotti, C. (2018). Proteins and chemical chaperones involved in neuronal nicotinic receptor expression and function: An update. *British Journal of Pharmacology*, *175*(11), 1869–1879. https://doi.org/10.1111/bph.13777.

Danthi, S., & Boyd, R. T. (2006). Cell specificity of a rat neuronal nicotinic acetylcholine receptor α7 subunit gene promoter. *Neuroscience Letters*, *400*(1–2), 63–68. https://doi.org/10.1016/j.neulet.2006.02.067.

Deneris, E. S., Francis, N., McDonough, J., Fyodorov, D., Miller, T., & Yang, X. (2000). Transcriptional control of the neuronal nicotinic acetylcholine receptor gene cluster by the β43' enhancer, Sp1, SCIP and ETS transcription factors. *European Journal of Pharmacology*, *393*(1–3), 69–74. https://doi.org/10.1016/S0014-2999(99)00883-3.

Du, Q., Tomkinson, A. E., & Gardner, P. D. (1997). Transcriptional regulation of neuronal nicotinic acetylcholine receptor genes. *Journal of Biological Chemistry*, *272*(23), 14990–14995. https://doi.org/10.1074/jbc.272.23.14990.

Durek, T., Shelukhina, I., Tae, H.-S., Thongyoo, P., Spirova, E., Kudryavtsev, D., Kasheverov, I., Faure, G., Corringer, P.-J., Craik, D., Adams, D., & Tsetlin, V. I. (2017). Interactions of synthetic human SLURP-1 with nicotinic acetylcholine receptors. *Scientific Reports*, *7*.

Flores, C. M., Rogers, S. W., Pabreza, L. A., Wolfe, B. B., & Kellar, K. J. (1992). A subtype of nicotinic cholinergic receptor in rat brain is composed of α4 and β2 subunits and is up-regulated by chronic nicotine treatment. *Molecular Pharmacology*, *41*(1), 31–37.

Fornasari, D., Battaglioli, E., Flora, A., Terzano, S., & Clementi, F. (1997). Structural and functional characterization of the human α3 nicotinic subunit gene promoter. *Molecular Pharmacology*, *51*(2), 250–261. https://doi.org/10.1124/mol.51.2.250.

Gallego, X., Cox, R., Laughlin, J., Stitzel, J., & Ehringer, M. (2013). Alternative CHRNB4 3'UTRs mediate the allelic effects of SNP rs1948 on gene expression. *PLoS ONE*, *8*(5).

George, A. A., Bloy, A., Miwa, J. M., Lindstrom, J. M., Lukas, R. J., & Whiteaker, P. (2017). Isoform-specific mechanisms of a3β4*-nicotinic acetylcholine receptor modulation by the prototoxin lynx1. *FASEB Journal*, *31*(4), 1398–1420. https://doi.org/10.1096/fj.201600733R.

Gu, S., Matta, J. A., Lord, B., Harrington, A. W., Sutton, S. W., Davini, W. B., & Bredt, D. S. (2016). Brain α7 nicotinic acetylcholine receptor assembly requires NACHO. *Neuron*, *89*(5), 948–955. https://doi.org/10.1016/j.neuron.2016.01.018.

Henderson, B. J., Grant, S., Wong, B. K., Shahoei, R., Huard, S. M., Saladi, S. S. M., Tajkhorshid, E., Dougherty, D. A., & Lester, H. A. (2018). Menthol stereoisomers exhibit different effects on α4β2 nAChR upregulation and dopamine neuron spontaneous firing. *eNeuro*, *5*(6). https://doi.org/10.1523/ENEURO.0465-18.2018.

Henderson, B. J., & Lester, H. A. (2015). Inside-out neuropharmacology of nicotinic drugs. *Neuropharmacology*, *96*, 178–193. https://doi.org/10.1016/j.neuropharm.2015.01.022.

Henderson, B. J., Srinivasan, R., Nichols, W. A., Dilworth, C. N., Gutierrez, D. F., Mackey, E. D., ... Lester, H. A. (2014). Nicotine exploits a COPI-mediated process for chaperone-mediated up-regulation of its receptors. *The Journal of General Physiology*, *143*(1), 51–66.

Henderson, B. J., Wall, T. R., Henley, B. M., Kim, C. H., Mckinney, S., & Lester, H. A. (2017). Menthol enhances nicotine reward-related behavior by potentiating nicotine-induced changes in nAChR function, nAChR upregulation, and DA neuron excitability. *Neuropsychopharmacology*, *42*(12), 2285–2291. https://doi.org/10.1038/npp.2017.72.

Henderson, B. J., Wall, T. R., Henley, B. M., Kim, C. H., Nichols, W. A., Moaddel, R., Xiao, C., & Lester, H. A. (2016). Menthol alone upregulates midbrain nAChRs, alters nAChR subtype stoichiometry, alters dopamine neuron firing frequency, and prevents nicotine reward. *Journal of Neuroscience*, *36*(10), 2957–2974. https://doi.org/10.1523/JNEUROSCI.4194-15.2016.

Hogan, E. M., Casserly, A. P., Scofield, M. D., Mou, Z., Zhao-Shea, R., Johnson, C. W., Tapper, A. R., & Gardner, P. D. (2014). MiRNAome analysis of the mammalian neuronal nicotinic acetylcholine receptor gene family. *RNA*, *20*(12), 1890–1899. https://doi.org/10.1261/rna.034066.112.

Jeanclos, E. M., Lin, L., Treuil, M. W., Rao, J., DeCoster, M. A., & Anand, R. (2001). The chaperone protein 14-3-3η interacts with the nicotinic acetylcholine receptor α4 subunit. *Journal of Biological Chemistry*, *276*(30), 28281–28290. https://doi.org/10.1074/jbc.m011549200.

Jones, A. K., Buckingham, S. D., & Sattelle, D. B. (2010). Proteins interacting with nicotinic acetylcholine receptors: Expanding functional and therapeutic horizons. *Trends in Pharmacological Sciences*, *31*(10), 455–462. https://doi.org/10.1016/j.tips.2010.07.001.

Kuryatov, A., Luo, J., Cooper, J., & Lindstrom, J. (2005). Nicotine acts as a pharmacological chaperone to up-regulate human α4β2 acetylcholine receptors. *Molecular Pharmacology*, *68*(6), 1839–1851. https://doi.org/10.1124/mol.105.012419.

Lester, H. A., Xiao, C., Srinivasan, R., Son, C. D., Miwa, J., Pantoja, R., Banghart, M. R., Dougherty, D. A., Goate, A. M., & Wang, J. C. (2009). Nicotine is a selective pharmacological chaperone of acetylcholine receptor number and stoichiometry. Implications for drug discovery. *AAPS Journal*, *11*(1), 167–177. https://doi.org/10.1208/s12248-009-9090-7.

Lippi, G., Steinert, J. R., Marczylo, E. L., D'Oro, S., Fiore, R., Forsythe, I. D., Schratt, G., Zoli, M., Nicotera, P., & Young, K. W. (2011). Targeting of the Arpc3 actin nucleation factor by miR-29a/b regulates dendritic spine morphology. *Journal of Cell Biology*, *194*(6), 889–904. https://doi.org/10.1083/jcb.201103006.

Lyukmanova, E. N., Shulepko, M. A., Kudryavtsev, D., Bychkov, M. L., Kulbatskii, D. S., Kasheverov, I. E., Astapova, M. V., Feofanov, A. V., Thomsen, M. S., Mikkelsen, J. D., Shenkarev, Z. O., Tsetlin, V. I., Dolgikh, D. A., & Kirpichnikov, M. P. (2016). Human secreted Ly-6/uPAR related protein-1 (SLURP-1) is a selective allosteric antagonist of α7 nicotinic acetylcholine receptor. *PLoS ONE*, *11*(2). https://doi.org/10.1371/journal.pone.0149733.

Lyukmanova, E. N., Shulepko, M. A., Shenkarev, Z. O., Bychkov, M. L., Paramonov, A. S., Chugunov, A. O., Kulbatskii, D. S., Arvaniti, M., Dolejsi, E., Schaer, T., Arseniev, A. S., Efremov, R. G., Thomsen, M. S., Dolezal, V., Bertrand, D., Dolgikh, D. A., & Kirpichnikov, M. P. (2016). Secreted isoform of human Lynx1 (SLURP-2): Spatial structure and pharmacology of interactions with different types of acetylcholine receptors. *Scientific Reports*, *6*. https://doi.org/10.1038/srep30698.

Mao, D, Perry, DC, Yasuda, RP, Wolfe, BB, & Kellar, KJ. (2008). The α4β2α5 nicotinic cholinergic receptor in rat brain is resistant to up-regulation by nicotine in vivo. *Journal of Neurochemistry*, *104*, 446–456.

Marks, M. J., Burch, J. B., & Collins, A. C. (1983). Effects of chronic nicotinic infusion on tolerance development and nicotinic receptors. *Journal of Pharmacology and Experimental Therapeutics*, *226*(3), 817–825.

Marks, M. J., McClure-Begley, T. D., Whiteaker, P., Salminen, O., Brown, R. W. B., Cooper, J., Collins, A. C., & Lindstrom, J. M. (2011). Increased nicotinic acetylcholine receptor protein underlies chronic nicotine-induced up-regulation of nicotinic agonist binding sites in mouse brain. *Journal of Pharmacology and Experimental Therapeutics*, *337*(1), 187–200. https://doi.org/10.1124/jpet.110.178236.

Marks, M. J., Pauly, J. R., Gross, S. D., Deneris, E. S., Hermans-Borgmeyer, I., Heinemann, S. F., & Collins, A. C. (1992). Nicotine binding and nicotinic receptor subunit RNA after chronic nicotine treatment. *Journal of Neuroscience*, *12*(7), 2765–2784. https://doi.org/10.1523/jneurosci.12-07-02765.1992.

Marks, M. J., Rowell, P. P., Cao, J. Z., Grady, S. R., McCallum, S. E., & Collins, A. C. (2004). Subsets of acetylcholine-stimulated 86Rb+ efflux and [125I]-epibatidine binding sites in

C57BL/6 mouse brain are differentially affected by chronic nicotine treatment. *Neuropharmacology*, *46*(8), 1141–1157. https://doi.org/10.1016/j.neuropharm.2004.02.009.

Matta, J. A., Gu, S., Davini, W. B., & Bredt, D. S. (2021). Nicotinic acetylcholine receptor redux: Discovery of accessories opens therapeutic vistas. *Science*, *373*(6556). https://doi.org/10.1126/science.abg6539.

Matta, J. A., Gu, S., Davini, W. B., Lord, B., Siuda, E. R., Harrington, A. W., & Bredt, D. S. (2017). NACHO mediates nicotinic acetylcholine receptor function throughout the brain. *Cell Reports*, *19*(4), 688–696. https://doi.org/10.1016/j.celrep.2017.04.008.

Mazzaferro, S., Whiteman, S., Alcaino, C., Beyder, A., & Sine, S. M. (2021). NACHO and 14-3-3 proote expression of distinct subunit stoichiometries of the 42 acetylcholine receptor. *Cellular and Molecular Life Sciences*, *78*(4), 1565–1575.

Merlie, J. P., & Lindstrom, J. (1983). Assembly in vivo of mouse muscle acetylcholine receptor: Identification of an α subunit species that may be an assembly intermediate. *Cell*, *34*(3), 747–757. https://doi.org/10.1016/0092-8674(83)90531-7.

Merlie, J. P., & Smith, M. M. (1986). Synthesis and assembly of acetylcholine receptor, a multisubunit membrane glycoprotein. *The Journal of Membrane Biology*, *91*(1), 1–10. https://doi.org/10.1007/BF01870209.

Miwa, J. M., Anderson, K. R., & Hoffman, K. M. (2019). Lynx prototoxins: Roles of endogenous mammalian neurotoxin-like proteins in modulating nicotinic acetylcholine receptor function to influence complex biological processes. *Frontiers in Pharmacology*, *10*. https://doi.org/10.3389/fphar.2019.00343.

Moretti, M., Mugnaini, M., Tessari, M., Zoli, M., Gaimarri, A., Manfredi, I., Pistillo, F., Clementi, F., & Gotti, C. (2010). A comparative study of the effects of the intravenous self-administration or subcutaneous minipump infusion of nicotine on the expression of brain neuronal nicotinic receptor subtypes. *Molecular Pharmacology*, *78*(2), 287–296. https://doi.org/10.1124/mol.110.064071.

Nagavarapu, U., Danthi, S., & Boyd, R. T. (2001). Characterization of a rat neuronal nicotinic acetylcholine receptor α7 promoter. *Journal of Biological Chemistry*, *276*(20), 16749–16757. https://doi.org/10.1074/jbc.M009712200.

Nashmi, R., Xiao, C., Deshpande, P., McKinney, S., Grady, S. R., Whiteaker, P., Huang, Q., McClure-Begley, T., Lindstrom, J. M., Labarca, C., Collins, A. C., Marks, M. J., & Lester, H. A. (2007). Chronic nicotine cell specifically upregulates functional α4* nicotinic receptors: Basis for both tolerance in midbrain and enhanced long-term potentiation in perforant path. *Journal of Neuroscience*, *27*(31), 8202–8218. https://doi.org/10.1523/JNEUROSCI.2199-07.2007.

Nichols, W. A., Henderson, B. J., Yu, C., Parker, R. L., Richards, C. I., Lester, H. A., & Miwa, J. M. (2014). Lynx1 shifts α4β2 nicotinic receptor subunit stoichiometry by affecting assembly in the endoplasmic reticulum. *Journal of Biological Chemistry*, *289*(45), 31423–31432. https://doi.org/10.1074/jbc.M114.573667.

Pauly, J. R., Marks, M. J., Robinson, S. F., van de Kamp, J. L., & Collins, A. C. (1996). Chronic nicotine and mecamylamine treatment increase brain nicotinic receptor binding without changing α4 or β2 mRNA levels. *The Journal of Pharmacology and Experimental Therapeutics*, *278*, 361–369.

Peng, X., Gerzanich, V., Anand, R., Whiting, P. J., & Lindstrom, J. (1994). Nicotine-induced increase in neuronal nicotinic receptors results from a decrease in the rate of receptor turnover. *Molecular Pharmacology*, *46*(3), 523–530.

Perry, D. C., Dávila-García, M. I., Stockmeier, C. A., & Kellar, K. J. (1999). Increased nicotinic receptors in brains from smokers: Membrane binding and autoradiography studies. *Journal of Pharmacology and Experimental Therapeutics*, *289*(3), 1545–1552.

Puddifoot, C. A., Wu, M., Sung, R. J., & Joiner, W. J. (2015). Ly6h regulates trafficking of alpha7 nicotinic acetylcholine receptors and nicotine-induced potentiation of glutamatergic signaling. *Journal of Neuroscience*, *35*(8), 3420–3430. https://doi.org/10.1523/JNEUROSCI.3630-14.2015.

Rauthan, M., Gong, J., Liu, J., Li, Z., Wescott, S. A., Liu, J., & Xu, X. Z. S. (2017). Micro-RNA regulation of nAChR expression and nicotine-dependent behavior in *C. elegans*. *Cell Reports*, *21*(6), 1434–1441. https://doi.org/10.1016/j.celrep.2017.10.043.

Rosenberg, M. M., Yang, F., Giovanni, M., Mohn, J. L., Temburni, M. K., & Jacob, M. H. (2008). Adenomatous polyposis coli plays a key role, in vivo, in coordinating assembly of the neuronal nicotinic postsynaptic complex. *Molecular and Cellular Neuroscience*, *38*(2), 138–152. https://doi.org/10.1016/j.mcn.2008.02.006.

Sallette, J., Bohler, S., Benoit, P., Soudant, M., Pons, S., Le Novère, N., Changeux, J. P., & Corringer, P. J. (2004). An extracellular protein microdomain controls up-regulation of neuronal nicotinic acetylcholine receptors by nicotine. *Journal of Biological Chemistry*, *279*(18), 18767–18775. https://doi.org/10.1074/jbc.M308260200.

Schwartz, R., & Kellar, K. J. (1983). Nicotinic cholinergic binding sites in the brain: Regulation in vivo. *Science*, *220*.

Simon, D. J., Madison, J. M., Conery, A. L., Thompson-Peer, K. L., Soskis, M., Ruvkun, G. B., Kaplan, J. M., & Kim, J. K. (2008). The MicroRNA miR-1 regulates a MEF-2-dependent retrograde signal at neuromuscular junctions. *Cell*, *133*(5), 903–915. https://doi.org/10.1016/j.cell.2008.04.035.

Song, C., Shi, J., Xu, J., Zhao, L., Zhang, Y., Huang, W., Qiu, Y., Zang, R., Chen, H., & Wang, H. (2021). Post-transcriptional regulation of α7 nAChR expression by miR98-5p modulates cognition and neuroinflammation in an animal model of Alzheimer's disease. *FASEB Journal*, *35*(6), e21658.

Tekinay, A. B., Nong, Y., Miwa, J. M., Lieberam, I., Ibanez-Tallon, I., Greengard, P., & Heintz, N. (2009). A role for LYNX2 in anxiety-related behavior. *Proceedings of the National Academy of Sciences of the United States of America*, *106*(11), 4477–4482. https://doi.org/10.1073/pnas.0813109106.

Thomsen, M. S., Arvaniti, M., Jensen, M. M., Shulepko, M. A., Dolgikh, D. A., Pinborg, L. H., Härtig, W., Lyukmanova, E. N., & Mikkelsen, J. D. (2016). Lynx1 and Aβ1–42 bind competitively to multiple nicotinic acetylcholine receptor subtypes. *Neurobiology of Aging*, *46*, 13–21. https://doi.org/10.1016/j.neurobiolaging.2016.06.009.

Wanamaker, C. P., & Green, W. N. (2005). N-linked glycosylation is required for nicotinic receptor assembly but not for subunit associations with calnexin. *Journal of Biological Chemistry*, *280*(40), 33800–33810. https://doi.org/10.1074/jbc.M501813200.

Wang, Z. Z., Fuhrer, C., Shtrom, S., Sugiyama, J. E., Ferns, M. J., & Hall, Z. W. (1996). The nicotinic acetylcholine receptor at the neuromuscular junction: Assembly and tyrosine phosphorylation. In *Vol. 61. Cold Spring Harbor symposia on quantitative biology* (pp. 363–371). Cold Spring Harbor Laboratory Press. https://doi.org/10.1101/sqb.1996.061.01.039.

Wang, F., Nelson, M. E., Kuryatov, A., Olale, F., Cooper, J., Keyser, K., & Lindstrom, J. (1998). Chronic nicotine treatment up-regulates human but not α3β4 acetylcholine receptors stably transfected in human embryonic kidney cells. *Journal of Biological Chemistry*, *273*(44), 28721–28732. https://doi.org/10.1074/jbc.273.44.28721.

Wonnacott, S. (1990). The paradox of nicotinic acetylcholine receptor upregulation by nicotine. *Trends in Pharmacological Sciences*, *11*(6), 216–219. https://doi.org/10.1016/0165-6147(90)90242-Z.

Wu, M., Puddifoot, CA, Taylor, P, & Joiner, WJ. (2015). Mechanisms of inhibition and potentiation of α4β2 nicotinic acetylcholine receptors by members of the LY6 protein family. *Journal of Biological Chemistry*, *290*(40), 24509–24518.

CHAPTER TEN

Nonmammalian models for studying nAChRs; zebrafish, fruit fly, and worm. What have we learned?

Most of this book focuses on human nAChRs and what we have learned about these from studying rats, mice, or humans. However, nAChRs are widely expressed in many species. Interestingly, several important observations have been made using model organisms such as *Caenorhabditis elegans* (worm), *Drosophila melanogaster* (fruit fly), and *Danio rerio* (zebrafish). These systems are amenable to a number of different assays that are also used in rodents, but that can be exploited on a larger scale.

10.1 *Danio rerio* (*D. rerio*)
10.1.1 Biology and nervous system

Zebrafish (*D. rerio*) have been used a model to study vertebrate development since the 1930s. Breeding and maintenance costs are relatively inexpensive compared with other vertebrates (Boyd, 2013; Westerfield, 2000). Zebrafish are raised in aquatic systems that can hold thousands of fish at a reasonable cost or in smaller systems suitable for individual labs (Boyd, 2013). Crosses are easily done with 100–300 embryos produced from each mating. While adult zebrafish are 3–4 cm long, embryonic and larval fish are much smaller with large numbers (50 or more) able to fit in a 60-mm Petri dish and are free swimming. No feeding is necessary due to the presence of the yolk sac until around day 6. The small size of zebrafish makes them ideal for use in 96- or 384-well plates for high-throughput screening (Boyd, 2013). One advantage of studying nAChRs and cholinergic systems in zebrafish is that nervous system development takes place rapidly with the earliest stages of development outside the female. Many features of the nervous system are visible after 24 hours postpofertilization (hpf). The first somites appear at about 10 hpf, while in rat, this occurs at 9–10 days. At 72 hpf embryogenesis is complete. Most organs are fully developed at around 3–5 days postfertilization (dpf) (Ackermann & Paw, 2003; Delvecchio, Tiefenbach, & Krause, 2011). Zebrafish from 3 dpf to 29 dpf are referred to as larvae and as juveniles from 30 to 89 dpf (Westerfield, 2000). Zebra are considered mature at 90 dpf and can reproduce and live from 3 to 5 years. Zebrafish are also transparent through about 14 dpf, and larvae can be maintained in that state after treatment with phenyl-2-thiourea (PTU) in the water to prevent formation of melanocytes. This allows for the visualization of cells expressing nAChRs and nAChR RNAs (Ackerman et al., 2009). A mutant strain, Caspar, can be used to visualize fluorescent signals in adults (White et al., 2008). Nicotine can be added to the water or injected intramuscularly to produce a dose-dependent increase in brain nicotine (Ponzoni et al., 2014). Zebrafish are active during the day as are humans and in contrast to rodents. This makes many behavioral tests easier to administer. Many lines of zebrafish have been developed that express fluorescent proteins in various cells that allow visualization of specific neuron populations in living animals. Electrical activity while the fish are performing various tasks can be monitored using voltage or Ca^{2+}-sensitive dyes (Boyd, 2013). Knockdown and knockout zebrafish can easily be developed using CRISPR. All of these features make zebrafish an excellent model to study nAChRs.

The zebrafish ventral telencephalic population is homologous to the mammalian cholinergic basal forebrain system and the isthmic superior reticular cell population is a homolog of the mammalian peduncul-opontine/laterodorsal tegmental system (Boyd, 2013). The CNS cholinergic system has been mapped in adult zebrafish (Clemente et al., 2004; Mueller et al., 2004). Acetylcholinesterase (AChE) was detected in the retina, dorsal telencephalic area, ventral telencephalic area, multiple areas of the diencephalon including the ventral thalamus, and the mesencephalon with high expression in the optic tectum (Clemente et al., 2004). AChE was also present in multiple areas of the rhombencephalon including nuclei reticularis (Clemente et al., 2004). Choline acetyltransferase (ChAT) was detected in the retina, lateral nucleus of the ventral telencephalic area, multiple areas of the diencephalon including the dorsal thalamus, habenula, and posterior tuberculum (Mueller et al., 2004). ChAT-positive cells were detected in many areas of the mesencephalon including rostral tegmental nucleus and isthmic region. Rhombencephalon labeling was seen in multiple populations of motor nerve nuclei, reticular formation, and the central gray area (Mueller et al., 2004). For more detailed localization information, see Clemente et al. (2004) and Mueller et al. (2004).

The zebrafish nervous system develops a forebrain, midbrain, and hindbrain as in other vertebrates. Zebrafish don't have a necortex, but the telencephalon mediates higher functions as in other nonmammalian vertebrates. Zebrafish motor, sensory, and autonomic nervous systems are similar to other vertebrate models. Similar locations of cholinergic neurons as in other vertebrates were seen in the telencephalon, tegmentum, cerebellum retina, spinal cord, and olfactory bulb (Boyd, 2013; Clemente et al., 2004). Vesicular acetylcholine transporters (VAChT), the high-affinity choline transporter, and AChE are present in zebrafish early in development (Rima et al., 2020). As can be seen, though different than in mammals, zebrafish have a highly developed cholinergic nervous system. As will be described below, many aspects of cholinergic signaling and nAChRs can be explored in zebrafish.

10.1.2 nAChRs Genes

The zebrafish genome has been sequenced (Howe et al., 2013), and the nAChR gene sequences are very similar to those of other vertebrates including humans (Pedersen et al., 2019; Boyd unpublished, Tables 10.1 and 10.2). Zebrafish nAChR genes are mostly syntenic with human and rodent nAChR genes (Barbazuk et al., 2000).

Zebrafish neuronal nAChR genes as well as the muscle nAChR genes have been cloned (Ackerman et al., 2009; Ackerman & Boyd, 2016; Mongeon et al., 2011; Zirger et al., 2003). The neuronal genes include orthologs of human, rats, and mice (Pedersen et al., 2019). Two $\alpha2$, one $\alpha3$, two $\alpha4$, $\alpha5$, $\alpha6$, two $\alpha7$, $\alpha8$, two $\alpha9$, two $\alpha10$, and an $\alpha11$ gene have been identified.(Table 10.1). One $\beta2$, two $\beta3$, one $\beta4$, and two $\beta5$ genes also exist in the genome. Many of the zebrafish nAChR protein sequences are between 60% and 81% identical to their human orthologs (Pedersen et al., 2019) (Table 10.2). For example, $\beta2$ has an 81% identity to the human $\beta2$, while zebrafish $\alpha7$ is 76% identical, $\alpha6$ is 71% identical, $\beta3$ is 75%, and $\alpha3$ is 74% (Table 10.2). Sequence similarities with nAChRs from other species are highest in extracellular domains and in transmembrane region 2 (Boyd, 2013). The most diverse region between zebrafish nAChRs is in the intracellular loop between transmembrane regions 3 and 4. This feature is conserved with other human, rodent, and other species nAChRs in that this is also the most diverse region when comparing sequence of different subunits within a species. The similarity of the zebrafish nAChRs with those of other

Table 10.1 Zebrafish neuronal nAChR subunit genes (not in table format in ELSA).

Name	Chr.	Gene ID ENSDARG	Transcript ID ENSDART	Protein ID ENSDARP	ZFIN
α2a*	17	00000006602	00000033947	00000028704	ZDB-GENE-040108-2
α2b	20	00000057025	00000136317	00000119522	ZBD-GENE-041001-99
α3*	18	00000100991	00000160975	00000131497	ZBD-GENE-070822-1
α4a	6	00000087071	00000153740 0	0000128324	ZBD-GENE-130530-903
α4b*	11	00000070724	00000018614	00000022694	ZDB-GENE-090505-3
α5	18	00000003420	00000021372	00000002331	ZBD-GENE-050417-440
α6*	1	00000055559	00000031546	00000038141	ZBD-GENE-090312-91
α7a*	7	00000101702	00000171463	00000139906	ZDB-GENE-040108-3
α7b	25	00000101522	00000163134	00000140988	
α8*	1	00000087224	00000176849	00000144302	ZBD-GENE-090313-161
			00000127506	00000108882	
α9a	1	00000054680	00000189921	00000154815	ZBD-GENE-090312-63
α9b	14	00000099181	00000158505	00000139294	
α10a	15	00000011113	00000152531	00000127070	ZBD-GENE-060503-725
			00000012872	00000015717	
α10b	21	00000108118	00000178123	00000143851	
α11	19	00000037427	00000138971	00000120968	ZDB-GENE-060503-606
			00000141649		
β2	16	00000017790	00000041625	00000041624	ZDB-GENE-070821-3
β3 a*	1	00000052764	00000074678	00000069164	ZDB-GENE-040108-1

β3b	14	00000038508	00000050037	00000050036	ZDB-GENE-030102-4
β4*	18	00000101677	000000161429	00000138900	ZDB-GENE-120809-1
β5a	1	00000115778	00000187834	00000149406	ZDB-GENE-090312-169
β5b*	14	00000021392	00000173319	00000142507	ZDB-GENE-120809-5
			00000039985	00000039984	

*cDNA cloned from zebrafish embryos or larvae by Boyd RT.

Gene, transcript, protein IDs and ZFIN ID from Sanger Ensembl Zv11 Gene Build.

α4a is noted as (XP_021333008) in Pedersen et al. (2019).

α7b is noted as (XP_005174278) in Pedersen et al. (2019).

α8 is noted as si:ch73-380n15.2 in Pedersen et al. (2019).

α9a is noted as XP_021326868 in Pedersen et al. (2019).

α9b noted as XP_00133964 in Pedersen et al. (2019).

α10b noted as α10l (XP_021324910) in Pedersen et al. (2019).

α11 is also noted as si:ch211-39a7.1.

β5a also noted as β2a.

β5b also noted as β2like.

Table 10.2 Comparison of protein sequence identity between human and zebrafish nAChRs.

Zebrafish	α2a	α2b	α3	α4a	α4b	α5	α6	α7a	α7b	α8	α11	α9a	α9b	α10a	α10b	β2	β5a	β5b	β3a	β3b	β4
Human																					
α2	72	72	59	61	66	54	59	41	42	40	38	33	32	34	34	50	51	47	54	53	49
α3	60	61	74	57	57	47	71	39	38	37	35	34	36	33	34	50	51	45	48	50	48
α4	70	68	58	60	66	54	56	38	38	36	35	32	35	32	35	48	51	48	52	53	51
α5	53	51	47	48	52	72	49	35	36	36	35	36	34	37	35	42	43	40	65	65	41
α6	57	56	66	52	55	50	71	36	36	35	33	34	35	33	34	49	49	43	49	50	48
α7	40	41	39	38	40	37	36	76	76	72	67	37	39	39	37	37	40	38	35	35	37
α9	36	36	36	33	34	35	36	38	38	38	36	66	69	61	62	33	34	33	35	35	34
α10	35	37	35	33	36	36	37	40	42	41	40	56	57	59	60	37	38	35	36	34	37
β2	53	52	50	47	50	43	48	37	38	35	36	33	33	33	34	81	67	64	41	44	69
β3	54	54	47	48	53	65	50	35	36	36	34	36	35	36	36	43	43	41	75	75	44
β4	53	50	48	47	49	45	45	38	38	36	36	33	32	34	35	64	64	58	44	43	65

animals that have been used to study nAChRs supports using zebrafish as a model to understand nAChR function.

10.1.3 Localization and expression

The developmental expression patterns of several zebrafish nAChR subunits have been determined including α2a, α4b, α6, α7a, β2, and β3 (Ackerman et al., 2009; Ackerman & Boyd, 2016; Welsh et al., 2009; Zirger et al., 2003). PCR has also been used to determine the timing of expression of multiple nAChR subunit RNAs (Table 10.3). Three were detected in maternal RNA (α2a, β5b, and β3a). This indicates that nAChRs may be involved in very early developmental events, even before zygotic transcription begins. α4b appears at 3 hpf, α7a is detected at 8 hpf, and α6 by 10 hpf (see Table 10.3). While elsewhere in this book, the expression patterns of nAChRs in other species are summarized, the zebrafish patterns also reveal

Table 10.3 Expression of zebrafish nAChR RNAs during early development.

Zf gene	α2a	α3	α4b	α6	α7a	α8	β5b	β3a	β4
Stage									
M	(+)	NT	(−)	(−)	(−)	(−)	(+)	(+)	(−)
2 h	(+)	NT	NT	NT	NT	NT	NT	(+)	NT
3 h	NT	NT	(+)	NT	NT	(−)	(−)	NT	NT
4 h	NT	NT	NT	NT	NT	NT	NT	(+)	(−)
5 h	(+)	NT	NT	NT	NT	NT	NT	(+)	NT
6 h	NT	NT	(+)	NT	NT	(−)	(+)	NT	NT
8 h	(+)	NT	NT	NT	(+)	(−)	(−)	(+)	NT
10 h	NT	NT	(+)	(+)	NT	(−)	NT	(+)	(−)
12 h	(+)	(−)	NT	(+)	(+)	(−)	(−)	(+)	NT
14 h	NT	(−)	NT	(+)	NT	(−)	(−)	(+)	(−)
16 h	NT	(−)	(−)	(+)	NT	(−)	NT	(+)	(−)
18 h	(+)	(−)	(+)	(+)	(+)	(−)	(+)	(+)	NT
24 h	(+)	(+)	(+)	(+)	(+)	(−)	(+)	(+)	(+)
36 h	(+)	(+)	(+)	(+)	(+)	(+)	(+)	(+)	NT
48 h	(+)	(+)	(+)	(+)	(+)	(+)	(+)	(+)	NT
60 h	NT	(+)	(+)	(+)	NT	(+)	(+)	NT	NT
72 h	(+)	(+)	(+)	(+)	(+)	(+)	(+)	(+)	(+)
96 h	NT	(+)	(+)	(+)	(+)	(+)	(+)	NT	NT
7 d	NT	(+)	(+)	(+)	(+)	(+)	(+)	NT	NT

M, maternal RNA.
h, hours postfertilization.
d, days postfertilization.
NT, not tested.
(−), not detected.
(+), detected.

some aspects that support a role for early expression of nAChRs in development. It also provides the background information for the use of zebrafish as an important model for studying nAChRs.

The α6 gene was expressed at around 10 hpf near the onset of neurogenesis (Ackerman et al., 2009). α6 is expressed in Rohon-Beard (RB) and spinal cord neurons by 24 hpf and was also expressed in the hindbrain, pineal gland, trigeminal ganglion, and diencephalon. Spinal cord expression was transient as α6 was not detected at 48 hpf, but continued to be expressed in the pineal gland, trigeminal ganglion and now expressed in the locus coeruleus and diencephalic catecholaminergic cluster (Ackerman et al., 2009). α6 was also expressed in the tectum and retina at 48 hpf. At 72 hpf, pineal and trigeminal expression continued with now higher expression in the tectum, initial expression in cranial sensory neurons and in non-catecholamingergic cells in the midbrain and hindbrain in addition to continued expression in the diencephalic catecholaminergic cluster (Ackerman et al., 2009). Retinal expression was high at both 72 and 96 hpf.

α4 expression was first detected at 3 hpf near the time when zygotic transcription begins (Kimmel et al., 1995) making it one the earliest transcribed genes. There are two α4 genes identified, and this one is α4b (Pedersen et al., 2019). α4 expression was different from that of α6. By 24 hpf, α4b was expressed in rhombomeres 4–7 and in cranial neural crest cells. α4b was not detected in the retina and was detected at low levels in the forebrain and midbrain. Since α4 is expressed in the forebrain and retina in rodents and humans, this could reflect that the α4b gene has a different expression pattern than that of α4a. By 48 hpf, midbrain and hindbrain expression increased in reticulospinal neurons and the nucleus of the medial longitudinal fascicle, but no spinal cord expression was detected (Ackerman et al., 2009). High expression in specific areas in midbrain and hindbrain was still present at 72 and 96 hpf (Ackerman et al., 2009).

There are two β3 genes, and β3a is surprisingly expressed in unfertilized eggs as part of the maternal RNA population (Ackerman & Boyd, 2016). This is an indication of a possible early role in development. β3a expression continued throughout development up to 72 hpf and was not examined after that (Zirger et al., 2003). β3a was detected in the retina at 18 and 24 hpf and from 36 to 72 hpf was detected in an area of the retina consistent with retinal ganglion cells with the highest expression in an area near where the optic nerve exits the eye (Zirger et al., 2003). In rat, β3 is detected at embryonic day (ED)15 retinal ganglion cells (Zirger et al., 2003). There are two α7 genes, and α7a is expressed early as are the others, being detected at 8 hpf. In situ hybridization detected a limited pattern in hindbrain at 72 and 96 hpf and in the retina (Zirger et al., 2003). This α7 gene may have a more limited expression than α7b whose pattern of expression has not been identified to our knowledge.

α2a, β5b, and β3a are also surprisingly detected in unfertilized eggs. The α2 expression pattern is complex, and the early localization and transient expression are indicative of a possible developmental role. α2a RNA is detected by in situ hybridization from 10 to 96 hpf (Ackerman & Boyd, 2016). At 10 hpf, some expression is seen in somites and by 18 hpf is also expressed in the ventral midbrain, hindbrain, and somites. At 24 h, forebrain expression is limited, with diffuse expression in the ventral midbrain and with high expression in the anterior hindbrain neurons (Ackerman & Boyd, 2016). High expression also is present in the anterior spinal cord bilaterally at 24 hpf. At 48 hpf, expression is reduced in the midbrain and with reduced spinal cord expression. By 72 hpf, spinal cord expression is not detected with increased midbrain and limited hindbrain expression (Ackerman & Boyd, 2016). At 96 midbrain expression is low and no spinal cord expression is detected.

Closer examination showed that α2 was expressed dorsally to motoneurons (islet-1 positive cells), perhaps in interneurons, in the spinal cord in the first seven anterior segments, and in a location consistent with motoneurons down the rest of the spine at 24 hpf (co-expressed with islet −1) (Ackerman & Boyd, 2016). Expression in motoneurons was low and punctate with most of the labeling at the level of the somites and rostral to the motor nuclei. α2a was not detected at 24 hpf in RB cells in anterior spine by in situ hybridization (Ackerman & Thomas Boyd, 2016), but was detected at some level of the spine in 20–22 hpf RB cells by in situ hybridization and α2a antibody labeling (Menelaou et al., 2014). α2a was also detected in RB cells at 30–33 hpf using an α2a antibody (Menelaou et al., 2014). This may not have been the same level of the spine as was examined in Ackerman & Boyd, 2016. At 48 hpf, α2a expression was in spine, but dorsal to spinal motor neurons (Ackerman & Boyd, 2016). β2 subunits were detected in 36 hpf zebrafish RB neurons using antibody (Welsh et al., 2009). β2 was also detected in ventral secondary motoneurons that project dorsally at 72 hpf, but not in ventrally projecting motoneuron axons (Welsh et al., 2009).

Medial habenula-interpeduncular nucleus (MHb-IPN) pathway was shown in mice to be important for control of the aversive effects of nicotine (Fowler et al., 2011). Multiple nAChR subunit genes are expressed in the zebrafish MHb, making this another model to investigate mechanisms of nAChR signaling in this pathway. α2a, α2b, and β4 are expressed in the IPN at 4 dpf (Hong et al., 2013). These subunit RNAs and also α5 and α7 RNAs are detected in adult IPN (Hong et al., 2013). α3 RNAs were not detected in the adult IPN.

α7 was also detected in 22–24 hpf embryos and was located in a position consistent with motoneurons while α3 and α2a were in a position consistent with interneurons. α3 and α7 were also expressed in the brain (Rima et al., 2020). At 6 dpf, larvae α3 and α7 were present along the length of the spinal cord (Rima et al., 2020). α4 and α7 nAChRs are detected in adult zebrafish brain (Viscarra et al., 2020). α9, but not α10, nAChRs are detected in the zebrafish lateral line organ hair cells (similar to human cochlear hair cells). This model is useful to study mammalian hearing loss. Zebrafish express α9 nAChR homomeric receptors as do mammalian cochlear hair cells although in mammals, α9/α10 heteromers may also play a role (Freixas et al., 2021). In summary, multiple nAChR genes homologous to those expressed in humans are present in the zebrafish.

Early expression of nAChR RNAs points to a role in development. Assembled nAChRs have also been detected in developing zebrafish (Zirger et al., 2003). Epibatidine, which binds to assembled receptors and not just subunits, binds to several nAChRs subtypes with high affinity (see Chapter 3). High-affinity binding of epibatidine was detected using homologous competition binding in 2 dpf and 5 dpf zebrafish. In 2 dpf zebrafish, two binding sites were detected with IC_{50}s of 28.4 pM and 8.9 nM and in 5 dpf zebrafish two sites with IC_{50}s of 28.6 pM and 29.7 nM (Zirger et al., 2003). Specific subtypes can't be assigned at this age, but assembled receptors are present with affinities comparable with those of human nAChRs (Zirger et al., 2003). The higher-affinity site may represent α4β2* nAChRs. Epibatidine binding in adult zebrafish (6–12 mo.) also revealed two types of sites with K_ds of 20 pM and 5 nM (Ponzoni et al., 2014). Incubation with α-Bgt eliminated the lower-affinity site indicating that this is formed of α7 nAChRs (Ponzoni et al., 2014). Zebrafish α-Bgt-bound receptors were almost completely precipitated with an anti-α7 nAChR antibody. Surprisingly, an α4 antibody didn't precipitate the epibatidine-labeled population as it does in rat. This could indicate that the epitope was not available or some unrecognized differences exist in the C-terminus between zebrafish and rat α4 subunits. Zebrafish possess two major populations of nAChRs, epibatidine-bound and α-Bgt-bound. Competition binding showed that varenicline bound epibatidine-labeled receptors with the highest affinity followed by other agonists cytisine, CC26 (cytisine derivative), nicotine, and CC4 (another cytisine derivative (Ponzoni et al., 2014). Among antagonists, the conotoxin MII had the highest affinity followed by MLA, α-Bgt, DHβE, and mecamylamine (Ponzoni et al., 2014). Binding to the α-Bgt-labeled population also showed that among agonists varenicline had the highest affinity followed by cytisine, CC26, nicotine, and CC4. As expected among antagonists, α-Bgt had the highest affinity followed by MLA, MII, DHβE, and mecamylamine (Ponzoni et al., 2014). These results are generally in line in what is observed for other vertebrates, and the binding selectivity is conserved, supporting a role for zebrafish in studying nAChR function (Ponzoni et al., 2014). Keeping with studies in other vertebrates, chronic nicotine exposure increased epibatidine binding, indicating nAChR upregulation, but not α-Bgt binding (Ponzoni et al., 2021).

10.1.4 Pharmacology of zebrafish nAChRs

Zebrafish have been used in a number of behavioral studies regarding the effects of nicotine on development, gene expression, and addiction. These studies using cholinergic drugs with zebrafish have to be considered with regard to whether these compounds act in the same way and at similar nAChR subtypes as in rodents and humans. The pharmacology of multiple nAChR subtypes in zebrafish has now been examined, and there are many similarities and a few differences (Papke et al., 2012). Zebrafish α4β2, α2β2, α3β4, and α7 homomeric neuronal nAChRs were expressed in Xenopus oocytes. Muscle α1β1βεδ nAChRs were also examined (Papke et al., 2012).

Nicotine showed low potency and efficacy at zebrafish muscle nAChRs. All five of the receptor subtypes tested were activated by 3 μM ACh or higher (Papke et al., 2012). ACh was the most potent for α4β2 nAChRs (13 μM) and lowest for the α3β4 nAChRs (73 μM). α2β2, muscle, and α7 ranged from 39 to 48 μM. α4β2 nAChRs showed smaller currents than the mammalian counterparts (Papke et al., 2012). Zebrafish nAChR kinetics were similar to mammalian nAChRs with the heteromeric subtypes demonstrating concentration-dependent ACh responses. Mammalian α7 nAChRs desensitize quickly, and zebrafish nAChRs also showed the same concentration-dependent desensitization. Nicotine was a full agonist for α7 nAChRs compared with ACh, but only a partial agonist for α4β2, α2β2, and α3β4 nAChRs. Nicotine was efficacious and potent for α4β2 nAChRs.

Cytisine is described elsewhere in this book and has been used a platform to produce varenicline (Chantix), used for smoking cessation. To determine if zebrafish are suitable for studying the effects of cytisine on nAChRs, the efficacy of cytisine on zebrafish nAChRs was determined. Cytisine was a weak partial agonist for zebrafish $\alpha3\beta4$ nAChRs in contrast to being a full agonist for mammalian nAChRs (Papke et al., 2012). Side effects of cytisine on rat and mouse are thought to be due to actions at ganglionic receptors that express $\alpha3\beta4$ subtypes. This may indicate that a drug modeled after cytisine identified in a zebrafish screen may have ganglionic effects in humans (Boyd, 2013). Cytisine is a full agonist for zebrafish $\alpha7$ nAChRs as it is for mammalian $\alpha7$ nAChRs. Cytisine has the highest potency for $\alpha4\beta2$ nAChRs, more so than for mammalian $\alpha4\beta2$ nAChRs (Papke et al., 2012).

$\alpha7$ agonists choline, 4OH–GTS-21, and tropane were tested against zebrafish $\alpha7$ nAChRs (Papke et al., 2012). None of these compounds were $\alpha7$ selective in zebrafish and activated $\alpha4\beta2$ nAChRs. 4OH–GTS-21 inhibited all heteromeric receptors. Testing compounds based on these structures may not be advisable in zebrafish due to some lack of subtype specificity (Boyd, 2013; Papke et al., 2012).

Mecamylamine has been used a nonselective antagonist for nAChRs in numerous studies. In zebrafish, it was most potent at $\alpha3\beta4$ nAChRs with a rank order of potency of $\alpha3\beta4 > $ muscle $ > \alpha2\beta2 > \alpha4\beta2 > \alpha7$ (Boyd, 2013; Papke et al., 2012). This is similar to the activity of mecamylamine in mammals. Mecamylamine activity in zebrafish at muscle receptors was similar to that in mice, inhibiting the receptors with a similar EC_{50}s (Boyd, 2013). Mecamylamine reversibly blocked muscle, $\alpha7$, and $\alpha3\beta4$ nAChRs. This is in contrast to rat nAChRs, whereby mecamylamine exposed to $\alpha3\beta4$ nAChRs produced a 60% residual inhibition (Boyd, 2013). ACh and mecamylamine together produced long-term inhibition in $\beta2$ containing nAChRs. Activity at muscle nAChRs was reversible, supporting drug studies in zebrafish.

Mecamylamine is not a selective antagonist for zebrafish CNS neuronal nAChRs. However, mecamylamine and nicotine can be used to target nAChR in zebrafish homologous to their mammalian counterparts (Papke et al., 2012). DHβE, an antagonist for $\alpha4\beta2$ nAChRs in mammals, also blocked the responses of $\alpha4\beta2$ zebrafish nAChRs (but not $\alpha7$) expressed in oocytes (Ponzoni et al., 2014). MLA, an $\alpha7$ nAChR antagonist in mammals, surprisingly blocked both zebrafish $\alpha7$ nAChR and $\alpha4\beta2$ AChR responses (Ponzoni et al., 2014). MLA's effect on blockage of zebrafish $\alpha4\beta2$ nAChRs differs from that of mammals. CC4 and CC26 blocked the zebrafish $\alpha4\beta2$ response to ACh in oocytes, but had little effect on $\alpha7$ n AChRs (Ponzoni et al., 2014). Given the conservation of receptors demonstrated by ligand-binding studies described above, zebrafish are a good model for studying the effects of drugs on nAChRs . Much of the pharmacology appears to be conserved; however, some differences in pharmacology between mammalian and zebrafish nAChRs exist.

10.1.5 Zebrafish nAChRs in development

Because of the ease of access to developing zebrafish embryos, they can be used to examine the effects of nicotine or other drugs on development. While in some cases, nicotine appears to be neuroprotective, application of 50 μM nicotine to zebrafish from 5 to 96 hpf produced apoptosis in the 96 hpf embryos (Matta et al., 2007). When the embryos were co-incubated with 50 μM nicotine and 20 μM DHβE, apoptosis was not observed, indicating a role for α4β2* nAChRs (Matta et al., 2007).

In addition, 20 μM nicotine exposure to 1 dpf embryos caused reduced length of the notochord and eye diameter (Parker & Connaughton, 2007). Exposure to 40 μM produced reduced swimming behavior, startle response, and increased death. Maternal smoking is correlated with reduced birth weight in humans, and zebrafish exposure to nicotine also produced reduced weight (Parker & Connaughton, 2007).

Exposure of zebrafish to nicotine was shown to delay development of secondary spinal motoneurons and produce some muscle paralysis (Svoboda et al., 2002). Embryos exposed from 22 to 66 hpf to 33 μM nicotine examined at 66 hpf had a drastic reduction in the number of ventral secondary motor neurons. The nicotine exposure caused a delay in motoneuron development rather than death because if nicotine was removed at 66 hpf; by 120 hpf, the number of motoneurons was restored (Svoboda et al., 2002). However, the surviving motoneurons still had altered trajectories, and thus, the transient nicotine exposure produced a long-term effect. The effects of transient nicotine exposure (up to 3 dpf, but then removed) were still evident at up to 86 dpf (Menelaou & Svoboda, 2009).

Since nicotine can also affect muscle function, it is possible that the paralysis was caused at the muscle level, not the motoneuron. However, the effect of nicotine also was shown in a mutant zebrafish (sofa potato) that lacked muscle nAChRs, where the motoneuron pathfinding errors still occurred (Welsh et al., 2009). DHβE blocked the action of nicotine indicating the role of α4β2 nAChRs, but MLA at 2 μM didn't. Treatment with epibatidine in order to stimulate α4β2 nAChRs, in a manner similar to nicotine, produced pathfinding errors (Welsh et al., 2009). Interestingly, MLA at 1 μM reduced ACh-induced currents in $\alpha4\beta2$ nAChRs expressed in oocytes by 79%(Ponzoni et al., 2014), but apparently didn't affect motoneuron development. DHβE may affect other zebrafish subtypes or other subtypes in addition to $\alpha4\beta2$ nAChRs may be involved. Nicotine has an EC_{50} for both $\alpha2\beta2$ and $\alpha4\beta2$ nAChRs of 6 μM (Papke et al., 2012), so it is possible that some of the effects on pathfinding may be mediated by $\alpha2\beta2$ nAChRs.

Nicotine can desensitize nAChRs after prolonged exposure, and it is possible that nicotine's actions are due to reduction of nAChR signaling. Preliminary studies (Ackerman et al., 2006) show that knockdown of zebrafish $\alpha2$a RNA also alters motoneuron axonal pathfinding of dorsally extending axons in 72 and 96 hpf fish Fig. 10.1. The morphology was normal at 24 hpf and 48 hpf, but multiple defective phenotypes manifested by 72 hpf. This $\alpha2$a knockdown produced fish with various motor function abnormalities. Reintroduction of $\alpha2$a RNA rescued much of the motor phenotype (Ackerman et al., 2006). Another study using antisense morpholinos designed to knock down $\alpha2$a RNA expression blocked the nicotine-induced bending (but not spontaneous activity) in 20–22 hpf embryos with these effects rescued by reintroduction of $\alpha2$a RNA (Menelaou et al., 2014). The effects of the $\alpha2$a knockdown were consistent with a role for $\alpha2$a nAChRs in upstream signaling within the spinal cord and not on muscle nAChRs (Menelaou et al., 2014). Primary motoneurons exhibited normal trajectories at 30 hpf (Menelaou et al., 2014), in contrast to other studies at 72 and 96 hpf showing deficits in secondary motoneurons after antisense $\alpha2$a knockdown (Ackerman et al., 2006). RB neurons were not affected by the $\alpha2$a knockdown treatment. (Menelaou et al., 2014) At later stages, 23–29 hpf, $\alpha2$a knockdown reduced but didn't eliminate nicotine-induced motor output, indicating the additional role of non-$\alpha2$-containing nAChRs.

Larval zebrafish were exposed to imidacloprid, a pesticide, and nicotine to determine the effects on the developing zebrafish. Fish were treated from 4 hpf to 5 dpf. Larval swimming in the dark was reduced by both treatments. Fish treated as larvae and allowed to grow for 1.5 months (adolescents) or 3 months (adults) showed decreased novel tank exploration and an increased response to startle stimuli after imidacloprid treatment (Crosby et al., 2015). Nicotine-treated fish showed an increased startle response as adolescents. Many of the developmental effects of maternal nicotine exposure seen in rodents (see Chapter 7) can also be examined in zebrafish with more control of drug application and access to multiple developmental stages.

Nicotine possibly produces its effects through desensitization of nAChRs at a crucial time in development or stimulating receptors that

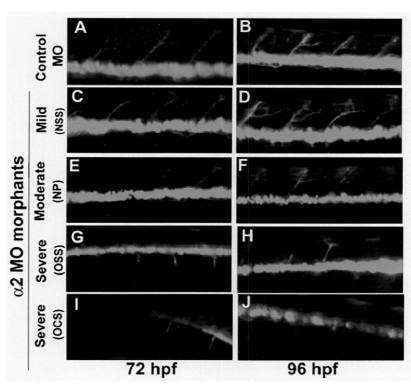

Fig. 10.1 α2 nAChR morphants display abnormal pathfinding of dorsal motor neurons. Islet 1 GFP transgenic zebrafish that label (MiP) dorsal extending axons were examined after α2 nAChR gene knockdown at 72 hpf and 96 hpf. A and B: Control Morpholino (Control MO) injected embryos displayed a normal pattern of axon extension. (A) At 72 hpf, the axons first projected ventrally along the common path and then at the first checkpoint grew dorsally with a single branch point at the dorsal edge of the extending axon. (B) At 96 hpf, there was more extensive branching, but the pattern remained with the single branch point at the dorsal myotome. (C) and (D) Mild phenotype displayed both normal patterns of axon extension and axons that extended fully to the dorsal surface but had extra branching along the axon. (C) At 72 hpf, 81.75% were normal while 18.25% of the axons extended fully to the dorsal surface but did not branch properly. (D) At 96 hpf, 82.67% of the axons were normal while 17.33% displayed extra branching that began to innervate adjacent somites. E and F: Moderate Phenotype displayed normal axonal extension, truncated axons, axons that extend fully to the dorsal surface but did not take a branch point, and finally, axons were missing. (E) At 72 hpf, 66.4% were normal, 14.11% were truncated, 10.18% extend fully to the dorsal surface but did not take a branch point, and 9.31% of the axons were missing. (F) At 96 hpf, 86.37% of the axons were normal, 9.04% displayed extra branching that extended into adjacent somites, and 4.59% extended to the dorsal surface with no branch points. Unlike at 72 hpf, there were not truncated or missing axons. G and H: Severe Phenotype (OSS) had truncated and missing axons. (G) At 72 hpf, 77.35% of the axons were truncated and 22.65% were missing. I and J: Severe Phenotype (OCS) had missing and truncated axons. (I) At 72 hpf, 48.51% were truncated and 51.49% were missing. (J) At 96 hpf, 82.49% of the axons were truncated and 17.51% were missing.

would normally be activated by ACh, but to an inappropriate extent or at the wrong time. Many studies have been done exposing zebrafish to nicotine or other chemicals. Due to the ease of working with zebrafish, large numbers of drugs can be tested at once and zebrafish at multiple ages can be easily examined. Given the effects of nicotine on development due to maternal smoking, an additional model to study nicotine's effects on development or gene expression is desirable. The use of zebrafish has been important in understanding the role of nAChRs in development, and they are an important model that can be added to rodents, nonhuman primates, and the other species described in this chapter as important contributors to understanding nAChRs.

10.1.6 Applications of zebrafish for behavioral studies

The cholinergic system of zebrafish has many conserved elements compared with other vertebrates. Many CNS effects of nicotine are mediated via the dopaminergic system as can be seen for smoking addiction and locomotor effects. While different from mammals, the zebrafish also has a complex dopaminergic system that performs many of the functions of the VTA, SN, NAc, etc., in mammals. Zebrafish have orthologs of two tyrosine hydroxylase genes, dopamine transporters, dopamine β-hydroxylase, and four dopamine receptor genes similar to those present in humans (Boyd, 2013). The receptor genes are expressed in the tegmentum and diencephalon and in a pattern consistent with what is seen in mammals. The zebrafish dopamine transporter gene, slc6a3, is co-expressed with tyrosine hydroxylase in 13 dopamine neuron clusters (Boyd, 2013). Zebrafish dopaminergic neurons are located in the pretectum, ventral telencephalon, ventral diencephalon, hypothalamus, and retina (Boyd, 2013). Zebrafish have been used to study effects of nicotine and other drugs on the dopaminergic system.

Zebrafish also demonstrate numerous higher functions such as memory, social behaviors, conditioned responses, and drug-seeking behavior (Dooley & Zon, 2000; Guo, 2004; Klee et al., 2011; Norton & Bally-Cuif, 2010). Many behavioral tests are used with rodents to test memory, anxiety, etc. However, behavioral assays for zebrafish have also been developed to test anxiety, depression, withdrawal from nicotine, memory, and others. A few studies using these will be described to demonstrate the utility of zebrafish to study behaviors dependent on nAChR signaling.

10.1.6.1 Anxiety

Tests used to model anxiety in zebrafish include place preference (edge or bottom of tank), tank diving tests, locomotor activity, and light/dark preference (Norton & Bally-Cuif, 2010). Anxiety can be evaluated in zebrafish using a novel tank test (Viscarra et al., 2020). Zebrafish express anxiety by spending more time on the bottom of a tank. Nicotine and a cholinergic drug UFR2709, a nAChR antagonist for $\alpha4\beta2$, (Faundez-Parraguez et al., 2013) reduce time spent on the bottom of the tank, which is interpreted as reduced anxiety.

Another study using the novel tank test (NTT) in zebrafish showed that nicotine, cotinine, anatabine, and methylanatabine all reduced anxiety in adult fish (Alzualde et al., 2021). Computational docking showed that these compounds interacted with $\alpha4\beta2$ nAChRs. Multiple clinically applied anxiolytics such as fluoxetine and buspirone have also been proven in zebrafish to reduce anxiety determined by the NTT (Alzualde et al., 2021).

A place preference assay similar to the NTT was used to test the anxiolytic effects of nicotine in zebrafish (Levin, Bencan, & Cerutti, 2007). Nicotine was also anxiolytic, and mecamylamine blocked this effect if given with nicotine, but not before. $\alpha4\beta2$ and $\alpha7$ nAChRs may be involved since MLA and DHβE blocked the anxiolytic effects of nicotine (Bencan & Levin, 2008). However, given that MLA also affects zebrafish $\alpha4\beta2$ nAChR, $\alpha7$ nAChR involvement is not clear.

10.1.6.2 CPP

Nicotine-induced conditioned place preference (CPP) also occurs in zebrafish as in rodents (Boyd, 2013). Intramuscular nicotine induced CPP in zebrafish as did application of nicotine to the water (Ponzoni et al., 2014). Varenicline, CC4, CC26, and cytisine also induced CPP and abolished the reinforcing effects of nicotine when applied together (Ponzoni et al., 2014). The reinforcing effects of CC4 and cytisine with zebrafish are similar to that seen in rats. The reinforcing effects of nicotine were blocked by high doses of mecamylamine, DHβE, and MLA. The conotoxin MII, an $\alpha6\beta2^*$ antagonist, didn't block the effects of nicotine (Ponzoni et al., 2014). Thus, CPP to nicotine in zebrafish appears to be mediated by $\alpha4\beta2$ nAChRs. The MLA effects make sense since MLA antagonizes both $\alpha4\beta2$ and $\alpha7$ zebrafish nAChRs (Papke et al., 2012). The zebrafish results supporting a role of $\alpha4\beta2$ in nicotine -induced CPP are consistent with that seen in mice (Ponzoni et al., 2014). UFR 2709 treatment before nicotine exposure blocked the CPP response produced by nicotine

(Viscarra et al., 2020). UFR 2709 decreased ethanol consumption in rats, (Viscarra et al., 2020), supporting zebrafish as a model to study addiction and drugs of abuse. Nicotine-induced CPP was also produced in a biased protocol, with increased levels of phosphorylated cAMP response element binding protein (pCREB) detected in zebrafish brain regions thought be involved in reward following CPP (Kedikian, Faillace, & Bernabeu, 2013). This increase is also consistent with what has been observed in mice.

10.1.6.3 Withdrawal

Withdrawal from nicotine has been examined in adult zebrafish (Ponzoni et al., 2021). Mecamylamine-induced and spontaneous withdrawal were examined. Mecamylamine-induced nicotine withdrawal produced increased anxiety in two zebrafish tests, increased freezing, and decreased tank exploration (Ponzoni et al., 2021). Spontaneous withdrawal produced anxiety as well (tested by increased bottom dwelling), decreased memory attention, and reduced motivation (Ponzoni et al., 2021). These zebrafish behaviors model those seen in other animals after nicotine withdrawal and indicate that zebrafish might be used to test drugs that may mitigate drug withdrawal symptoms.

10.1.6.4 Epigenetics

Epigenetic effects have been seen in human and rodent studies (see Chapter 7). Zebrafish demonstrate epigenetic changes after nicotine exposure. Adult zebrafish were exposed to nicotine during the day, but not at night for 14 days (D/N) to mimic human consumption (Pisera-Fuster et al., 2020). These zebrafish demonstrated CPP, but not fish that were chronically treated for 14 days. Interestingly, inhibition of histone deacetylation blocked development of nicotine-induced CPP (Pisera-Fuster et al., 2020). The same inhibitor (4-phenylbutyrate) also prevented nicotine-induced CPP in rats (Pisera-Fuster et al., 2020). Examination of dopaminergic regions of the zebrafish brain (analogous to VTA, NAc, and dorsal habenula) showed increased expression of histone deacetylase I RNA in chronic and D/N nicotine-exposed fish. Expression of lysine acetyltransferase CREB-binding protein RNA was also increased in D/N nicotine-exposed fish. The levels of DNA-methylating enzyme RNAs were also increased as well as methyl-CpG-binding protein (MeCP2) RNA after nicotine treatment (Pisera-Fuster et al., 2020). Higher levels of acetylated histone 3 and pCREB were also seen in D/N-treated zebrafish (Pisera-Fuster et al., 2020).

10.1.6.5 Transgenerational assays

Transgenerational effects of nicotine exposure have been noted in rodent studies (see Chapter 7). In a recent study, FO zebrafish were exposed to the environmental toxins polychlorinated biphenyls (PCB) and poly-brominated diphenyl ethers (PBDE). The F2-generation fish demonstrated anxiety, and DNA methyltransferase (dnmt3ba) expression was increased in F2 and F4 fish (Alfonso et al., 2019). F3 and F4 larvae were hypoactive after light to dark transition. FO zebrafish embryos exposed to methylmercury showed visual deficits and hyperactivity in unexposed F2 fish (Carvan et al., 2017).

Zebrafish are an ideal model to study mechanisms by which transgenerational transmission may occur due to the ease of rearing and accessibility of many ages for molecular and behavioral assays. Effects of nicotine and of signaling through specific nAChR subtypes in transgenerational and epigenetic effects should be evaluated using the zebrafish model.

10.1.6.6 Learning and memory

Zebrafish exhibit learning and memory, although the cells involved are not completely understood. Several tests have been used to examine cholinergic signaling in zebrafish related to memory. A T maze has been used for spatial learning, and tests for delayed spatial alternation and active avoidance conditioning are available (Boyd, 2013; Norton & Bally-Cuif, 2010). Zebrafish have been trained in a delayed spatial alternation (DSA) task. Zebrafish exposure to low levels of nicotine improved performance, but higher levels led to impaired performance (Levin & Chen, 2004). This is typical in mammalian studies as well, with the inverted U response curve typical of responses to nicotine. Nicotine produced this response with a delay (Levin et al., 2006). Mecamylamine blocked these effects if given 5 min before the task, but not if given 40 min prior to the test (Levin et al., 2006. Nicotine may be producing its effects on learning by acting on high-affinity nAChRs or through desensitization (Levin et al., 2006).

Nicotine in addition to producing improved choice accuracy in a DSA task also increased the levels of the dopamine metabolite dihydroxyphenylacetic acid (DOPAC) (Eddins et al., 2009). This increase in DOPAC was mediated by nAChRs in that mecamylamine blocked this effect. No changes were detected in dopamine, serotonin, or norepinephrine levels (Eddins et al., 2009). A change in dopamine might have been expected, but DOPAC is an indicator of synaptic dopamine indicating the level of dopamine turnover/reuptake. In zebrafish, nicotine may affect dopamine metabolism into DOPAC

(Eddins et al., 2009). Zebrafish can be used to assess learning and the role of neurotransmitter systems in the process.

Object memory recognition using three-dimensional objects has also been studied in zebrafish (May et al., 2016). In total, 50 mg/L nicotine enhanced the familiar object recognition, demonstrating another mammalian test that can be applied to zebrafish.

10.1.6.7 Other behaviors

Zebrafish have also been used to examine the role of the habenula in olfaction (Krishnan et al., 2014). Nicotine-induced cholinergic activity rendered an olfactant aversive. This effect was blocked by mecamylamine, an indication of the involvement of nAChRs (Krishnan et al., 2014). Zebrafish tests for locomotion, social preference, sleep, vision, and aggression are also available (Norton & Bally-Cuif, 2010).

10.1.7 Utility of zebrafish for high-throughput screening (HTS) of cholinergic drugs

Drug discovery work begins with molecular and cellular in vitro assays in cells or cell lines and then progresses to preclinical studies using animal models. Most of these models have been mice, rats, and some primates. Human clinical trials are the next step, but it takes 10 or more years to get to this point. Most drugs (99%) don't make it to market, but many millions of dollars are spent on each during the developmental stages (Bowman & Zon, 2010). Problems in animal studies usually involve absorption, distribution, metabolism, excretion, or toxicity problems, the so-named ADMET properties (Bowman & Zon, 2010).

Animals studies are expensive in part due to animal care costs and the limitations on the number of experiments done or drugs that can be tested in a given animal. Using a large number of vertebrate animals that can be easily manipulated genetically, molecularly and can be grown in large numbers inexpensively can speed up the drug discovery process immensely. Zebrafish provide an ideal model for HTS that can uncover problems or promising pathways early in the process. As shown above, zebrafish share nAChR gene homology with rodents and humans and have dopaminergic and cholinergic nervous systems that function in similar ways regarding learning, memory, anxiety, and addiction. Many cellular and molecular pathways are conserved with other vertebrates, and thus, zebrafish are ideal for assessing the effects of cholinergic drugs and treatment on molecular mechanisms as well as behavior. CRISPR has been used to modify the zebrafish genome and may be used to create fish expressing specific mutant nAChRs (Potekhina et al., 2020). Whole animal and large-scale screens are difficult with rodents, but much easier in zebrafish. Whole animal screens can be done with *D. melanogaster* and *C. elegans*, which also express a variety of nAChRs, but being a vertebrate gives zebrafish a significant advantage.

Zebrafish are currently being used for target validation, drug reprofiling, toxicological screens and to test drug efficacy (Boyd, 2013). Rather than using microtiter plates with cells, zebrafish can be used in 96- or 384-well plates. Embryos and larvae can live in fish water up until 6 dpf without feeding, and drugs can then be applied by multichannel pipettes or robots and absorbed through the skin or gills (Boyd, 2013). Candidate drugs often fail due to toxicity, and large toxicology screens are easy with zebrafish. Transgenic zebrafish are available labeling specific populations of neurons or other cells (Boyd, 2013, Potekhina et al., 2020). Zebrafish have been used to monitor electrical activity of circuits in vivo (Marsden & Granato, 2015), including in multiple populations of neurons (Potekhina et al., 2020).

Zebrafish nAChRs sequences are known and can be modified by CRISPR (Potekhina et al., 2020) or expression knocked out by antisense morpholino oligonucleotides targeted to specific RNAs. These can be used to test drugs in a high-throughput way on "knock out" animals much easier than in rodents. Structure-activity relationship (SAR) studies are often used in vitro to understand the molecular basis of ligand binding and to increase potency or efficacy of drugs. SAR studies can also be done in zebrafish to quickly examine the biological effect of a new drug variant (i.e., cytisine to varenicline), but in vivo and on a large scale. For example, SAR in vivo has already been used to develop inhibitors of bone morphogenetic proteins and VEGF (Hao et al., 2010).

In vitro systems can't model complex neurobiological systems. Thus, rodent and primate studies are used preclinically to test behaviors. As shown above, zebrafish can be used to test a number of complex behaviors in response to nicotine and other cholinergic compounds, similar to other animals, but on a larger, less expensive scale. Zebrafish behavior can be analyzed on a large scale using automated systems. For example 4000 small molecules were tested for effects on zebrafish sleep/wake cycles, and 14,000 were tested for effects on high-intensity light pulses (Rihel & Schier, 2012). Nicotine affects sleep (Borniger et al., 2017), and tests for drugs that may affect zebrafish sleep could eliminate drugs at an early step in a pipeline. HTS behavioral assays can utilize either larvae or adult zebrafish. Recently, an assay using 96 zebrafish larvae at a time was used to screen 374 compounds for sedative/hypnotic activity (Yang et al., 2018). Larvae are smaller and have developed motor and sensory systems and are capable of responding to environmental cues and performing simple motor tasks (Boyd, 2013). Locomotor assays and open field testing, standard tests used to study nAChRs in rodents, has been done with zebrafish larvae in multiwell plates and analyzed by automated systems (Boyd, 2013). Zebrafish also demonstrate an acoustic startle response that is altered by prepulse inhibition (PPI). Antipsychotic drugs increase PPI in zebrafish (Boyd, 2013).

Other larval behaviors such as learning, CPP, and feeding response can be adapted to HTS. A high-throughput assay for CPP in zebrafish has already been developed (Mathur et al., 2011). Tests of spatial learning are also being developed (Boyd, 2013). A learning and memory task has been developed that can be scaled up to 400 fish a day (Gerlai, 2010). Changes in gene expression induced by specific drugs such as nicotine and varenicline can be quickly analyzed in small groups of zebrafish by using RNA-Seq. Various ages can be used in a multiwell format to quickly get results from exposure to various doses and exposure times.

In summary, new molecules characterized in zebrafish will need to be validated in humans. However, given the utility of zebrafish for multiple biochemical, genetic, and behavioral assays in a high-throughput format make using the zebrafish model worth the effort.

10.2 *Caenorhabditis elegans (C. elegans)*

In addition to zebrafish, the worm *C. elegans* is another non-mammalian model useful for studying nAChRs. *C. elegans* is a free-living nematode with 302 neurons (Jones & Sattelle, 2004) out of a total of around 900 cells. About 60%–80% of the all *C. elegans* genes are present in humans Taki, Pan, Lee, & Zhang, 2014). Approximately 120 neurons are cholinergic with nAChRs localized to NMJ and multiple neuron populations. *C. elegans* has been used as a model organism to study development and aging. The worms are about 1 mm long, develop in 2–3 days, and have a life span of 2–3 weeks (Taki, Pan, & Zhang, 2014). The nervous system structure in *C. elegans* has been well mapped. Knockout animals can be developed easily. Transgenic worms have been produced with various population of neurons labeled with GFP including AChR:GFP strains (Philbrook & Francis, 2016). Many lines with specific mutations have been developed including in multiple nAChR genes. CRISPR can also be applied to *C. elegans* AChRs, and RNAi can be used to knock down gene expression.

The *C. elegans* genome contains at least 29 nAChR-like genes (Jones & Sattelle, 2004; Jones et al., 2007; Philbrook & Francis, 2016). These genes have various levels of homology to rodent and human nAChRs with up to 60% identity in the second transmembrane region (Philbrook & Francis, 2016). Both homomeric (similar to mammalian α7) and heteromeric nAChRs are present. The genes were placed into five groups: ACR-16, UNC-38, UNC29, ACR-8, and DEG-3 (Jones et al., 2007). Members of the families are homologous to alpha or nonalpha nAChR subunits in other species The ACR-16 group has homology to mammalian/chick α7, α8, α9, and α10 genes.. For example, deg-3, des-2, and unc-38 have been identified as alpha subunits, while lev-1 and unc-29 are more similar to nonalpha genes (Jones & Sattelle, 2004). *C. elegans* subunits have distinct patterns of expression in muscle and various neurons. Multiple *C. elegans* nAChRs have been expressed in heterologous systems and their pharmacology defined. In general, they form cation channels such as nAChRs in other species and are often antagonized by mecamylamine (Jones & Sattelle, 2004).

Gain-of-function mutations have also been produced in *C. elegans* nAChRs as in mice (Tapper et al., 2004) and can used to study function, and the effects of nicotine. *C. elegans* neurons can be patch-clamped and Ca^{2+} indicators can be used to monitor function (Matta et al., 2007). The half-life of nicotine is not known, but nicotine can be applied at controlled doses to the worm medium easily. *C. elegans* have cholinergic and dopaminergic neurons and can respond to nicotine (Engleman et al., 2016). *C. elegans* behavior can be monitored easily and with a short lifetime and accessibility to molecular tools, it provides an excellent system to study nAChRs and mechanisms of nicotine's effects on neurons. Overall, *C. elegans* represent an excellent model expressing multiple nAChR genes with similar functions to those in rodents and humans.

10.2.1 Ric-3 and chaperoning

A significant contribution to the field of nAChRs was made with the discovery of RIC-3, a molecule used for assembly, trafficking, and chaperoning of nAChRs. RIC-3 was identified in a screen seeking to identify suppressors of deg-3-induced cell death. The deg-3 mutation was in the pore-forming domain of the AChR subunit DEG-3. Ric-3 was shown to be required for the function of four nAChRs (Halevi et al., 2002). Ric-3 was expressed in most *C. elegans* neurons including sensory neurons, interneurons, and motor neurons (Halevi et al., 2002). Ric-3 is required for cholinergic transmission in the pharyngeal muscle (Halevi et al., 2002). RIC-3 is in the soma and is also a membrane protein, possibly associated with the endoplasmic reticulum (ER). DEG-3 was expressed normally in Ric-3 mutants, but the location at axons was reduced indicating that RIC-3 was required for proper folding, assembly, or trafficking (Halevi et al., 2002). Mutations in Ric-3 in *C. elegans* lowered surface expression of the LEV-1 nAChRs (Millar., 2008). RIC-3 also functions with mammalian nAChRs increasing the expression of α7 nAChRs in Xenopus oocytes. The human RIC-3 protein also increases expression. This is due to enhancement of folding and assembly (Millar., 2008) and appears to be specific mostly for nAChRs. Ric-3 is required for efficient expression of mammalian nAChRs in many cell lines and ooctyes. RIC-3 physically interacts with nAChRs and can be co-precipitated with α7, α3, α4, β2, and β4 subunits (Millar., 2008). Ric-3 binds to unassembled nAChRs. While ric-3 has now been studied in multiple species, *C. elegans* has provided a good model to study trafficking of nAChRs.

10.2.2 Behavior

Work on *C. elegans* has also shown that some mechanisms involved in addiction are conserved in the worm including withdrawal and sensitization to nicotine. Several other behaviors in worms are modulated by nicotine including spicule ejection, egg-laying, and locomotor activity (Matta et al., 2007; Sellings et al., 2013), indicating that nicotine has behavioral effects in worm as in other model organisms. Responsiveness to taste or olfactory stimuli can also be modified by nicotine in *C. elegans* (Sellings et al., 2013). Young adult worms were attracted to high concentrations of nicotine. This effect was blocked by mecamylamine and varenicline. Worms tended to stay in areas of a plate with nicotine and not approach another rewarding substance, benzaldehyde (Sellings et al., 2013). Mutants for two nAChRs genes (acr-5 and acr-15) didn't approach nicotine, but did approach benzaldehyde, indicating a specific effect and not a general effect on appetitive behavior (Sellings et al., 2013). Acr-5 and acr-15 are similar to mammalian α7 subunits. Further experiments demonstrated that nicotine was acting as a reward, not a locomoter stimulant or secondary reinforcer (Sellings et al., 2013).

Worms were shown to respond to acute nicotine (1.5 μM) by increasing locomotion, but developed tolerance to chronic nicotine and responded as did untreated worms (Feng et al., 2006). After removal of nicotine from the worms for 16 h (withdrawal), the worms responded as did acutely exposed worms by increasing activity again (Feng et al., 2006). The worm motor response was also sensitized to repeated low-dose nicotine exposure. These effects depended on neuronal expression of nAChRs, but not in muscle (Feng et al., 2006). These behaviors are similar to that seen in rodents. Mutants in nAChRs (acr-16) didn't exhibit nicotine-dependent behaviors, and DHβE (α4β2 antagonist) blocked the effects of nicotine (Feng et al., 2006). Replacement of worm nAChRs with mouse α4β2 in a nAChR mutant background restored the response to nicotine, while expression of mouse α7 didn't (Feng et al., 2006). ACR-16 is also homologous to mammalian α7 nAChRs. Worm α7 orthologs may function as do mammalian α4β2 nAChRs (Sellings et al., 2013). Transient receptor potential canonical (TRPC) channels are involved in the worm response to nicotine (Feng et al., 2006).

nAChRs are involved in mammalian nociception, and *C. elegans* contain defined pain sensing neurons providing another model to study pain signaling through nAChRs. Pain and temperature-sensing neurons in *C. elegans* express multiple nAChR subunit combinations, some heteromeric and some homomeric (Cohen et al., 2014). Mutants in nAChR genes showed that specific nAChRs were required for responsiveness of specific neurons to high-threshold mechanical stimuli or cold (Cohen et al., 2014). Signaling appears to be mediated by ACh activation of nAChRs, Ca^{2+} entry, and subsequent downstream Ca^{2+} signaling (Cohen et al., 2014). At least some mechanisms of pain signaling are conserved between mammals and *C. elegans*.

10.2.3 Epigenetics

Effects of nicotine are thought to be modulated by epigenetic changes. One type of change is alteration in microRNA (miRNA) expression. Nicotine alters the expression of miRNAs in multiple systems including rodents and humans (Huang & Li, 2009; Taki, Pan, & Zhang, 2014) *C. elegans* has been used to study the effects of nicotine on miRNAs. Chronic nicotine treatment of postembryonic worms produced an increase in 40 miRNAs. The higher dosage produced the greatest changes (Taki, Pan, Lee, & Zhang, 2014), but three were increased by exposure to 20 µM nicotine. Five of the miRNAs have significant homology to human miRNAs. Multiple genes were targeted by these miRNAs including fos-1, a gene implicated in addictive pathways (Taki, Pan, Lee, & Zhang, 2014). Transient exposure of larval *C. elegans* to nicotine (F0 generation) also led to changes in miRNA expression in the unexposed F1 and F2 worms (Taki, Pan, & Zhang, 2014). Several miRNAs were affected in multiple generations. Target genes for these miRNAs were among those thought to play a possible role in addiction and nicotine-induced behavior (Taki, Pan, & Zhang, 2014). *C. elegans* may prove to be a good model to study epigenetic and transgenerational effects of nicotine at the cellular level.

In summary, *C. elegans* provides another model system to study nAChRs. The similarities in nAChRs are enough to warrant studies in a simple system of the effects of compounds on behaviors (locomotion, reward, and pain). The roles of *C. elegans* orthologs of human or rodent nAChRs can be quickly tested in an organism that is easy to grow, manipulate genetically, and label or remove specific cell populations.

10.3 *Drosophila melanogaster* (*D. melanogaster*)

D. melanogaster, commonly known as the fruit fly, has been the subject of thousands of publications. These studies have revealed important information about signaling, basic cell biology, and development. Some the first genes involved in development (homeobox) were discovered in the fly, and these genes are conserved in humans as well. Drosophila has around 300,000 neurons and exhibits a number of complex behaviors (Matta et al., 2007). Their life span is around 10 days, and their genetics has been studied for more than 100 years. Genes can easily be inserted, deleted, and specific cells labeled.

ACh is the major excitatory neurotransmitter in Drosophila in neurons. nAChRs are not present at the NMJ, typical of insects. nAChRs mediate fast synaptic transmission in the fly brain. The cholinergic loci (ChAT and VAChT genes) are clustered as they are in vertebrates (Matta et al., 2007). Flies also express AChE. Ten nAChR genes have been identified and are grouped as α-like (7) or β-like (3) (Gundelfinger & Hess, 1992; Sattelle et al., 2005). Dα1–Dα7 are alpha-like and Dβ1-3 are nonalpha like. Dα5–Dα7 are most similar to vertebrate α7 nAChRs (Sattelle et al., 2005). Comparison of the genes with those of other species indicates that Drosophila nAChRs possess an N-terminal extracellular domain, four transmembrane regions, and the short C-terminal domain (Rosenthal & Yuan, 2021). Alpha subunits have the characteristic vicinal cysteines, but reversion between alpha and beta subunits during evolution complicates the designation of ligand-binding subunits (Rosenthal & Yuan, 2021). Expression of Drosophila nAChR subunits (including Dα1–Dα4, Dβ1, and Dβ2) has been mapped to various regions of the nervous system including the brain, protocerebrum, deutocerebrum, ventral nerve, optic lobe, and medulla (Sattelle et al., 2005). Drosophila nAChRs are implicated in developmental events, plasticity, and synapse formation (Rosenthal & Yuan, 2021) with developmental control of subunit expression (Rosenthal & Yuan, 2021). Assembly of Drosophila nAChRs subunits has been demonstrated by co-immunoprecipitation and function by expression of some Drosophila subunits with vertebrates' nAChR subunits in oocytes or insect S2 cells (Sattelle et al., 2005). Drosophila subunits alone have yet to be expressed in heterologous systems, making the assignment of function to specific subtypes difficult. Knockout mutants for all 10 genes are available (Lu et al., 2022).

Nicotine can be delivered to flies by injection or headless thorax preparations (Matta et al., 2007). However, for more acute examinations of the effects of nicotine, volatilization is used. For chronic exposure, nicotine is incorporated into the food. Acute exposure induces grooming and locomotion (Matta et al., 2007).

10.3.1 Biogenic amine signaling and Parkinson's disease

Multiple components of the dopaminergic-related system are present in Drosophila (Fuenzalida-Uribe et al., 2013). nAChRs regulate the release of biogenic amines such as dopamine in the mammalian brain. nAChRs also have been shown to be involved in control of octopamine (fly homolog of norepinephrine/epinephrine) release in flies (Fuenzalida-Uribe et al., 2013. α-Bgt sensitive nAChRs were shown to be involved as PNU-282987, an α7 nAChR agonist, increased release of octopamine from a fly preparation, and α-Bgt blocked this effect (Fuenzalida-Uribe et al., 2013). nAChR control of octopamine release was important for control of a reduced startle response in the presence of nicotine.

Familial Parkinson's disease (PD) cases in humans are associated with loss-of-function mutations in the PARK2 gene (Chambers et al., 2013). Fruit flies also express a PARK2 gene, park, that is an ortholog of the human genes. Park loss-of-function mutations cause degeneration of neurons in Drosophila. A park heterozygous mutant was used that had a similar phenotype to the homozygous loss-of-function mutant and that shows a phenotype similar to human PD patients with neuronal degeneration, motor problems, and shortened life span (Chambers et al., 2013). Nicotine exposure reduced the mortality of the heterozygous mutants, but increased it in wild-type flies. Nicotine also improved the deficit in climbing seen in the heterozygous mutants, but inhibited it in control flies (Chambers et al., 2013). Nicotine also improved flight and prevented a loss of olfaction that occurs in the untreated mutants (Chambers et al., 2013). Drosophila may be useful to rapidly and, in a high-throughput manner, screen cholinergic compounds for treatment of PD.

10.3.2 Mechanisms of insecticide action

Perhaps the best application of Drosophila is for study of nAChR insecticide targets. Many insecticides have been developed that impair nAChR function. Neonicotinoids are a widely used insecticide with 24% of the world market (Lu et al., 2022). These include imidacloprid, thiacloprid, and acetamiprid and act as nAChR agonists. Insecticides produce hyperactivity and convulsion and death at higher doses (Lu et al., 2022). Mutations in specific Drosophila nAChRs subunits produced various levels of resistance to insecticides, indicating that activation of nAChRs by insecticides leads to death. Specific nAChR subtypes are activated by specific neonicotinoids (Lu et al., 2022). Detailed structure–function studies are being done to map specific residues responsible for activity of specific insecticides such as imidacloprid and thiacloprid (Ihara et al., 2021). The use of nAChR knockout or mutant Drosophila can be used to screen for insecticides with a high potency and efficacy at insect nAChRs and thus require lower levels of drug to be effective.

In summary, *D. melanogaster* possess nAChRs with structures similar to those expressed in vertebrates. The technical advantages of Drosophila make this another model to study nAChR function in development and for potential drug discovery.

References

Ackerman, K. M., & Boyd, R. T. (2016). Analysis of nicotinic acetylcholine receptor (nACHR) gene expression in zebrafish (*Danio rerio*) by in situ hybridization and PCR. *Neuromethods, 117*, 1–31.

Ackerman, K. M., Nakkula, R., Beattie, C. E., & Boyd, R. T. (2006). Zebrafish neuronal nicotinic acetylcholine receptors: Gene structures, expression, and effects of transient embryonic RNA knockdown. Program no. 526. In *Neuroscience meeting planner*Society for Neuroscience.

Ackerman, K. M., Nakkula, R., Zirger, J. M., Beattie, C. E., & Boyd, R. T. (2009). Cloning and spatiotemporal expression of zebrafish neuronal nicotinic acetylcholine receptor alpha 6 and alpha 4 subunit RNAs. *Developmental Dynamics*, *238*(4), 980–992. https:// doi.org/10.1002/dvdy.21912.

Ackermann, G. E., & Paw, B. H. (2003). Zebrafish: A genetic model for vertebrate organogenesis and human disorders. *Frontiers in Bioscience: A Journal and Virtual Library*, *8*, d1227–d1253. https://doi.org/10.2741/1092.

Alfonso, S., Blanc, M., Joassard, L., Keiter, S. H., Munschy, C., Loizeau, V., Bégout, M. L., & Cousin, X. (2019). Examining multi- and transgenerational behavioral and molecular alterations resulting from parental exposure to an environmental PCB and PBDE mixture. *Aquatic Toxicology*, *208*, 29–38. https://doi.org/10.1016/j.aquatox.2018.12.021.

Alzualde, A., Jaka, O., Latino, D. A. R. S., Alijevic, O., Iturria, I., de Mendoza, J. H., Pospisil, P., Frentzel, S., Peitsch, M. C., Hoeng, J., & Koshibu, K. (2021). Effects of nicotinic acetylcholine receptor-activating alkaloids on anxiety-like behavior in zebrafish. *Journal of Natural Medicines*, *75*(4), 926–941. https://doi.org/10.1007/s11418-021-01544-8.

Barbazuk, W., Korf, I., & Kadavi, C. (2000). The role of alpha7 and alpha4beta2 nicotinic receptors in the nicotine-induced anxiolytic effect in zebrafish. *Physiology and Behavior*, *10*(9), 408–412.

Bencan, Z., & Levin, E. D. (2008). The role of alpha7 and alpha4beta2 nicotinic receptors in thenicotine-induced anxiolytic effect in zebrafish. *Physiology and Behavior*, *95*(3), 408–412.

Borniger, J. C., Don, R. F., Zhang, N., Boyd, R. T., & Nelson, R. J. (2017). Enduring effects of perinatal nicotine exposure on murine sleep in adulthood. *American Journal of Physiology - Regulatory Integrative and Comparative Physiology*, *313*(3), R280–R289. https://doi.org/10.1152/ajpregu.00156.2017.

Bowman, T. V., & Zon, L. I. (2010). Swimming into the future of drug discovery: In vivo chemical screens in zebrafish. *ACS Chemical Biology*, *5*(2), 159–161. https://doi.org/ 10.1021/cb100029t.

Boyd, R. T. (2013). Using zebrafish for identification and development of new CNS drugs affecting nicotinic and dopaminergic systems. In *Vol. 2. CNS drug discovery* (pp. 381–406). Bentham Science Publishers.

Carvan, M. J., Kalluvila, T. A., Klingler, R. H., Larson, J. K., Pickens, M., Mora-Zamorano, F. X., Connaughton, V. P., Sadler-Riggleman, I., Beck, D., & Skinner, M. K. (2017). Mercury-induced epigenetic transgenerational inheritance of abnormal neurobehavior is correlated with sperm epimutations in zebrafish. *PLoS ONE*, *12*(5). https://doi.org/ 10.1371/journal.pone.0176155.

Chambers, R. P., Call, G. B., Meyer, D., Smith, J., Techau, J. A., Pearman, K., & Buhlman, L. M. (2013). Nicotine increases lifespan and rescues olfactory and motor deficits in a Drosophila model of Parkinson's disease. *Behavioural Brain Research*, *253*, 95–102. https://doi.org/10.1016/j.bbr.2013.07.020.

Clemente, D., Porteros, A., Weruaga, E., Alonso, J. R., Arenzana, F. J., Aijón, J., & Arévalo, R. (2004). Cholinergic elements in the zebrafish central nervous system: Histochemical and immunohistochemical analysis. *Journal of Comparative Neurology*, *474*(1), 75–107. https://doi.org/10.1002/cne.20111.

Cohen, E., Chatzigeorgiou, M., Husson, S. J., Steuer-Costa, W., Gottschalk, A., Schafer, W. R., & Treinin, M. (2014). *Caenorhabditis elegans* nicotinic acetylcholine receptors are required for nociception. *Molecular and Cellular Neuroscience*, *59*, 85–96. https:// doi.org/10.1016/j.mcn.2014.02.001.

Crosby, E. B., Bailey, J. M., Oliveri, A. N., & Levin, E. D. (2015). Neurobehavioral impairments caused by developmental imidacloprid exposure in zebrafish. *Neurotoxicology and Teratology*, *49*, 81–90. https://doi.org/10.1016/j.ntt.2015.04.006.

Delvecchio, C., Tiefenbach, J., & Krause, HM. (2011). The zebrafish: A powerful platform for in vivo, HTS drug discovery. *Assay and Drug Development Technologies*, *9*, 354–361.

Dooley, K., & Zon, L. I. (2000). Zebrafish: A model system for the study of human disease. *Current Opinion in Genetics and Development*, *10*(3), 252–256. https://doi.org/10.1016/S0959-437X(00)00074-5.

Eddins, D., Petro, A., Williams, P., Cerutti, D.T., & Levin, E.D. (2009). Nicotine effects on learning in zebrafish: The role of dopaminergic systems. Psychopharmacology, 202(1–3), 103–109. doi: https://doi.org/10.1007/s00213-008-1287-4.

Engleman, E. A., Katner, S. N., & Neal-Beliveau, B. S. (2016). *Caenorhabditis elegans* as a model to study the molecular and genetic mechanisms of drug addiction. In *Vol. 137. Progress in molecular biology and translational science* (pp. 229–252). Elsevier B.V. https://doi.org/10.1016/bs.pmbts.2015.10.019.

Faundez-Parraguez, M., Farias-Rabelo, N., Gonzalez-Gutierrez, J. P., Etcheverry-Berrios, A., Alzate-Morales, J., Adasme-Carreño, F., Varas, R., Bermudez, I., & Iturriaga-Vasquez, P. (2013). Neonicotinic analogues: Selective antagonists for α4β2 nicotinic acetylcholine receptors. *Bioorganic and Medicinal Chemistry*, *21*(10), 2687–2694. https://doi.org/10.1016/j.bmc.2013.03.024.

Feng, Z., Li, W., Ward, A., Piggott, B. J., Larkspur, E. R., Sternberg, P. W., & Xu, X. Z. S. (2006). A *C. elegans* model of nicotine-dependent behavior: Regulation by TRP-family channels. *Cell*, *127*(3), 621–633. https://doi.org/10.1016/j.cell.2006.09.035.

Fowler, C. D., Lu, Q., Johnson, P. M., Marks, M. J., & Kenny, P. J. (2011). Habenular α5 nicotinic receptor subunit signalling controls nicotine intake. *Nature*, *471*(7340), 597–601. https://doi.org/10.1038/nature09797.

Freixas, A. E. C., Moglie, M. J., Castagnola, T., Salatino, L., Domene, S., Marcovich, I., Gallino, S., Wedemeyer, C., Goutman, J. D., Plazas, P. V., & Elgoyhen, A. B. (2021). Unraveling the molecular players at the cholinergic efferent synapse of the zebrafish lateral line. *Journal of Neuroscience*, *41*(1), 47–60. https://doi.org/10.1523/JNEUROSCI.1772-20.2020.

Fuenzalida-Uribe, N., Meza, R. C., Hoffmann, H. A., Varas, R., & Campusano, J. M. (2013). NAChR-induced octopamine release mediates the effect of nicotine on a startle response in *Drosophila melanogaster*. *Journal of Neurochemistry*, *125*(2), 281–290. https://doi.org/10.1111/jnc.12161.

Gerlai, R. (2010). High-throughput behavioral screens: The first step towards finding genes involved in vertebrate brain function using zebrafish. *Molecules*, *15*(4), 2609–2622. https://doi.org/10.3390/molecules15042609.

Gundelfinger, E. D., & Hess, N. (1992). Nicotinic acetylcholine receptors of the central nervous system of Drosophila. *BBA—Molecular Cell Research*, *1137*(3), 299–308. https://doi.org/10.1016/0167-4889(92)90150-A.

Guo, S. (2004). Linking genes to brain, behavior and neurological diseases: What can we learn from zebrafish? *Genes, Brain and Behavior*, *3*(2), 63–74. https://doi.org/10.1046/j.1601-183X.2003.00053.x.

Halevi, S., McKay, J., & Palfreyman, M. (2002). The *C. elegans* ric-3 gene is required for maturation of nicotinic acetylcholine receptors. *EMBO Journal*, *21*(5), 1012–1020. https://doi.org/10.1093/emboj/21.5.1012.

Hao, J., Ho, J. N., Lewis, J. A., Karim, K. A., Daniels, R. N., Gentry, P. R., Hopkins, C. R., Lindsley, C. W., & Hong, C. C. (2010). In vivo structure—Activity relationship study of dorsomorphin analogues identifies selective VEGF and BMP inhibitors. *ACS Chemical Biology*, *5*(2), 245–253. https://doi.org/10.1021/cb9002865.

Hong, E., Santhakumar, K., Akitake, C. A., Ahn, S. J., Thisse, C., Thisse, B., Wyart, C., Mangin, J. M., & Halpern, M. E. (2013). Cholinergic left–right asymmetry in the habenulo-interpeduncular pathway. *Proceedings of the National Academy of Sciences of the United States of America*, *110*(52), 21171–21176. https://doi.org/10.1073/pnas.1319566110.

Howe, K., Clark, M. D., Torroja, C. F., Torrance, J., Berthelot, C., Muffato, M., Collins, J. E., Humphray, S., McLaren, K., Matthews, L., McLaren, S., Sealy, I., Caccamo, M., Churcher, C., Scott, C., Barrett, J. C., Koch, R., Rauch, G. J., White, S., … Stemple, D. L. (2013). The zebrafish reference genome sequence and its relationship to the human genome. *Nature*, *496*(7446), 498–503. https://doi.org/10.1038/nature12111.

Huang, W., & Li, M. D. (2009). Nicotine modulates expression of miR-140, which targets the 3-untranslated region of dynamin 1 gene (Dnm1). *International Journal of Neuropsychopharmacology*, *12*(4), 537–546. https://doi.org/10.1017/S146114 5708009528.

Ihara, M., Hikida, M., Matsushita, H., Yamanaka, K., Kishimoto, Y., Kubo, K., … Matsuda, K. (2021). Loops D,E, and G in the Drosophila Dα1 subunit contribute to high neonicotinoid sensitivity of Dα1-chicken β2 nicotinic acetylcholine receptor. *British Journal of Pharmacology*, *175*, 1999–2012.

Jones, A. K., Davis, P., Hodgkin, J., & Sattelle, D. B. (2007). The nicotinic acetylcholine receptor gene family of the nematode *Caenorhabditis elegans*: An update on nomenclature. *Invertebrate Neuroscience*, *7*(2), 129–131. https://doi.org/10.1007/s10158-007-0049-z.

Jones, A. K., & Sattelle, D. B. (2004). Functional genomics of the nicotinic acetylcholine receptor gene family of the nematode, *Caenorhabditis elegans*. *BioEssays*, *26*(1), 39–49. https://doi.org/10.1002/bies.10377.

Kedikian, X, Faillace, MP, & Bernabeu, R. (2013). Behavioral and molecular analysis of nicotine-conditioned place preference in zebrafish. *PLoS ONE*, *8*(7), e69453. https://doi.org/10.1371/journal.pone.0069453.

Kimmel, C. B., Ballard, W. W., Kimmel, S. R., Ullmann, B., & Schilling, T. F. (1995). Stages of embryonic development of the zebrafish. *Developmental Dynamics*, *203*(3), 253–310. https://doi.org/10.1002/aja.1002030302.

Klee, E. W., Ebbert, J. O., Schneider, H., Hurt, R. D., & Ekker, S. C. (2011). Zebrafish for the study of the biological effects of nicotine. *Nicotine and Tobacco Research*, *13*(5), 301–312. https://doi.org/10.1093/ntr/ntr010.

Krishnan, S., Mathuru, A. S., Kibat, C., Rahman, M., Lupton, C. E., Stewart, J., Claridge-Chang, A., Yen, S. C., & Jesuthasan, S. (2014). The right dorsal habenula limits attraction to an odor in zebrafish. *Current Biology*, *24*(11), 1167–1175. https://doi.org/10.1016/j.cub.2014.03.073.

Levin, E.D., Bencan, Z., Cerutti, D.T., 2007. Anxiolytic effects of nicotine in zebrafish. Physiology and Behavior, 90, (1):54-58.

Levin, E. D., & Chen, E. (2004). Nicotinic involvement in memory function in zebrafish. *Neurotoxicology and Teratology*, *26*(6), 731–735. https://doi.org/10.1016/j.ntt.2004.06.010.

Levin, E. D., Limpuangthip, J., Rachakonda, T., & Peterson, M. (2006). Timing of nicotine effects on learning in zebrafish. *Psychopharmacology*, *184*(3–4), 547–552. https://doi.org/10.1007/s00213-005-0162-9.

Lu, W., Liu, Z., Fan, X., Zhang, X., Qiao, X., Huang, J., & Palli, S. R. (2022). Nicotinic acetylcholine receptor modulator insecticides act on diverse receptor subtypes with distinct subunit compositions. *PLoS Genetics*, *18*(1), e1009920. https://doi.org/10.1371/journal.pgen.1009920.

Marsden, K. C., & Granato, M. (2015). In vivo Ca2+ imaging reveals that decreased dendritic excitability drives startle habituation. *Cell Reports*, *13*(9), 1733–1740. https://doi.org/10.1016/j.celrep.2015.10.060.

Mathur, P., Lau, B., & Guo, S. (2011). Conditioned place preference behavior in zebrafish. *Nature Protocols*, *6*(3), 338–345. https://doi.org/10.1038/nprot.2010.201.

Matta, S. G., Balfour, D. J., Benowitz, N. L., Boyd, R. T., Buccafusco, J. J., Caggiula, A. R., Craig, C. R., Collins, A. C., Damaj, M. I., Donny, E. C., Gardiner, P. S., Grady, S. R., Heberlein, U., Leonard, S. S., Levin, E. D., Lukas, R. J., Markou, A., Marks, M. J.,

McCallum, S. E., … Zirger, J. M. (2007). Guidelines on nicotine dose selection for in vivo research. *Psychopharmacology*, *190*(3), 269–319. https://doi.org/10.1007/s00213-006-0441-0.

May, Z., Morrill, A., Holcombe, A., Johnston, T., Gallup, J., Fouad, K., Schalomon, M., & Hamilton, T. J. (2016). Object recognition memory in zebrafish. *Behavioural Brain Research*, *296*, 199–210. https://doi.org/10.1016/j.bbr.2015.09.016.

Menelaou, E., & Svoboda, K. R. (2009). Secondary motoneurons in juvenile and adult zebrafish: Axonal pathfinding errors caused by embryonic nicotine exposure. *Journal of Comparative Neurology*, *512*(3), 305–322. https://doi.org/10.1002/cne.21903.

Menelaou, E., Udvadia, A. J., Tanguay, R. L., & Svoboda, K. R. (2014). Activation of α2A-containing nicotinic acetylcholine receptors mediates nicotine-induced motor output in embryonic zebrafish. *European Journal of Neuroscience*, *40*(1), 2225–2240. https://doi.org/10.1111/ejn.12591.

Millar. (2008). RIC-3 a nicotinic acetylcholine receptor chaperone. *British Journal of Pharmacology*, 153.

Mongeon, R., Walogorsky, M., Urban, J., Mandel, G., Ono, F., & Brehm, P. (2011). An acetylcholine receptor lacking both γ and ε subunits mediates transmission in zebrafish slow muscle synapses. *Journal of General Physiology*, *138*(3), 353–366. https://doi.org/10.1085/jgp.201110649.

Mueller, T., Vernier, P., & Wullimann, M. F. (2004). The adult central nervous cholinergic system of a neurogenetic model animal, the zebrafish *Danio rerio*. *Brain Research*, *1011*(2), 156–169. https://doi.org/10.1016/j.brainres.2004.02.073.

Norton, W., & Bally-Cuif, L. (2010). Adult zebrafish as a model organism for behavioural genetics. *BMC Neuroscience*, 11. https://doi.org/10.1186/1471-2202-11-90.

Papke, R. L., Ono, F., Stokes, C., Urban, J. M., & Boyd, R. T. (2012). The nicotinic acetylcholine receptors of zebrafish and an evaluation of pharmacological tools used for their study. *Biochemical Pharmacology*, *84*(3), 352–365. https://doi.org/10.1016/j.bcp.2012.04.022.

Parker, B., & Connaughton, V. P. (2007). Effects of nicotine on growth and development in larval zebrafish. *Zebrafish*, *4*(1), 59–68. https://doi.org/10.1089/zeb.2006.9994.

Pedersen, J. E., Bergqvist, C. A., & Larhammar, D. (2019). Evolution of vertebrate nicotinic acetylcholine receptors. *BMC Evolutionary Biology*, *19*(1). https://doi.org/10.1186/s12862-018-1341-8.

Philbrook, A., & Francis, M. M. (2016). Emerging technologies in the analysis of *C. elegans* nicotinic acetylcholine receptors. In *Vol. 117. Neuromethods* (pp. 77–96). Humana Press Inc. https://doi.org/10.1007/978-1-4939-3768-4_5.

Pisera-Fuster, A., Faillace, M. P., & Bernabeu, R. (2020). Pre-exposure to nicotine with nocturnal abstinence induces epigenetic changes that potentiate nicotine preference. *Molecular Neurobiology*, *57*(4), 1828–1846. https://doi.org/10.1007/s12035-019-01843-y.

Ponzoni, L., Braida, D., Pucci, L., Andrea, D., Fasoli, F., Manfredi, I., Papke, R. L., Stokes, C., Cannazza, G., Clementi, F., Gotti, C., & Sala, M. (2014). The cytisine derivatives, CC4 and CC26, reduce nicotine-induced conditioned place preference in zebrafish by acting on heteromeric neuronal nicotinic acetylcholine receptors. *Psychopharmacology*, *231*(24), 4681–4693. https://doi.org/10.1007/s00213-014-3619-x.

Ponzoni, L., Melzi, G., Marabini, L., Martini, A., Petrillo, G., Teh, M.-T., Torres-Perez, J. V., Morara, S., Gotti, C., Braida, D., Brennan, C. H., & Sala, M. (2021). Conservation of mechanisms regulating emotional-like responses on spontaneous nicotine withdrawal in zebrafish and mammals. *Progress in Neuro-Psychopharmacology and Biological Psychiatry*, *111*, 110334. https://doi.org/10.1016/j.pnpbp.2021.110334.

Potekhina ES, Bass DY, Kelmanson IV, Fetisova ES, Ivanenko AV, Belousov VV, Bilan DS (2020) Drug screening with genetically encoded fluorescent sensors: today and tomorrow. International Journal of Molecular Sciences 22(1):148. https://doi.org/10.3390/ijms22010148.

Rihel, J., & Schier, A. F. (2012). Behavioral screening for neuroactive drugs in zebrafish. *Developmental Neurobiology, 72*(3), 373–385. https://doi.org/10.1002/dneu.20910.

Rima, M., Lattouf, Y., Abi Younes, M., Bullier, E., Legendre, P., Mangin, J. M., & Hong, E. (2020). Dynamic regulation of the cholinergic system in the spinal central nervous system. *Scientific Reports, 10*(1). https://doi.org/10.1038/s41598-020-72524-3.

Rosenthal, J. S., & Yuan, Q. (2021). Constructing and tuning excitatory cholinergic synapses: The multifaceted functions of nicotinic acetylcholine receptors in Drosophila neural development and physiology. *Frontiers in Cellular Neuroscience, 15.* https://doi.org/10.3389/fncel.2021.720560.

Sattelle, D. B., Jones, A. K., Sattelle, B. M., Matsuda, K., Reenan, R., & Biggin, P. C. (2005). Edit, cut and paste in the nicotinic acetylcholine receptor gene family of *Drosophila melanogaster. BioEssays, 27*(4), 366–376. https://doi.org/10.1002/bies.20207.

Sellings, L., Pereira, S., Qian, C., Dixon-McDougall, T., Nowak, C., Zhao, B., Tyndale, R. F., & van der Kooy, D. (2013). Nicotine-motivated behavior in *Caenorhabditis elegans* requires the nicotinic acetylcholine receptor subunits acr-5 and acr-15. *European Journal of Neuroscience, 37*(5), 743–756. https://doi.org/10.1111/ejn.12099.

Svoboda, K., Vijayaraghavan, S., & Tanguay, R. (2002). Nicotinic receptors mediate changes in spinal motoneuron development and axonal pathfinding in embryonic zebrafish exposed to nicotine. *The Journal of Neuroscience, 22,* 10731–10741.

Taki, F. A., Pan, X., Lee, M. H., & Zhang, B. (2014). Nicotine exposure and transgenerational impact: A prospective study on small regulatory microRNAs. *Scientific Reports, 4.* https://doi.org/10.1038/srep07513.

Taki, F. A., Pan, X., & Zhang, B. (2014). Chronic nicotine exposure systemically alters microRNA expression profiles during post-embryonic stages in *Caenorhabditis elegans. Journal of Cellular Physiology, 229*(1), 79–89. https://doi.org/10.1002/jcp.24419.

Tapper, A. R., McKinney, S. L., Nashmi, R., Schwarz, J., Deshpande, P., Labarca, C., Whiteaker, P., Marks, M. J., Collins, A. C., & Lester, H. A. (2004). Nicotine activation of α4* receptors: Sufficient for reward, tolerance, and sensitization. *Science, 306*(5698), 1029–1032. https://doi.org/10.1126/science.1099420.

Viscarra, F., González-Gutierrez, J., Esparza, E., Figueroa, C., Paillali, P., Hödar-Salazar, M., Cespedes, C., Quiroz, G., Sotomayor-Zárate, R., Reyes-Parada, M., Bermúdez, I., & Iturriaga-Vásquez, P. (2020). Nicotinic antagonist UFR2709 inhibits nicotine reward and decreases anxiety in zebrafish. *Molecules, 25*(13). https://doi.org/10.3390/molecules25132998.

Welsh, L., Tanguay, R. L., & Svoboda, K. R. (2009). Uncoupling nicotine mediated motoneuron axonal pathfinding errors and muscle degeneration in zebrafish. *Toxicology and Applied Pharmacology, 237*(1), 29–40. https://doi.org/10.1016/j.taap.2008.06.025.

Westerfield, M. (2000). *The zebrafish book* (4th ed.). A guide for the laboratory use of zebrafish (Danio rerio): Univ. of Oregon Press.

White, R. M., Sessa, A., Burke, C., Bowman, T., LeBlanc, J., Ceol, C., Bourque, C., Dovey, M., Goessling, W., Burns, C. E., & Zon, L. I. (2008). Transparent adult zebrafish as a tool for in vivo transplantation analysis. *Cell Stem Cell, 2*(2), 183–189. https://doi.org/10.1016/j.stem.2007.11.002.

Yang, X., Jounaidi, Y., Dai, J. B., Marte-Oquendo, F., Halpin, E. S., Brown, L. E., … Forman, S. A. (2018). High-throughput screening in larval zebrafish identifies novel potent sedative-hypnotics. *Anesthesiology, 129*(3), 459–476. https://doi.org/10.1097/ALN.0000000000002281.

Zirger, J. M., Beattie, C. E., McKay, D. B., & Boyd, R. T. (2003). Cloning and expression of zebrafish neuronal nicotinic acetylcholine receptors. *Gene Expression Patterns, 3*(6), 747–754. https://doi.org/10.1016/S1567-133X(03)00126-1.

Index

Note: Page numbers followed by *f* indicate figures and *t* indicate tables.

Printed in the United States
by Baker & Taylor Publisher Services